ENGINEERING
MATERIALS
TECHNOLOGY

ENGINEERING MATERIALS TECHNOLOGY

JAMES A. JACOBS
School of Technology
Norfolk State University

THOMAS F. KILDUFF
Division of Engineering Technologies
Thomas Nelson Community College

PRENTICE-HALL, INC.
Englewood Cliffs, New Jersey 07632

Library of Congress Cataloging in Publication Data

Jacobs, James A. (date)
   Engineering materials technology

   Includes bibliographical references and index.
   1. Materials.   I. Kilduff, Thomas F. (date)
II. Title.
TA403.J26 1985   620.1'1   84-10679
ISBN 0-13-278045-3

Editorial/production supervision: Nancy Milnamow and Diana Drew
Interior design: Nancy Milnamow
Cover design: 20/20 Services, Inc.
Art production: Meg Van Arsdale
Manufacturing buyer: Anthony Caruso

Printed in the United States of America

10 9 8

ISBN 0-13-278045-3 01

Prentice-Hall International, Inc., *London*
Prentice-Hall of Australia Pty. Limited, *Sydney*
Editora Prentice-Hall do Brasil, Ltda., *Rio de Janeiro*
Prentice-Hall Canada Inc., *Toronto*
Prentice-Hall Hispanoamericana, S.A., *Mexico*
Prentice-Hall of India Private Limited, *New Delhi*
Prentice-Hall of Japan, Inc., *Tokyo*
Prentice-Hall of Southeast Asia Pte. Ltd., *Singapore*
Whitehall Books Limited, *Wellington, New Zealand*

As authors we are indebted to the many people whose constructive suggestions and cooperation helped bring this book to completion. However, most of all, we dedicate this book to our wives, Martha and Virginia, and daughters, Sherri, Tammi, Jeanene, Liz, and Suzy, to all of whom we owe a major debt for their understanding, unfailing support, optimism, and tangible assistance as we toiled for longer than six years.

CONTENTS

Module 5   *Metallic Materials*   **155**

Module 7   Ceramics   *306*

Module 8   Composite Materials   *346*

Contents

Module 9   *Electronic-Related Materials*   *398*

Module 10   Appendix of Tables   **435**

PREFACE

Rapid technological advances coupled with a growing scarcity of vital natural resources cause frequent changes in engineering and industrial materials. An indication of such change comes in the estimate that an engineer's half-life is four years (i.e., four years after graduation much of the knowledge learned in schooling is outdated). So, too, with the engineering technician and numerous other materials users. This textbook focuses on the need to give the proper preparation so the reader can deal with these inevitable changes.

The decision to produce this book resulted from several factors. First, the experience of nearly two decades of teaching in programs of engineering technology, industrial technology, and industrial education, plus teaching workers from business, industry and government, brought a frustration over the textbooks for industrial materials. Materials science textbooks focus on traditional engineering programs with a very theoretical treatment that requires a strong mathematics and science background. At the other end of the spectrum, one finds texts with a superficial treatment of most materials, with often only in-depth coverage of metals. Such limited coverage does not suffice in the highly competitive and changing climate of modern industry.

Pursuing a better way to communicate the subject of materials technology to students other than those enrolled in traditional engineering programs resulted in a two-year grant from the National Aeronautics and Space Administration. NASA felt our concern reflected a national problem. The research brought us into contact with many projects dealing with the improved teaching and understanding of technical subjects. Such projects ranged from secondary schools, to community colleges and technical institutes, to universities and industrial training. Many of the efforts aimed at teaching materials and processes technology or materials science. Some National Science Foundation (NSF) grants focused on industrial arts and science students and joint NSF and American Society for Engineering Education support developed into Education Modules of Materials Science and Engineering (EMMSE) that employed effective instructional strategies. We participated in the EMMSE project also. Still, all of our research failed to turn up a materials technology textbook suitable for our audience. So we began the development of such a text.

AUDIENCE

The design of *Engineeering Materials Technology* focuses on that large group of people who require an understanding of industrial materials but lack the preparation for a traditional engineering course in materials science. These readers may or may not possess a background in physics or chemistry. As materials users in such roles as manufacturing technicians, designers, purchasing agents, field engineers, sales representatives, engineering technologists, quality controllers, quality assurance inspectors, or technical educators, they need a sound understanding of available materials to meet immediate job requirements. They also must possess mastery of the basic concepts common to all materials, which prepares them to adapt to the rapid developments in materials technology. Memorization of long lists of tool steels or plastics serves limited value since one consults references such as handbooks, standards, or catalogs for that type of data. However, these materials users need to learn basic principles for problem solving and know how to use the references. The mastery of concepts, a higher level of learning, allows one to synthesize one's knowledge of materials facts to select, reject, order, design, fabricate, maintain, test, and process. The approach that we seek to foster in this text has a lasting effect upon which the reader can build as new technology develops. Radical changes to many industries resulting from the advantages offered by such materials as optical fibers, advanced composites, and structural adhesives require a full range of materials technology knowledge if a person wishes to stay current.

DESIGN

We sought a careful blending of the whys with the hows and whats, because we recognize that most of our readers need job entry understanding that focuses more on the applications of materials rather than their development. We present a brief overview of some basic principles in chemistry and physics of materials in an early unit in order to develop a comprehension of structure and properties common to most materials, which aids in the prediction of how materials react to given stresses and environments. Then in units on specific materials such as plastics, the reader does not deal with page after page of materials (plastics) and properties; rather the most common materials (plastics) are compared with materials of the same group and of other groups (metals, ceramics, and other polymers).

We wish to develop a breadth of knowledge of the range of properties of various materials groups while counting on this foundation to make the reader competent to develop the depth of knowledge when dealing with specific job requirements.

SPECIAL FEATURES

An application of modular instruction units such as that used by EMMSE seeks to foster a degree of self-instruction. The self-contained modules are divided into four major segments: Pause and Ponder, instructional concepts, Applications and Alternatives, and Self-Assessment. Pause and Ponder presents case histories of familiar products that prompt the reader to contemplate the familiar, while also posing some questions that establish a rationale for concepts presented in the module and the opportunity for problem solving. Next, instructional concepts are presented with simplified figures, illustrations of applications, and sufficient tables and diagrams to develop proficiency in reading these tools of the trade. Applications and Alternatives are used to present unique methods of how industry uses the principles covered in the module. Self-Assessment at the end of each module provides questions and problems so the reader gains practice and deals with points for class interaction. This section employs more of a problem-solving approach than regurgitation of facts.

Our colleagues using this text may find coverage of certain topics beyond that which time allows. The numbering used on each section provides an easy means to assign reading to students. Your program objectives may exclude topics, but as material users, your graduates will have access to important topics to refer back to in this book when on their jobs.

The use of both Systeme International d'Unites (SI) and U.S. customary units in the text recognizes that the dual system of measurement seems a reality for the foreseeable future. Appendices include measurement conversion factors, standard symbols and abbreviations, and selected tables of properties.

To ensure consistency with industrial practice, we maintain close relationship with contacts in industry and government. They generously provided illustrations, data, and examples of current and projected applications. Each module received evaluation from at least one specialist on the topic covered.

The *Instructor's Manual* for *Engineering Materials Technology* provides a variety of aids including answers to the Self-Assessment sections at the end of each module. The *Manual* offers sections on supplementary activities with sources of instructional aids, activities for students, laboratory manuals, free periodicals, and ASTM standards. Faculty may wish to order copies of the *Manual* for bookstore sale so students can check their solutions to Self-Assessment questions.

The authors would like to obtain instructor feedback and recommendations for subsequent editions. Please send this information to the authors in care of Prentice-Hall.

ACKNOWLEDGMENTS

The authors wish to acknowledge valuable assistance from the following evaluators: Samuel J. Scott, Structural Engineer for NASA; Louis C. Nenninger, III, Ceramic Engineer for Owens-Illinois; Dr. George G. Marra, Deputy Director, U.S. Forest Products Laboratory, Madison, Wisconsin; Charles V. White, Associate Professor, General Motors Institute; Cindy Neilson, Ceramic Engineer, IBM; Walter E. Thomas, Western Carolina University, Frank J. Rubino, Middlesex County College; Howard Hull, New York City Technical College; Charles Flanders, Texas A & M University; Mario J. Restive, Mohawk Valley Community College; Richard I. Phillips, Southwest Missouri State University.

The following organizations and individuals provided valuable assistance in many forms, including manuscript review, technical advice, illustrations, and technical data: Prince Manufacturing Co.; International Business Machines; Bell Laboratories; National Aeronautics and Space Administration; Bethlehem Steel Corporation; Mobay Chemical Corp.; Dr. Wilfred A. Côte, College of Environmental Science and Forestry, State University of New York; U.S. Forest Products Laboratory, Madison, Wisconsin; Bureau of Mines, U.S. Department of the Interior; Howmedica, Inc.; Ford Motor Co.; Morrison Molded Fiber Glass Co.; General Electric; T. L. Brown, *Chemistry,* Prentice-Hall, Inc., 1981; D. C. Giancoli, *Physics,* Prentice-Hall, Inc., 1980; International Nickel Co.; American Society for Metals; Howmet Turbine Components Corp.; Portland Cement Assoc.; Tinius Olsen Testing Machine Co.; John Deere Co.; American Ceramics Society; Wilson Instrument Division of ACCO; Chrysler Corp.; Corning Glassworks; American Can Co.; General Motors; United States Steel Corp.; Buehler Ltd.; American Iron and Steel Institute; C. O. Smith, *The Science of Engineering Materials,* Prentice-Hall, Inc., 1977; Republic Steel Corp.; Reynolds Metal Co.; Hoeganaes Corp.; Dr. Phillip H. Geil, University of Illinois; Monsanto Plastics and Resins Co.; Shell Chemical Co.; E. I. Du Pont de Nemours; PPG Industries; Measurement Group, Inc.; Firestone Tire and Rubber Co.; B. F. Goodrich; American Plywood Association; Dr. Alan A. Marra, University of Massachusetts; Champion Spark Plug; Hercules, Inc.; Permali, Inc.; Solar Energy Research Institute, U.S. Department of Energy; Sargent-Welch; Chemical Rubber Co.; Cummins Engine Co.

1

THE STUDY OF ENGINEERING MATERIALS TECHNOLOGY

1.1 PAUSE AND PONDER

Advances in technology result from the constant search for improvements. Since prehistoric times the desire to make a task easier or do a better job has given birth to innovations. Very often a designer in one field may observe a new development outside of his or her field and then apply the principle of that development for an innovation in his or her own field.

One such innovator who has made several significant breakthroughs in athletic equipment is Howard Head, who made revolutionary changes in snow skis and tennis racquets. Head turned the art of tennis racquet design into a science when he applied the principles of mechanics to produce a new shaped racquet. Figure 1-1a shows the comparison of the conventional racquet to the new Prince racquet that Head used to obtain a U.S. patent. Through careful analysis, he theorized that an enlarged area for the racquet head would increase the "sweet spot" or "center-of-percussion" which is the spot on a tennis racquet, ball bat, or golf club that yields the cleanest hit of the ball. The principle applied in the design increased the polar moment of inertia, which means the increased width of the racquet distributed some of the racquet's weight farther out from the center. The result is less twisting of the racquet when a ball is hit slightly off center.

The next step in this innovation was to follow a recent trend in sports equipment that involved using newly developed engineering materials and materials processing. A number of racquets were made using the new design. Each racquet employed different materials to produce its own distinct feel. Figure 1-1b shows the variety of materials and their different

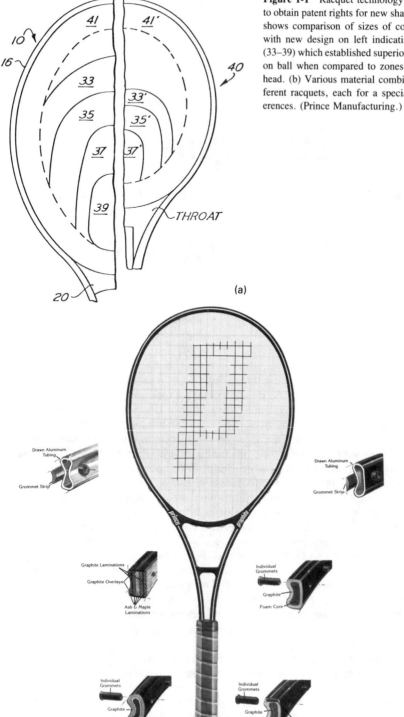

Figure 1-1 Racquet technology. (a) Patent drawing used to obtain patent rights for new shaped Prince tennis racquet; shows comparison of sizes of conventional head on right with new design on left indicating "zones of restitution" (33–39) which established superior ability for placing power on ball when compared to zones (33'–37') on older style head. (b) Various material combinations produced six different racquets, each for a special "feel" for player preferences. (Prince Manufacturing.)

cross sections. The I beam extrusion is a 7046 aluminum alloy developed for automotive bumper supports and has good strength and flexibility, plus high stress corrosion resistance (to be explained later in the book). The tubular extrusion of the same 7046 alloy is made by first hot extruding a tube and then cold drawing it to increase strength while producing a thinner, lighter tubing for a stiffer racquet than the I beam construction. Other racquets were composites since they use several different materials. The graphite composite provides very high stiffness which is achieved by using epoxy-impregnated graphite fibers wrapped around a mixture of foam plastic and cork that undergo several processing steps to achieve a final product. The laminated wood racquet also employs graphite fibers with layers of ash and maple that provide the player with the new racquet design, yet still keep the "feel of wood." Another composite employs 75% graphite and 25% fiber glass as reinforcing fibers for a lower-priced stiff racquet. On the other end of the composite cost spectrum, their most exotic racquet uses thin bands of boron fibers to reduce weight and increase strength, but it also carries a very high price. The handles for all six racquets are made of polyurethane plastic foam. Foaming plastic provides air bubbles that absorb shock and reduce vibration.

Some of the terms used to describe these innovations in tennis racquets may be new to you, but as you study this book the concepts will become clear. As you read about engineering materials, consider how these new tennis racquets evolved. This module mentions other developments related to materials technology. Throughout this module, the book, and in your daily life, be observant for uses of materials. Ask yourself why a specific material was used and see if you can determine how another material may be better suited for certain products. When you buy a product, study the labels to see what materials go into the product. Make decisions on purchases based on the suitability of the materials for your intended use. The concepts, examples, tables, and information in this book will help you to make good choices of materials, both as a private consumer and in your career. Your knowledge will have many applications and you will be prepared with alternatives for material selection.

1.2 WHY STUDY ENGINEERING MATERIALS TECHNOLOGY?

Engineering Materials technology is the term used to cover industrial fields dealing with the science and engineering of materials, materials testing and materials processing. This study encompasses many engineering and scientific specialties involving research, development, design, production, and maintenance of materials, products, and systems. The underlying purpose of this book is to present a broad and simplified understanding of the nature and properties of engineering materials while stressing the linkage between the structure of materials and their properties. By presenting information on the general characteristics of the major members of the family of industrial materials, it is hoped that you will retain a working knowledge of the properties of the major materials systems as an aid in carrying out your different roles in our technological world.

The ideal engineering material in the eyes of most people engaged in the technology of materials would have the following characteristics:

1. Endless source of supply
2. Cheap to refine and produce
3. Energy efficient
4. Strong, stiff, and dimensionally stable at all temperatures
5. Lightweight
6. Corrosion resistant
7. No harmful effects on the environment or people
8. Biodegradable
9. Numerous secondary uses

We know that such a material, particularly a single solid material, does not exist. In fact, many applications require varying degrees of one property, such as stiffness, in order to carry out the designed role.

The history of civilization reveals the constant need to design engineering materials that contain a select few of the preceding characteristics to fit a particular application. The quest for new materials that exhibit desired properties can be seen in the abundance of metallic alloys (around 70,000), nearly infinite number of plastics and rubbers, and thousands of ceramics used to help satisfy our present-day technological needs. The construction of mud and straw huts by ancient tribes could be the first signs of humanity's use of composite materials, which today are finding more applications in aircraft, automobile, sports and recreation industries for their great strength-to-weight ratios. In fact, this constant research and development effort in materials has produced a multitude of different materials that can be used for the same purpose. This has created competition among different materials industries in the marketplace.

Concurrent with this ongoing quest is the fact that the source of supply of most materials on our earth is limited. The *Global 2000 Report to the President* of the United States projects the limit to our mineral resources in the third millennium. This only recently recognized fact adds emphasis to the importance of designing secondary uses for materials and/or their ultimate recycling. Such decisions take place during the design process at that point when the material is first selected.

Module 1 provides a broad overview of materials technology and explains the value of studying materials in terms of their relation to communication, transportation, and production systems. It provides a rationale for why people in all walks of life should gain a knowledge of industrial materials so that they can deal with these many and varied materials in terms of the environment, energy, and society and also obtain the technological literacy necessary for today's consumer and intelligent citizen. The module also provides examples of recent and ongoing material development that exemplify the never ending quest for new materials in new products. Finally, basic materials concepts are set forth that provide a fundamental understanding for the study of subsequent modules.

1.2.1 Effects of Materials on Humanity

Whatever people build, they use materials. These construction materials are many and varied since each possesses different types of properties to fit a particular application. Running the gamut from dried mud and straw for huts, flint for tools and arrowheads, wood for ships, and petroleum, coal, and uranium for energy to composite materials for space vehicles, materials have been and will continue to be used by humanity to meet its needs for construction, energy, and weapons. Regardless of the age, humans have been directly involved with materials. They have shaped cultures and civilizations. Those peoples who possess an abundance of resources combined with an ability to use them have prospered. To the present day, pressing social issues concern the rapidity with which many of these sources of materials are being depleted, as well as the accessibility of materials sources in times of political upheaval and war.

1.2.2 Our Technological Society

Almost everything people have ever done has involved materials. Materials are directly tied to our very existence. The major epochs of our history have been labeled after materials. The Stone Age, Bronze Age, and Age of Iron all convey the intimate relationship of people to materials. It is interesting to ponder the "Age" that future historians may label our short period of years from the end of the twentieth century into the early twenty-first century. Perhaps the "Age of the Semiconductor" or the "Age of Composite Materials" or, better still, the "Age of Materials." Suffice it to say that materials are a basic resource of humanity. Not only do they play a crucial role in our way of life but also in the well-being and security of the nations of the world. The third millennium promises many changes in materials and materials usage.

1.2.3 Materials in Communications, Transportation, and Production Systems

It is possible to group much of the technological support of our society into three systems: communications, transportation, or production. Each system relies heavily on industrial materials. Throughout their careers, scientists, craftsmen, technologists, engineers, technicians, and other workers in the three systems deal with materials decisions. Often these decisions relate to the selection of materials to meet specific needs. The interrelation of materials and the methods of processing the materials always exist; so when a designer selects a material, a consideration for processing it must accompany the selection. The following examples of materials developments illustrate their integral existence in our technological systems.

Both the communication and transportation industries have made significant gains due to the rapid developments in miniaturized circuitry. Through ingenious selection of materials, clever design of circuitry, and highly sophisticated and clean manufacturing processes, the composites of ceramic and metallic memory chips have not only reduced size and weight but also lowered cost and energy demands. The

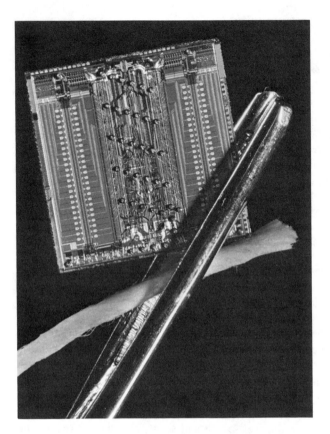

Figure 1-2 Memory chip. Through advances in materials technology, computer technology has made great strides to serve humanity. This miniature chip is small enough to slip through the eye of a needle yet so dense it can store more information than computers that filled whole rooms. As a result of refining photographic techniques (ultraviolet lithography) and other processes, silicon chips advanced from having about 100 elements in 1970 to over 100,000 in the early 1980s. Smaller and denser chips not only reduce overall size of equipment but also permit more rapid processing of data. (IBM.)

chips, often small enough to slip through the eye of a needle (Figure 1-2), allow such devices as computers, calculators, radios, televisions, and watches to offer greater capabilities with minute electrical energy needed to power them. Employed as microprocessors, these chips provide automotive designers with new flexibility in the control of the engine and accessories in a car.

The field of telephone and telegraph communications is undergoing a revolutionary change brought about by breakthroughs in materials science, specifically in fiberoptic technology. One such development was the production of high-purity glass, which is then processed into optical fibers with diameters as small as human hair (Figure 1-3). It has been recognized since the late 1800s that light could be used to transmit information. In 1966, glass was proposed as the medium to carry light, but it took some ten years of effort to produce glass that was sufficiently pure (free from impurities and defects) to allow light to travel through the glass fiber for a distance of over one kilometer without suffering any significant loss. Adaptions of this new technology by the television and computer industries promise great changes to our society.

The transportation industry has long depended on aluminum alloys with their favorable strength-to-weight ratios and good corrosion resistance. Although these alloys cost more per pound than steel, their weight per volume makes them cheaper;

Figure 1-3 Optical fibers. Loops of a hair-thin fiber, illuminated by laser light, represent the transmission medium lightwave systems. Typically, twelve fibers are embedded between two strips of plastic in a flat ribbon, and as many as twelve ribbons are stacked in a cable that can carry more than 40,000 voice channels. (Bell Laboratories.)

however, the problems created by heavy dependence on petroleum as an energy source have created the need for materials that are also energy efficient. In addition to recycling of some existing materials, the never-ending quest for strong, lightweight, dimensionally stable, corrosion-resistant materials that will retain their properties throughout an expanded range of temperatures under severe service conditions has been singled out for greater emphasis. One avenue of approach is to continue the development of composite materials, which to date have made great inroads in aerospace and aircraft industries and the sports and recreation industries. Figure 1-4 is based on a presentation NASA engineers used to depict the nature and advantages of composites in commercial aircraft.

The high cost of many of the new materials, such as graphite-fiber-reinforced composites, results from the production systems required to generate the materials and to fabricate parts, products, and systems that use these materials. Because production systems developed over the centuries around the dominant materials of wood and metal, with plastics joining the other two materials this century, industry produces its goods with certain processes dictated by the nature of wood, metal, and plastics. For example, the large foundries and mills used to make steel evolved over two centuries and represent high investments of money and technological development. To begin from scratch and put together the steel production system would be prohibitive. In fact, the United States has lost its world's leadership in iron and steel production to Japan and the Soviet Union because it has not adequately invested in modernization of its foundries and mills. In addition to the processes needed to produce steel, many others follow for shaping and fabricating, including welding, casting, and machining. Each involves special equipment whose development followed an evolution similar to foundries and steel mills. The rapid development of plastic production technology resulted from several factors, including a plentiful supply of

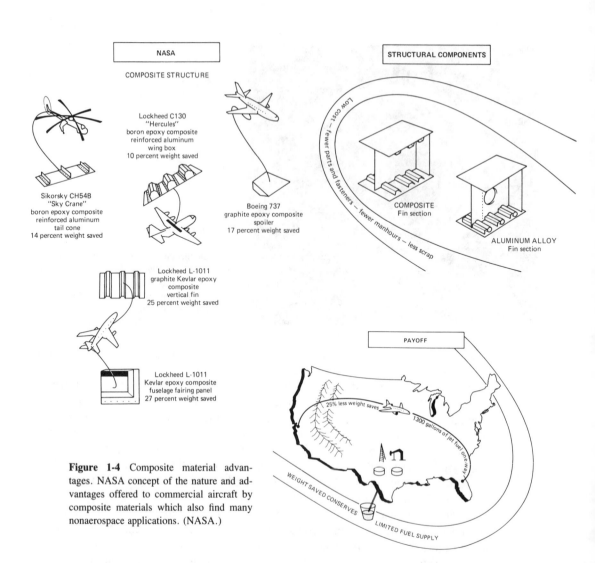

NASA

COMPOSITE STRUCTURE

STRUCTURAL COMPONENTS

Sikorsky CH54B
"Sky Crane"
boron epoxy composite
reinforced aluminum
tail cone
14 percent weight saved

Lockheed C130
"Hercules"
boron epoxy composite
reinforced aluminum
wing box
10 percent weight saved

Boeing 737
graphite epoxy composite
spoiler
17 percent weight saved

Lockheed L-1011
graphite Kevlar epoxy
composite
vertical fin
25 percent weight saved

Lockheed L-1011
Kevlar epoxy composite
fuselage fairing panel
27 percent weight saved

Low cost — fewer parts and fasteners — fewer manhours — less scrap

COMPOSITE
Fin section

ALUMINUM ALLOY
Fin section

PAYOFF

25% less weight saves

1300 gallons of jet fuel one way

WEIGHT SAVED CONSERVES

LIMITED FUEL SUPPLY

Figure 1-4 Composite material advantages. NASA concept of the nature and advantages offered to commercial aircraft by composite materials which also find many nonaerospace applications. (NASA.)

the raw material (oil) and the fact that production methods for plastics employ many of the processes used on wood and metal.

On the other hand, production systems for certain space-age materials are radically different from those for conventional materials. They often require considerable energy to produce, as in the case of magnesium and virgin aluminum. Some require new and elaborate production technology, such as for graphite fibers. To justify the replacement of steel, iron, or wood, the new materials must have tremendous advantages so that they become competitive for reasons other than production costs. The need for strong, lightweight materials to meet the demands for fuel-efficient ground, air, and sea vehicles provides such justification. As the auto industry adopts more graphite-fiber-reinforced composites, the increase in production resulting from the demand from this large industry should drive down the cost. Advances in the

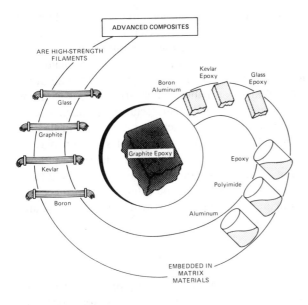

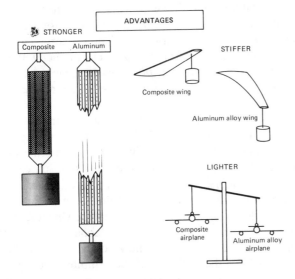

Figure 1-4 (continued).

production of materials used with electronic circuitry made rapid strides because of the tremendous advantages provided through miniaturization. The research and development phases of these electronic materials were expensive, involving high technology, but the new production systems are relatively small, very clean, and highly automated. Coupling these advantages with the large demand resulted in lower cost for computers, microprocessors, and control devices that use the materials of these new production systems. Low-cost, highly sophisticated calculators, watches, and automotive controls typify these developments. The strong surge toward automated manufacturing and robotics focuses on "design for manufacturing" which favorably views those materials processed, like chemicals rather than traditional processes. In other words, plastics, ceramics, and composites are often easier to automate than metals that may require stamping, machining, and mechanical fastening. The thrust toward "design for automation" and "design for manufacturability" focuses attention on better integration of design, development, and manufacturing through use of computer-aided design (CAD), computer-aided manufacturing (CAM), and flexible manufacturing systems (FMS). Throughout each phase, material selection plays a major role. Some goals sought are to eliminate threaded fasteners, keep the number of components to a minimum, allow continuous processing, reduce materials handling, and facilitate use of robots for materials handling and assembly. If you become involved in manufacturing for automation, many questions will arise on the nature, properties, processes, and applications of materials.

1.2.4 Energy and Society

The concept of energy was first formulated using mechanical problems. In so doing, the term *work* was used to describe the production of motion against some opposing force. *Energy* is defined, then, as the capacity to do work. We know that energy encompasses more than simple mechanical situations, such as a lifted weight or a compressed spring. First, we know that all matter has energy. Second, energy takes many additional forms: chemical energy, heat (thermal) energy, electromagnetic energy, biological energy, atomic energy, and nuclear energy. Third, energy can be converted from one form to another and transferred from one object to another without suffering loss. This last statement expresses the *principle of energy conservation*. Another term used in any discussion of energy is power. *Power* is defined as the time rate use of energy.

Just as humanity competed for food and materials throughout history, energy was also an important ingredient to life. Prior to A.D. 1200 the principal sources of energy were solar energy, wood, wind, and water. Shortly after this time the discoveries of coal, natural gas, and petroleum proved to be more promising sources of power. In our age the development of nuclear power was added to this category. Today in the United States, power consumption is about 100 times the average of the underdeveloped nations of the world. What is remarkable is that the United States represents some 6% of the world's population, yet it consumes about one-third of the world's energy. Consequently, the need is great for new policy decisions that will directly reflect the need to slow down the depletion of these limited sources of the world's power by conserving energy wherever possible, developing new energy sources, shifting energy consumption to sources other than petroleum, recyling of materials wherever possible, and substituting energy-efficient materials for those requiring much energy in their production. Uppermost is the necessity to control the harmful side effects that accompany the generation of huge amounts of energy. Thus a balance must be struck between having sufficient energy to keep the economy sound with continual increases in the standard of living, particularly of the underdeveloped nations, and protection against serious health hazards to present and future generations of people.

The success in our battle to reduce waste of earth's known materials and energy supplies will be contingent in part on our materials scientists finding newer materials with which we can use newer methods of manufacture, thus increasing productivity. If we could step into the future for a few moments, we might see the use of new materials applications in the following areas. Energy-expensive metals might be used more efficiently if less expensive "filler" materials were added. Can this be done? We do this with polymeric materials now. Electrical transformers require over millions of tons of soft magnetic iron alloy. The eddy current and particularly the magnetic losses in the core continue to reduce transformer efficiency. New improved steel alloys are needed through better control of the chemistry of steels and the processing of the transformer cores to bring greater results. Transmission lines of copper and aluminum for transmitting electrical power suffer excessive joule heat loss. Fine filaments of alloys such as tin, vanadium, and niobium operating at low temperatures are being investigated to produce superconducting cables.

The key to successful development of the aircraft turbine engine has been the materials development of a nickel and titanium alloy steel, which resists fracture in the corrosive environment of the hot gas plasma. Nickel, being expensive, has not been considered for low-cost turbine automobile engines. Instead, ceramics materials (compounds containing metallic and nonmetallic compounds, such as glass, spark plug insulators, porcelain, cement) are being developed for use in turbine engines operating at 2500°F with fewer polluting emissions than in the present internal combustion engine. In addition to a concern for strength-to-weight ratios, in selection of the most energy-efficient material one must also consider the energy required to produce the material itself.

These are a few examples of the many advances that seem likely to generate tomorrow's structures, appliances, and products. They result from the control of the chemistry of alloys and polymers to produce precise physical arrangements of crystals and molecules, which in turn exhibit the desired behavior. With this knowledge, technicians, technologists, and engineers of the future can make maximum use of the improved characteristics of new materials to give us new energy-efficient products.

1.3 TECHNOLOGICAL LITERACY

Our society continues to become more complex. To avoid what Alvin Tofler labels "future shock," citizens should stay abreast of technological developments; and one must be technologically literate, that is, understand the language and concepts of technology, in order to understand new technological advances.

1.3.1 The Materials Consumer

Each year an increasing number of newly developed materials is substituted for more familiar materials with limited properties. Forty percent of a manufactured item's price on today's markets represents materials cost. To be an informed and intelligent consumer requires a basic understanding of materials. The selection of a product can be improved by a greater understanding of the nature and properties of the materials used in the product. After a product has been purchased, the problem of failure, sometimes by the *abuse* or *misuse* of the product, can be lessened by such knowledge. Learning about the structure of materials, hence how materials behave, should permit an intelligent analysis of a failure, and possibly pinpoint its source and cause. A knowledgeable consumer stands a much better chance of success in demanding remedial action by both manufacturers and retailers of faulty products than one with less knowledge of the behavior of materials and a poor technical vocabulary needed to explain such behavior.

Wood is our oldest building material and possesses unique structure, warmth, and beauty; but perhaps more importantly, the ability to renew this raw material makes it of equal importance to the newer materials (Figure 1-5). In 1850 about 90% of U.S. energy came from wood, both in the raw state or refined into charcoal for fuel in glass and iron making. By 1900, wood accounted for only 20% of the nation's fuel and continued to decline. Wood usage as a fuel is now on the rise in both home and industry; those industries, such as lumber and paper, make extensive use of wood residues, including waste chemical by-products and bark. Wood does not figure as a major fuel for meeting the void left by depleted fossil fuels, but it holds promise as a substitute for more energy-intensive materials. In construction materials for similar phases of building, the comparison of fuel required for the preparation of building products shows wood requires the least energy, plastic nearly six times, and steel about eight times the amount of fuel needed for wood products. Concrete, glass, and other ceramics also show great promise as future materials for building and manufacture because the raw materials are as vast as the sands of the deserts and seas.

1.3.2 The Intelligent Citizen

The technology of materials also provides the individual with the necessary knowledge to make decisions based on personal values relating to political, social, and ecological issues. *A better informed citizen is a better citizen,* who is much needed in today's changing technological society in which the great issues over energy and materials resources are being debated and voted on. As an example, the present technology of jet engine design mandates the use of the metal chromium. An artificial shortage of chromium developed in the 1970s as a result of the politically inspired Rhodesian chromium embargo. Such a shortage produced by domestic and/or foreign political action tends to increase prices. Coming at times when the metals industry is under increasing economic pressure, such shortages could threaten shortages in the many grades of stainless steels on which we as a nation are most dependent.

Long-range industrial research and development of new alloys and nonmetals, intensification of programs to conserve and reclaim metals, enlargement of the search for and development of new domestic sources, and the utilization of ocean resources are all affected by issues that find a source in the political, social, economic, or ecological spheres within our society. Knowledge of the fact that the United States is dependent on foreign sources for most metals with the exception of iron and copper,

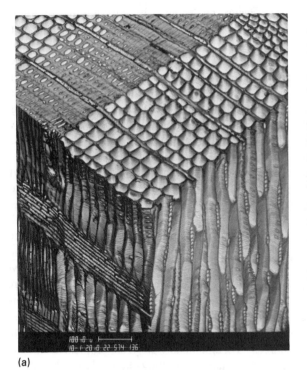

(a)

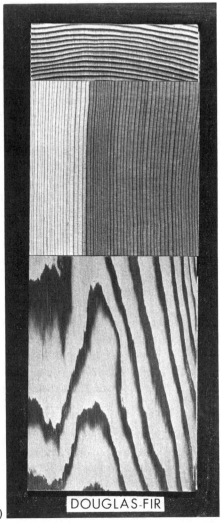

(b)

Figure 1-5 Wood's complex structure. (a) Electron microscope reveals microstructure. (Dr. Wilfred A. Côté, SUNY College of Environmental Science and Forestry.) (b)Macrostructure of wood seen in many different grain patterns offers warmth and beauty. (U.S. Forest Products Laboratory, Madison, Wisconsin.)

DOUGLAS-FIR

plus the negative effects of our excessive consumption rate of the world's known reserves, places greater emphasis on a good working knowledge of the technology of materials so that citizens can guide their governmental representatives in making the correct decisions on matters that will have a lasting effect on our lives and standards of living for years.

The Congressional Office of Technology Assessment developed 11 options for reducing waste in metals. Among these were (1) major redistribution of metals, as done during World War II, (2) metal substitution, (3) use of nonmetallic coatings, (4) recycling of scrap metal, and (5) product remanufacturing and reuse. Some of these, if not all, will require a major effort in the research and development fields. The report points out that there is waste not only on the consumer side but in the manufacturing of the product, where excess metals are used in the manufacture of the products. Also, engineers in the design of products must take a good look at secondary uses of their products and devices once wear and/or obsolescence mandates a change in the primary use of a product.

1.3.3 The Engineering Materials Technology Team

The complexities of our modern technological society characterized by ongoing explosions of knowledge and the rapid transmittal of such information to the distant corners of the world have contributed to the need for occupational specialization. The spectrum of technological activities in early times was sufficiently small to enable one person to possess the necessary knowledge and skills to function as a craftsperson and scientist. Today the depth of knowledge gained in an academic setting and the degree of skill needed to function as a craftsperson do not permit one person with the composite interests and aptitudes to function in more than one role. The obsolescence of both knowledge and jobs mandates a lifetime of learning. Consequently, specialization is the rule. Today's new or improved devices, designs, and systems are produced by a collective effort of a team known as the technology team or engineering team. This team consists of craftspeople, technicians, associate engineers, engineers, and scientists.

While certain of these members specialize in areas such as nondestructive testers, materials scientists, or metallurgical technicians, all members of the technology team deal with materials in varying degrees. To understand the technological team, one can define each member in general terms and then study their relationship to the major activities in engineering, such as research and development, design, manufacture, and maintenance. There are no clear-cut descriptions for the members of the technological team and jobs held in industry by them.

With our dependence on materials, it is not surprising that sooner or later a science of materials developed, along with its attendant engineering to stand beside other long-established disciplines. Thus *materials science and engineering,* as defined by a National Academy of Science study in 1971–1973, is the generation and application of knowledge relating the composition, structure, and processing of materials to other properties and uses. The science focuses on an understanding of the nature of materials, which in turn leads to theories or descriptions that explain how structure relates to composition, properties, and behavior. *Materials engineering,* on the other hand, deals with the synthesis and use of this knowledge in order to develop, prepare, modify, and apply materials to meet specified needs. The distinction between the two domains is one of viewpoint or emphasis as both domains merge and overlap one another. This helps explain why this new discipline is often called *engineering science* because of its rather applied nature. As an interdisciplinary or multidisciplinary science, materials science embraces some disciplines (e.g., metallurgy and ceramics) and some subdisciplines (e.g., solid-state physics and polymer chemistry), and also overlaps several engineering disciplines from mechanical to aerospace.

Engineering materials technology covers fields of applied science related to the science of materials, materials processing, and the many engineering specialties dealing with materials, such as research, development, design, manufacture, construction, and maintenance.

To provide a definition of materials, we can say they are a part of the matter of the universe. A more useful definition is that materials are substances whose properties make them useful in structures, machines, devices, or products. Basically, the word properties is used to describe the behavior of materials when subjected to some external force or condition. For example, the tensile strength of a metallic alloy is a measure of the material's resistance to an externally applied load or force. The family of materials includes such materials as metals, polymers (plastics), ceramics, semiconductors, wood, sand, stone, and composite materials, such as fiber glass, which are man-made materials composed of fibers embedded in a polymer resin matrix.

1.4 MATERIALS SYSTEMS

A modern vacuum cleaner is an example of a relatively complex product that requires a multitude of different materials for its manufacture and successful use by a consumer. To cite a few examples, there are metals such as in the motor housing, various plastics for handles, gears, hose, and the cleaner shell, plus rubber wheels and electrical plug. Such materials make up a materials system. Each separate material component must be compatible with every other component, while at the same time contributing its distinct properties to the overall characteristics of the system of which it is a part. Material components are selected on the basis of their properties, appearance, and cost. Another example is the materials system for an automobile or truck tire. Steel filaments in belted or radial tires must not separate from their surrounding rubber matrix material as they carry the tensile stresses in the tire under varying service conditions. Other types of reinforcing fillers such as glass or ceramic filaments in composite materials must be compatible with their companion matrix materials if these man-made materials are to function as designed.

1.4.1 Product Liability

Products, including equipment, machinery, and materials, are designed to be safe for the operator and the people who are in close proximity. In the United States over 160 safety standards are in use to cover a wide range of devices, products, and materials. The Occupational Safety and Health Act (OSHA) of 1970 sets forth the general conditions of safety that must be met by those involved in the manufacture or use of goods and services in this country.

Product liability is civil (as opposed to criminal) liability to an ultimate user for injury resulting from using a defective product. *Caveat emptor* (buyer beware) was the rule in the day when the manufacturer (the seller) and the buyer lived in the same community. Today the assembly of a product made of parts manufactured anywhere in the world and possibly sold by someone other than the manufacturer who completes the final assembly has made the buyer-beware rule obsolete. Liability law is taking its place, which is being redefined in court cases and through legislative action.

For those involved in materials, particularly the selection of materials in the design process, the present trend by the courts to identify members of a design team as being responsible for some fault in a product liability action mandates that all possess and use the latest and best information about the materials used, particularly their long-term characteristics.

APPLICATIONS AND ALTERNATIVES

Materials systems in bioengineering. Dramatic developments occur weekly in medicine due to advances of technology. The field of bioengineering combines biology and medicine with many specialities in engineering and materials science to foster technological developments to aid humans and animals. Bioengineering produces artificial organs, limbs, and related anatomic structures. The materials aspect of these devices becomes a critical concern because of the need for physiological compatibility between the materials and the human or animal systems. The materials system described next introduces you to topics that you will study in the following module.

Many examples exist of materials joined into systems to utilize complementary properties of the various materials used. Prosthetic devices used to replace defective joints in knees and hips provide an excellent case of such a materials system. These artificial joint implants are often necessary in patients suffering from accidents or disease. They consist of metal alloys, plastics, composites, and adhesives. Figure 1-6a shows a PCA™ (porous-coated anatomic) knee replacement made of a cobalt alloy, sintered cobalt alloy spheres, ultrahigh molecular weight polyethylene (UHMWPE),* and a copolymer adhesive of polymethylmethacrylate and styrene (PMMA-PS). Figure 1-6b shows a view from the front of the human leg with the PCA™ in place and a lag screw rejoining the fibular bone that was cut as a part of the operation. Figure 1-6c shows an x-ray taken 6 months after the operation with the PCA™ in place.

Such orthopedic surgical operations, known as total joint orthoplasty, date back to 1940 when Vitallium™ first served as the metal alloy in hip replacement devices. The alloy Vitallium™ consists of 65% cobalt with chromium and molybdenum and provides a hard, tough, noncorroding metal. The UHMWPE that bonds to the lower half of the PCA™ provides a self-lubricating bearing surface on which the highly polished Vitallium™ part hinges. You can picture the type of environment that the PMMA-PS adhesive must endure because of body fluids present during implantation, thus bearing witness to the value of this adhesive to achieve a good bond,

*American Society for Testing and Materials (ASTM) and American National Standards Institute (ANSI) abbreviations, acronyms, and symbols used throughout this book.

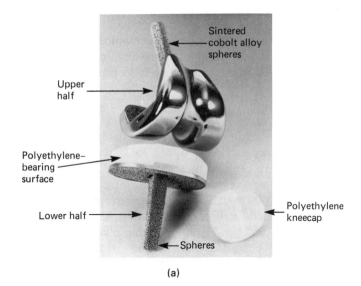

(a)

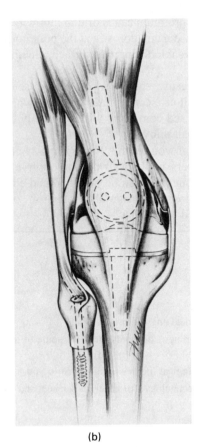

(b)

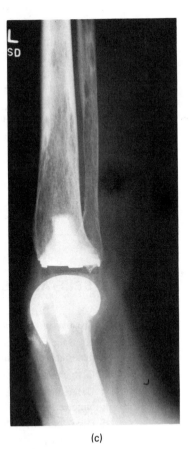

(c)

Figure 1-6 Human knee joint replacement. (a) P.C.A.™ total knee joint. (b) Frontal view drawing of total knee joint in human leg. (c) Postoperative 6-month follow-up X-ray; note polyethylene bearing surface and kneecap do not show up on X-ray). (Howmedica, Inc.)

which most other adhesives could not accomplish. This copolymer adhesive cures rapidly while providing a free radical reaction (the mixing of ingredients do not require close control for a proper bond). Since the operation requires cutting open the patient, the fast setting nature of PMMA-PS is required to speed along the operation. Unfortunately, PMMA cures into a brittle structure prone to fatigue fracture. The adhesive often deteriorates by loosening after years of service. The addition of a composite surface made of chrome-cobalt sintered beads (see Figure 1-6a) provides a porous surface designed to increase the surface area contacting the adhesive and improves the fatigue life of the PMMA. Tests revealed that cemented composite surfaces increase the ultimate tensile strength to 1,733,000 pounds per square inch (psi), compared to cemented surfaces of waffled Vitallium^{TM} at 670,000 psi and cemented surfaces of UHMWPE at 175,000 psi. Dynamic testing revealed an improved fatigue life of the porous coating 2.7 times longer than the waffled metal surface with a tension–compression load at 625 pounds.

Because of a concern for long-term failure with the PMMA, some orthopedic surgeons implant the total joint prosthesis without an adhesive. Instead, they use either cobalt-chromium beads or titanium wire mesh that joins to the metal parts to increase metal porosity at the bone-to-prosthesis interface. Bone tissue then grows into the porous surface, creating a stable interface.

Problems still remain because some people exhibit sensitivity to cobalt, plus studies reveal cobalt as a cancer-causing agent when injected into rats. Perhaps a polymer-based composite will eventually become an alternative material for the cobalt alloy.

What materials systems can you recall? Watch for uses of materials systems as you read newspapers and magazines or observe products, because often an idea used in one field can be modified and employed in another. Can you see applications for materials systems in other fields that use the principles discussed here?

1.5 SELF-ASSESSMENT

1-1 List five of the nine characteristics of the ideal engineering material.

1-2 From the examples in this module and from news accounts, describe some of the effects of materials on humanity.

1-3 Give explanations for three reasons technological literacy is important to you.

1-4 Define the terms (a) engineering materials technology, (b) materials science, and (c) materials engineering.

1-5 Give your own example of a materials system.

1-6 What does *caveat emptor* mean?

1-7 Who can be held liable in the event a person is harmed due to faulty design in a product used?

1-8 Cite examples of at least two materials each and their applications in systems of (a) transportation, (b) communications, and (c) production.

1-9 Name three periodicals that publish information on industrial materials, their processing, and applications.

1.6 REFERENCES AND RELATED READING

ATTENPOHL, D. G. *Materials in World Perspective.* New York: Springer-Verlag, 1980.

BEAKLEY, GEORGE C., and H. W. LEACH. *Careers in Engineering and Technology.* New York: Macmillian, 1981.

"Body Parts As Sculpture," *Discover*, June 1984, pp. 62–67.

BRADY, GEORGE S. *Materials Handbook.* New York: McGraw-Hill, 1977.

"The Bumpy Road to Submicron Lithography," *High Technology*, March 1983, pp. 26–29.

COOK, ALBERT M., and JOHN G. WEBSTER. *Therapeutic Medical Devices: Applications and Design.* Englewood Cliffs, N.J.: Prentice-Hall, 1982.

DIETER, GEORGE E. *Engineering Design—A Materials and Processes Approach,* New York: McGraw-Hill, 1983, from Materials Science and Engineering Series.

The Global 2000 Report to the President: Volume 1, *Entering the Twenty-First Century,* Volume II *Technical Report,* Volume III *The Governments Global Model.* Washington, D.C.: U.S. Government Printing Office, 1980.

GORDON, J. E. *The New Science of Strong Materials or Why You Don't Fall Through the Floor,* 2nd ed. New York: Penguin Books, 1978.

————. *Structures or Why Things Don't Fall Down.* New York: Penguin Books, 1978.

Materials, A Scientific American Book. San Francisco: W. H. Freeman, 1967.

"Materials in the News," *Machine Design*, Apr. 16, 1984, pp. 2–5.

RUSKINS, ARNOLD M. *Materials Considerations in Design.* Englewood Cliffs, N.J.: Prentice-Hall, 1967.

WERT, CHARLES. *Opportunities in Materials Science.* New York: Universal Publishing, 1973.

Periodicals

Discover	*Materials Engineering*
High Technology	*Plastic World*
Machine Design	*Popular Science*

2

NATURE AND FAMILY
OF MATERIALS

2.1 PAUSE AND PONDER

If you were given the responsibility of designing an automobile that had to be very fuel efficient, how would you begin? Of course, many factors must receive consideration, such as the body shape, power system, and total weight. What are some methods you would use to reduce weight? How could you improve the shape of the car to provide a smooth aerodynamic shape? What types of improvements can you conceive to power the car with a minimum of energy?

Automobile manufacturers approach fuel efficiency through various avenues many of which involve materials considerations. In an attempt to explore the advantages composites might offer the auto industry, Ford Motor Company invested $3.5 million to manufacture an experimental replica of a 1979 Ford LTD of almost entirely graphite-fiber-reinforced composites. The experimental car weighs 1250 pounds less than its conventional counterpart (Figure 2-1). From an average weight of 3760 pounds on 1977 models, Ford chopped off huge amounts of weight down to around 2500 pounds average weight for the 1985 models. Many material and design changes allowed the huge reduction aimed at meeting the federally mandated 27.5 miles-per-gallon average fleet fuel consumption for 1985.

Reinforced plastic composites are not new. The 1953 Chevrolet Corvette was the first production car with a glass-reinforced plastic body (see Figure 2-2). However, the advantages of glass-fiber composites were not sufficient at that time to persuade the auto industry to adopt this material

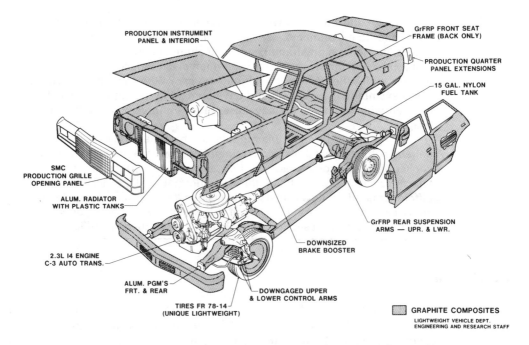

Figure 2-1   Ford lightweight concept vehicle. (Ford Motor Company.)

to replace sheet steel bodies. General Motors' use of fiber-glass SMC (sheet molding compound) body panels to replace sheet steel bodies on production cars marks another materials revolution in automaking and part of their effort to reduce vehicle weight by half by 1995. Fiber-reinforced plastic (FRP) offers advantages in the areas of total vehicle mass reduction, manufacturing energy saved, and total energy savings. The FRP cost equals that of steel, while being less than aluminum, but presently causes increases in the cost of labor, new equipment, and other factors resulting in the changeover from steel stamping to fiber-glass molding.

Figure 2-2   Composite car. 1953 Chevrolet Corvette was first fiber-glass production body. (Morrison Molded Fiber Glass Co.)

In one effort to lower weight with metals, Chrysler Corporation produced the XL experimental car that employed lightweight, high-strength metals, including high-strength steel and aluminum. Coupled with design and production innovations, the experimental car reduced weight by 630 pounds under their conventional mid-size car of the same group. These cases exemplify the research and development in the auto industry that, coupled with technological advances, bring improvements in materials usage.

As you read through this module and become acquainted with the general nature of materials, try to remember how each group of materials is used in a car; some materials from each of the groups find use in a typical car. In the early 1900s the first automobiles used less than 100 different materials; now cars use over 4000 material combinations. As you study each member of the family of materials, imagine how you would select materials for designing a car that would improve fuel efficiency.

2.2 NATURE

Through an understanding of the nature and structure of materials one can predict how materials should behave when exposed to certain forces and environments. The nature of materials results from their composition and structure, which determine their properties. This module presents basic concepts of atomic and molecular structure that lay a foundation for the discussion throughout the book on the nature, properties, and uses of materials. The ability to grasp these fundamental materials technology concepts not only allows you to deal with today's materials and processes, but also prepares you for new developments in materials.

2.2.1 Internal Structure

The structure of materials follows the structure of a building in that it develops through the joining of smaller units: particles of earth form bricks, bricks are stacked together to form walls, walls form rooms, and rooms make a building. Materials are forms of matter generally in the solid state. Subatomic particles form atoms, different atoms possess distinct characteristics and are categorized as elements. Molecules, which consist of atoms, combine into polymer chains and/or crystals to form solid materials. In their formation and processing, materials may exist in the four forms of matter: solid, liquid, gas, or plasma.

2.2.1.1 Atomic structure. An *atom* is the smallest particle of an element that possesses the physical and chemical properties of that element. It is the smallest particle of an element that can enter into a chemical change. Atoms are the basic building blocks of matter. The average diameter of an atom is only about 10^{-10} meter. It takes more than 10^6 atoms edge-to-edge to make the thickness of this page.

What the atom really looks like or how the charged particles distribute themselves and move within the atom is not exactly known. Consequently, scientists resort to various models of an atom to help explain its characteristics. One model, the

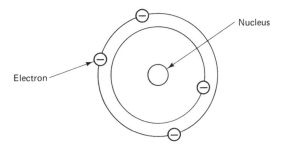

Figure 2-3 Planetary model of an atom.

planetary model, appears to be the most popular in explaining the nature of an atom (see Figure 2-3). However, this model is far from being the most accurate. The atom is pictured as a sphere with a very dense center, the nucleus, containing the protons and neutrons around which negatively charged particles (the electrons) are moving in circular or elliptical orbits much like the motion of the planets around the sun. The diameter of the nucleus is about 1/10,000 the size of the atom, but it contains more than 90% of the mass of the atom. If one represented the nucleus as a golf ball, the electrons would revolve in a sphere with a radius of about $1\frac{1}{2}$ miles. Another model of the atom, sometimes sketched in three dimensions, attempts to show the spaces within which the electrons are confined as fuzzy volumes.

For our purposes the nucleus consists of protons and neutrons. A *proton* is a particle of matter that carries a positive electrical charge equivalent to the negative charge on an electron. *Neutrons* are uncharged particles in the nucleus with a mass nearly equal to the proton's mass. The *atomic number* of an element is equal to the number of protons in the nucleus. The number of protons in the nucleus equals the number of electrons in the atom in the balanced, neutral, or equilibrium state. Elements in the periodic table (see Appendix 10.1) are in order according to their atomic numbers. The total number of protons and neutrons in the nucleus is the *mass number of the atom*. Protons and neutrons are referred to as *nucleons*.

All atoms of an element contain the same number of protons, but these atoms of the same element can contain different numbers of neutrons. In other words, these atoms are different, but their positive and negative charges are identical. Such atoms with different numbers of neutrons are *isotopes*. For example, hydrogen exists in three isotopic forms (see Figure 2-4).

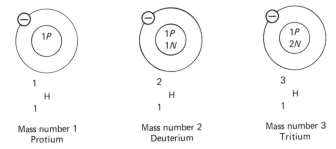

N.B. 99.985% of hydrogen is in the form of Protium.
Deuterium represents only 0.015%.

Figure 2-4 Isotopes of hydrogen.

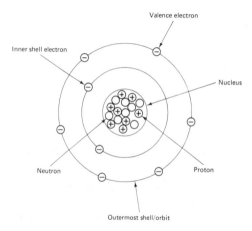

Figure 2-5 Subatomic parts of an atom. This atom is shown in an equilibrium, balanced, or electrically-neutral state.

In summary, an atom (see Figure 2-5) of an element is made up of a dense core, the *nucleus,* consisting of protons and neutrons surrounded by even smaller particles called electrons. The electrons carry a negative (−) electrical charge much like the charge at the negative (−) terminal of a battery. Those electrons that occupy the outermost ring or shell from the nucleus are *valence electrons.* If the atom is in a balanced or equilibrium state (see Figure 2-5), it possesses the same number of protons as electrons, and the electrical charge carried by the atom is zero. The negative (−) charge of the electrons is balanced by the positive (+) charge of the protons in the nucleus.

The process of pulling away (removing) or adding valence electrons from a balanced or neutral atom is called *ionization*. The atoms with unbalanced electrical charge are called *ions* (see Figures 2-6 and 2-7). The tendency of atoms to lose their valence electrons and the energy required to remove these electrons will be discussed in the treatment of the periodic table and its underlying concepts.

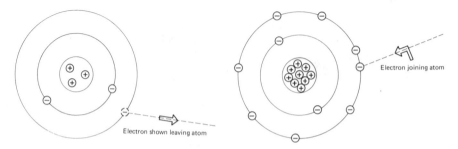

Figure 2-6 Positive ion or cation.

Figure 2-7 Negative ion or anion.

2.2.1.1.1 Electronic Structure. The energy levels in atoms have been given letter names starting with K, the first and lowest energy level (closest to the nucleus), and proceeding outward to Q the seventh level where the electrons have the highest energy. Each level can hold only a certain number of electrons at any one time. To determine the maximum number of electrons at any energy level, the relationship $2n^2$ can be used. The number of the energy level is substituted for n. For example, the third or M energy level can contain a maximum of (2×3^2) or 18 electrons. Figure 2-8a contains a simple sketch of all energy levels drawn concentric with the nucleus of an atom. Some energy levels are listed along with the maximum number of electrons permitted at each level in Figure 2-8b.

The manner in which the electrons of an atom distribute themselves in the ground or lowest energy state is the *electron configuration* of the atom. Learning the electron configuration of atoms permits us to predict what the atoms will do in the presence of other atoms. In other words, the number of electrons in an atom's outermost energy level (the valence electrons) is the key that controls the chemical properties of an element. As an example, hydrogen (H) with an atomic number of 1 in the ground state has its one electron in the K level. Carbon (C) has six electrons; two fill the K level and the remaining four are in the L level. Atoms that have a completed energy level are chemically inert; that is, their atoms do not react with other atoms. Neither do these atoms attract other electrons, nor do they readily lose their outer electrons. Hence, these atoms possess atomic stability. Helium (He) has a complete K energy level with its two electrons. Some atoms have the tendency to build up the number of electrons in their levels to eight. Once they arrive at eight, the next electron goes into the next higher level; furthermore, these atoms that only have one electron in their outermost energy level are extremely reactive. They tend to lose their electron. Examples of such elements are lithium (Li) and sodium (Na).

Probing deeper into the electron world, scientists have determined that the energy levels are divided into sublevels that also contain a maximum number of electrons. Table 2-1 contains a listing of these sublevels which are given small letter names. Table 2-2 summarizes the data presented in Figure 2-8b and Table 2-1.

The K level has only one sublevel, the *s* sublevel. Using a block diagram, the energy levels could be represented in a manner as shown in Figure 2-8b. The electrons

TABLE 2-1 ELECTRONIC SUBLEVELS AND THEIR ELECTRONS

Sublevels	Maximum number of electrons in each sublevel
s	2
p	6
d	10
f	14
g	18
h	22
i	26

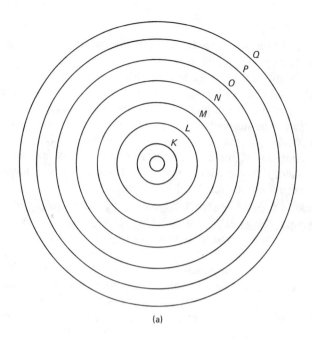

(a)

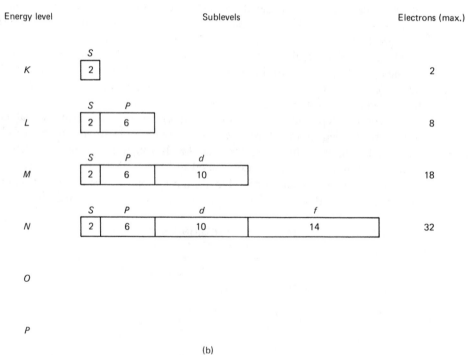

(b)

Figure 2-8 (a) Electron energy levels. (b) Block diagram of the first four energy levels showing their respective sublevels.

TABLE 2-2 ELECTRONIC ENERGY LEVELS, SUBLEVELS, AND THEIR ELECTRONS

Energy levels	Sublevels	Maximum number of electrons per level
K	*s* only	2
L	*s, p*	8
M	*s, p, d*	18
N	*s, p, d, f*	32
O	*s, p, d, f, g*	50
P	*s, p, d, f, g, h*	72
Q	*s, p, d, f, g, h, i*	98

fill these sublevels in a particular way. The first 18 elements of the periodic table build up their authorized number in a routine manner. Once the K level reaches its maximum of two electrons, the electrons begin filling up the L level until a maximum of eight electrons is reached, as with neon (Ne). Then the third energy level from the nucleus (M level) with its three sublevels *(s, p, d)* begins filling to its maximum of 18. This point is reached with the element argon (Ar) (atomic no. 18). Note that both neon (Ne) and argon (Ar) have a maximum possible number of electrons in their outermost energy levels. We know that these elements are classified as chemically inert gases; that is, they are very stable elements and do not react with other elements.

On the other hand, *let us look at some elements that have only one electron in their outermost energy level.* Lithium (Li), atomic no. 3, with two electrons in the K level and only one electron in the L level is extremely reactive. The same can be said of sodium (Na) and potassium (K). For elements with atomic numbers greater than 18, starting with potassium (K), the routine followed by the next 18 elements in filling their electron levels is somewhat different.

Through an understanding of the electronic structure of atoms the properties of elements can be ascertained and the bonding of atoms of elements together to form molecules, compounds, or alloys can be explained. Additionally, these electrons determine the size of atoms, the degree of electrical and thermal conductivity of materials, and the optical characteristics of materials.

When atoms take part in chemical reactions, there is an energy change or transfer, usually resulting in the release of energy in the form of heat. This energy change brings about a rearrangement of the electrons whose end result is an electron configuration with lower energy than any of the reactant elements. Only the outermost or valence electrons are rearranged. The electrons in the inner energy levels (or shells) are too stable or shielded from this activity. More will be said about this phenomenon when chemical bonding of atoms is discussed later in this module.

2.2.1.1.2 Periodic Table. The modern periodic table, as seen in Appendix 10.1, contains a wealth of information for the student who wishes to learn more about the chemistry of the elements. The table is made up of horizontal rows called *periods* and vertical columns called *groups*. Elements with the same number of valence electrons are placed in the same group. The groups (or families) are given number/letter designations based on differing schools of thought. Tables 10.1A and 1B designate titanium (At. No. 22) as a Group IVA element. Other designations used are IVB or 4. Transition elements (Sc thru Zn and Y thru Cd) are filling their 3d or 4d sublevels in their second energy levels counting from the outside. Period 1 contains two elements, H and He. Periods 6 and 7 having thirty-two elements each require the two inner transition element series (La thru Lu and Ac thru Lr) to be written below the rest of the table. To paraphrase some of the preceding, one can say the elements are arranged in order of increasing atomic number (at. no.). The elements with similar chemical properties recur at definite intervals, or *periodically*. This periodicity is produced by the number of electrons in the outer energy levels of the atoms of the elements. Such elements, having the same valence electrons, will have similar chemical properties. These are grouped together into families of elements known as *groups*. For example, carbon (C) with six electrons has four valence electrons. In the same Group (IVB), silicon (Si) with 14 electrons also has four valence electrons.

The atomic radius of an atom (size of atom) decreases as one moves from left to right across a period. As one moves down a particular group of elements, the size of the atoms increases. Ionization, as previously defined, involves the loss of a valence electron by an atom. The energy required to accomplish this is known as the *ionization potential*. This binding energy increases as the size of the atom decreases. In other words, the farther away an electron is from its nucleus, the smaller the force of attraction between it and its positively charged nucleus. Metallic elements on the far left of the table readily ionize, whereas atoms of nonmetals located on the right of the table more readily share or accept additional electrons and become negative ions. The observation to be made is that the tendency of atoms to ionize depends on their relative positions in the periodic table.

2.2.1.2 Molecular structure and bonding.

We mentioned previously that atoms join together to form molecules and that molecules combine to form various compounds. In some cases the atoms may be of the same element. Fluorine (F) exists as a diatomic molecule under normal conditions because two fluorine atoms combine to form one fluorine molecule. all the Group VIIB elements form diatomic molecules. In this section we will gain further understanding as to how the 100 or so elements combine or chemically bond to form not only more than 3 million organic compounds but the various inorganic and bulk metals such as iron or copper.

Molecules and compounds are formed or joined together by chemical bonds. *Chemically bonding* can simply be explained as the end product of the interaction of the electrical forces of attraction and repulsion between oppositely charged or similarly charged particles of matter. Chemical bonds are formed by the electrons in the atom's outermost energy levels or regions. In our study of these electron energy levels, we have learned that, generally, atoms that have eight electrons in their outermost orbit are very stable (in equilibrium state). Atoms with less than eight outermost electrons attempt to seek this stable condition by bonding with each other so that each atom can attain this stable configuration of eight electrons (valence electrons) in its outermost energy level. One way atoms can achieve this stability is by sharing electrons with other atoms that also need eight electrons in their outermost energy levels. For purposes of study, we can divide the various types of bonding into two groups, strong or *primary bonding* and weak or *secondary bonding*. In the first group we include covalent, ionic, and metallic bonding. The second group includes weak atomic bonds, electric dipoles, polar molecules, and the hydrogen bond. These secondary bonds are also known as van der Waals forces or bonds. It is emphasized that atoms of one material do not bond via one particular bonding mechanism only. One type of bond may predominate in a material, but in many materials the various bonding types are represented and produce mixed-bond types of materials.

2.2.1.2.1 Covalent or Electron-Pair Bonding.

The sharing of electrons between two or more atoms is known as *covalent or shared electron-pair bonding*. As an example, fluorine (F) atoms have a total of nine electrons; two in the inner energy level and seven in the outermost level. Two fluorine atoms can share one of the electrons with the other to form a single covalent bond, depicted with a single dash representing the bond.

Covalent bonding gets its name because a pair of valence electrons is a part of the electron structure of each atom bonding the two atoms together. In general, organic compounds (complex compounds containing carbon and/or hydrogen) form covalent bonds.

A *triple covalent* bond is one in which two atoms *share* three of their electrons with each other. Figure 2-9 illustrates this with nitrogen (N). Nitrogen needs eight electrons to fill the outer energy level. It only has five. Hence, it shares three with another nitrogen atom, forming a triple bond. A basic molecule of ethylene (C_2H_4) for producing polythylene is written in Figure 2-10. The simple hydrocarbon (hydrogen and carbon) methane is the natural molecule that forms the methane series of gases: methane, ethane, propane, butane, ethylene, and acetylene. Shown in Figure 2-11, the C represents carbon and the H represents hydrogen. Bonding of the elements is indicated with the small bar ($-$). Each bar indicates a covalent bond. Methane, ethane, and propane in Figure 2-12 are classified as saturated hydrocarbons since all linkages are single covalent bonds. Ethylene and acetylene are unsaturated hydrocarbons because their double or triple covalent bonds can react with other molecules of other elements. As the chains lengthen to around the molecular weight of 50, the state of matter of the hydrocarbon changes to a liquid, and at a molecular weight around 200 soft, waxy solids form and the solids become harder as more molecules join the chain. Hydrogen and carbon also form rings known as benzene rings with

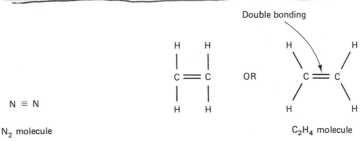

Figure 2-9   Triple covalent bond.

Figure 2-10   Covalent bonding of ethylene.

CH$_4$ molecule   **Figure 2-11** Covalent bonding of methane.

six carbon atoms and six hydrogen atoms. Figure 2-13 shows several ways of representing these rings. Covalent bonding occurs between atoms of nonmetallic elements such as hydrogen, fluorine, chlorine, and carbon and in molecular compounds such as water, carbon dioxide, ammonia, and ethylene. By sharing electrons the atoms achieve stability through the creation of interatomic electrical forces, which bond the atoms together to form molecules and compounds. Research has shown that double or triple bonds are stronger than single bonds between the same two atoms. Carbon forms four single covalent bonds.

    2.2.1.2.2 Ionic Bonding.   Instead of sharing electrons, some elements actually *transfer* electrons to other elements. This type of bonding is called ionic. In general, inorganic compounds bond ionically. Elements having outer electron levels that are almost full or almost empty tend to gain or lose electrons in order to complete their outer energy levels. A classic example of ionic bonding of two elements is sodium (Na) and chlorine (Cl). From the periodic table we learn that neutral sodium has 11 electrons, one of which is in the outer energy level. This single valence electron is weakly attracted to the positive nucleus. Chlorine (Cl) has 17 electrons, leaving seven electrons in the outer or third energy level. As you recall, chlorine needs eight electrons in its third energy level to attain stability. Since each atom attempts to attain equilibrium (lowest energy state) by completing its required number of electrons for each energy level, the sodium atom gives up its one electron through the action of a strong driving force of the chlorine atom to form a stable outer shell (Figure 2-14). This results in a temporary unstable condition that is soon changed by the sodium atom, which has now become a positive ion (cation) due to the loss of one of its electrons. The positively charged sodium atom (cation) interacts with the chlorine atom, which is now a negative ion (anion) due to the gain of the electron from the sodium atom. This interaction is a strong electrical force (electrostatic) between oppositely charged ions forming the ionic bond, which produces the molecule of sodium chloride (NaCl), common salt. Each of these elements has its own characteristic properties. Sodium is an active metal and chlorine is a poisonous gas. Yet combined as an ionic molecule they have the well-known properties of common salt. Sodium and chloride ions are also different from their respective neutral substances (atoms). Chlorine, whose atoms have seven valence electrons, is very active, with greenish-yellow color. Chloride ions, on the other hand, with eight electrons in the outer energy level are quite stable. Therefore, they do not react with metals and are colorless. All these bonds are electrostatic in origin so that the chief distinction among the various ways in which atoms bond to form molecules is in the distribution of the electrons around the atoms and molecules. *Metals* can be characterized by the tendency

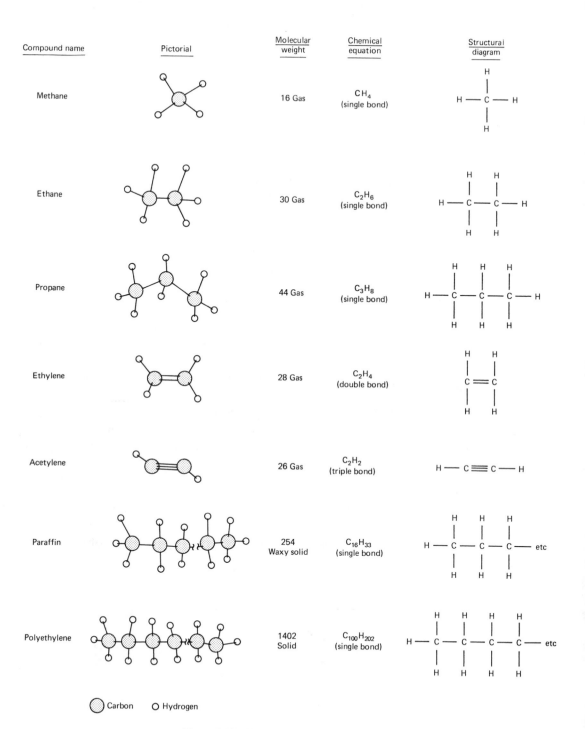

Compound name	Pictorial	Molecular weight	Chemical equation	Structural diagram
Methane		16 Gas	CH_4 (single bond)	
Ethane		30 Gas	C_2H_6 (single bond)	
Propane		44 Gas	C_3H_8 (single bond)	
Ethylene		28 Gas	C_2H_4 (double bond)	
Acetylene		26 Gas	C_2H_2 (triple bond)	
Paraffin		254 Waxy solid	$C_{16}H_{33}$ (single bond)	
Polyethylene		1402 Solid	$C_{100}H_{202}$ (single bond)	

Carbon   Hydrogen

Figure 2-12 Hydrocarbons: methane gas series.

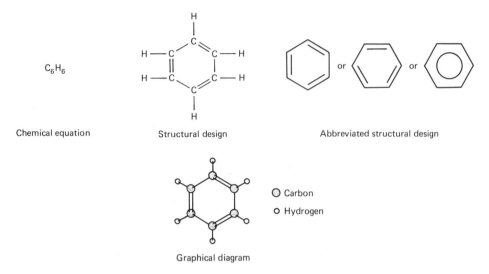

C_6H_6

Chemical equation   Structural design   Abbreviated structural design

○ Carbon
o Hydrogen

Graphical diagram

Figure 2-13 Benzene or benzene ring.

of their atoms to lose their valence electrons. An *active metal* is one that loses its electrons quite readily. Similarly, an *active nonmetal* is one whose atoms readily accept electrons. We also know that nonmetallic elements have a larger number of valence electrons than metallic elements.

2.2.1.2.3 Metallic Bonding. Metals contain one, two, or three valence electrons. These are shielded from the strong attractive forces of the positive nucleus by the inner electrons and, thus, they bond to the nucleus relatively weakly. Consequently, these metal atoms when in company with other metal atoms will lose their weakly held valence electrons, which, in turn, enter a common free-electron cloud, sea, or gas. These free or delocalized electrons can then move in three dimensions. Once these electrons leave their atoms, the atoms become positive ions (cations). An electrostatic balance is maintained between the cation and the *electron cloud*, which results in the cations arranging themselves in a regular three-dimensional pattern as

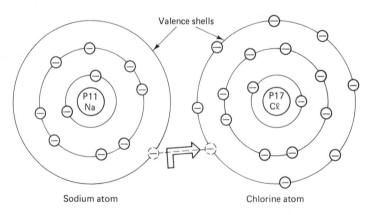

Figure 2-14 Ionic bonding of sodium chloride molecule.

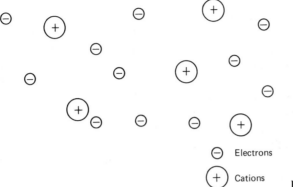

⊖ Electrons

⊕ Cations

Figure 2-15 Metallic bonding.

shown in Figure 2-15. This electrostatic balance is the glue that bonds the metallic structure. The ever-moving free-electron cloud acts like a matrix surrounding the positive metallic ions and provides rigidity. If these ions were represented as spheres, their space arrangement in bulk metal could be likened to a box full of ping-pong balls. Table 2.3 provides a summary of atomic bonding.

2.2.1.2.4 Secondary Bonding. Much weaker forces of attraction known as *van der Waals forces* produce bonding between atoms and between molecules. Group VIIIA elements (inert gases) have a full complement of valence electrons in

TABLE 2-3 SUMMARY OF ATOMIC BONDING*

Type of bond	Number of electrons "shared"	Kinds of atoms involved	Remarks
van der Waals (molecular)	0	same	Weak electrostatic attraction due to unsymmetrical electrical charges in electrically neutral (as a whole) atoms or molecules
Ionic	1 (or more) transferred	different	Strong electrostatic attraction
Covalent normal	2	same or different	Electron pair "revolves" in common orbit about both nuclei, one atom supplying one electron
coordinate	2	different	Electron pair "revolves" in common orbit about both nuclei, one atom supplying both electrons, the other atom supplying none (quite rare)
Metallic	∞	same	General attraction of a very large number of positive (metallic) ions for a dispersed cloud of electrons

*From C.O. Smith, *The Science of Engineering Materials, 2nd ed.*, Englewood Cliffs, NJ: Prentice-Hall, 1977.

their outer levels. Thus, they have no inclination to lose, gain, or share electrons. At ordinary temperatures they remain as single atoms (monatomic). Only at very low temperatures will these gases condense. It is the presence of van der Waals forces that permits this condensation. A similar situation occurs with covalently bonded molecules that have achieved equilibrium by sharing electrons. The van der Waals forces also permit these molecules to condense at low temperatures. All molecules consist of distributions of electrical charge. When two molecules are in close proximity, these electrical charge distributions interact and create intermolecular forces. The van der Waals forces of attraction between molecules have their origin in the forces of attraction of the nucleus (positive) of one molecule for the electron charge (negative) of a neighboring molecule, which form fluctuating dipoles in the two molecules. Such forces increase with an increase in the number of electrons per molecule because the electrons in the large molecule are more readily polarized by these forces.

In a *polar molecule* or *electric dipole,* the charges are polarized; that is, both the positive and negative charges are localized within the molecule. Figure 2-16 illustrates these two situations.

A hydrogen molecule is a simple example of a *nonpolar molecule.* This diatomic molecule, in which both atoms are of the same element and are covalently bonded together, is sketched in Figure 2-17. Each atom shares the bonding electrons equally, producing an electrical charge distribution that is symmetrical about a line joining the two nuclei.

Any diatomic molecule such as HCl in which the atoms are different from each other is polar (Figure 2-18a). The negative pole is located at the atom that is more electronegative (attracts electrons more strongly). In this case, the electrons are not shared equally. The chlorine atom, being more electronegative, attracts more electrons than the hydrogen atom does, which causes the shared electrons to be more closely associated with the chlorine nucleus. Figure 2-18b is a three-dimensional sketch of this molecule showing that, like the hydrogen molecule, it too has an electrical charge distribution that has symmetry around the line x-x.

A good example of a polar molecule containing more than two atoms is water (H_2O), sketched in Figure 2-19. Oxygen has a greater attraction for electrons than hydrogen. As a result, the four electrons being shared in the two covalent bonds remain closer to the oxygen atom. This unequal sharing of electrons has the effect

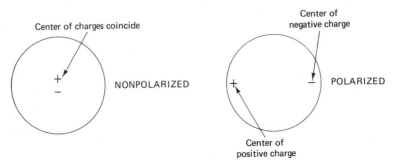

Figure 2-16 Schematic of a nonpolarized and a polarized molecule.

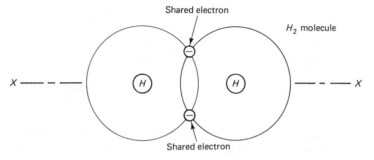

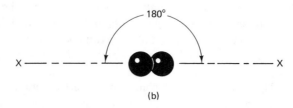

(a) Both atoms share two electrons equally forming a single covalent bond to complete each atom's outer energy level (K level).

(b)

Figure 2-17 (a) A nonpolar diatomic hydrogen molecule. (b) A linear hydrogen molecule.

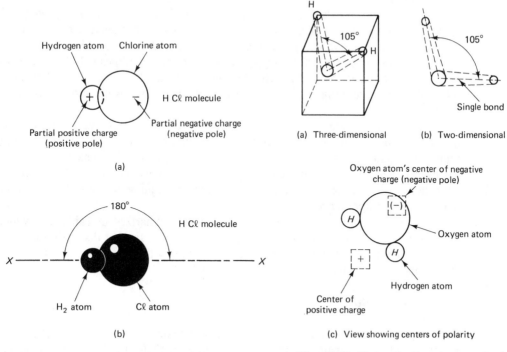

(a)

(b)

Figure 2-18 (a) A polar diatomic molecule of hydrogen chloride. (b) A linear hydrogen chloride molecule.

(a) Three-dimensional (b) Two-dimensional

(c) View showing centers of polarity

Figure 2-19 Various sketches of water molecule (H_2O).

Bond or bridge

H ——-◯————-⌁----H—◯

H   Water molecule

H   **Figure 2-20** Hydrogen bond (bridge) in liquid water.

of creating a partial positive charge between the hydrogen atoms and a partial negative charge on the oxygen, which produces a *polar covalent bond*. The three-dimensional sketch (Figure 2-19a) of the water molecule with the oxygen atom in the center of a cube with two hydrogen atoms occupying two of the eight corners shows a bent molecule (not symmetrical) with an angle between the two hydrogen atoms of about 105° (Figure 2-19b). Most covalent bonds are polar. The unsymmetrical charge distribution on the water molecule results in the formation of regions of positive (+) and negative (−) charges located at a maximum distance from each other (Figure 2-19c). If the charge distribution were symmetrical, the positive (+) and negative (−) charges would cancel each other, as both centers would be at the center of the molecule. Thus each center of charge can exert an attractive force on an adjacent charge of opposite sign. This statement explains, in part, how water molecules bond together to form the liquid water. As pictured, the hydrogen atoms bond on one side of the oxygen, leaving the protons of the hydrogen atom farther from the nucleus of the oxygen atom.

The covalent bonds between the hydrogen and oxygen atoms in a water molecule, being polar, result in an unsymmetrical charge distribution. The oxygen atom, having a slight negative charge, attracts a positive hydrogen atom belonging to an adjacent water molecule and forms a *hydrogen bond,* sometimes called a *hydrogen bridge*. This bond is the strongest of the secondary bonds. Figure 2-20 shows one schematic for representing the hydrogen bond between two water molecules in the liquid phase.

The hydrogen bond, as do all bonds, affects the properties and the behavior of materials, particularly polymeric materials. In thermoplastics, the hydrogen bond joins long chainlike molecules to each other. These relatively weak bonds can be easily loosened or broken by heating, permitting flow to take place. This explains why thermoplastics can be converted by heating to liquids and then back to solid plastic material upon cooling.

Our discussion of chemical bonding is well summarized in Table 2-3, which shows one secondary bond and the three primary bonds. Further discussion of these chemical bonding mechanisms comes up in dealing with the various groups within the family of materials.

2.2.2 Solid State

A solid is a sample of matter that has a fixed volume or size and a fixed shape, which it retains indefinitely without any need to confine it. Picturing the atom as a sphere of microscopic size, atoms that make up a solid are packed closer together than the same atoms making up a liquid or gas. We can say that solids do not flow like liquids nor do they expand like gases. The atoms of a solid do not possess the kinetic energy or the motion they would possess if they were in a liquid or gas. These atoms are quite subdued and do not act as independent units, but move with their neighboring atoms. Solids, as we will see, can be grouped into several categories, such as metals, plastics, ceramics, and composites. The attractive forces between atoms in a solid are much greater than between atoms in a liquid or gaseous state. As such, these forces may position the atoms in some orderly geometric arrangement to form a *crystalline structure* (as in most metals) in which the atoms vibrate, rotate, and oscillate around fixed locations, maintaining a minimum dynamic equilibrium between adjacent atoms. From this we can conclude that the geometric arrangement of atoms is related to its energy content. If the arrangement is orderly, this implies a minimum energy content. On the other hand, an unorderly random distribution of particles (producing an *amorphous structure* as in many plastics) is an indicator of motion. Hence, unbalanced forces are acting on the atoms and, therefore, an increase in internal energy by the system of particles is necessary to sustain such motion of particles from a small energy demand for vibration to a larger energy requirement for translational motion. Finally, another indicator of energy is temperature. If the temperature of the system of particles is high, the higher the disorder of the particles will be and the more energy that will be possessed by the system of particles. With solids we know that most of the motion of the atoms is confined to a vibrational type.

2.3 FAMILY OF MATERIALS

The wide variety of materials found in the automobile also exists in the entire family of materials. There is nearly a limitless variety of materials. With so many materials, how can one be expected to understand them? The method used in this text is to consider all materials as members of a big family. Materials that possess common characteristics are then placed into their own group within the family. Even though overlaps exist in the grouping system, it is easier to understand materials when relationships are identified.

As seen in Table 2-4, four groups, *metallics, polymers, ceramics,* and *composites,* comprise the main groups; a fifth group of *other materials* is used for materials that do not fit well into the four groups. Each group then divides into subgroups. Some examples of each subgroup are listed to illustrate common materials in the subgroup. Figure 2-21 illustrates the variety of materials used in one of the communications systems; telecommunications. Within any system such as transportation, communications, or construction, numerous materials systems exist. These systems must pull materials from the groups within the material's family to meet their special needs. The wise materials user recognizes the many options available because of the compatibility between groups and then proceeds to develop a materials system that meets a specific need.

TABLE 2-4 FAMILY OF MATERIALS

Group	Subgroup	Examples
Metallics (metals and alloys)	Ferrous	Iron
		Steel
		Cast iron
		Steel alloys
	Nonferrous	Aluminum
		Tin
		Zinc
		Magnesium
		Copper
		Gold
		Nonferrous alloys
	Powdered metal	Sintered steel
		Sintered brass
Polymerics	Man-made	Plastics
		Elastomers
		Adhesives
		Paper
	Natural	Wood
		Rubber
	Animal	Bone
		Skin
Ceramics	Crystalline compounds	Porcelain
		Structural clay
		Abrasives
	Glass	Glassware
		Annealed glass
Composites	Wood based	Plywood
		Laminated timber
		Impregnated wood
	Plastic based	Fiber glass
		Graphite epoxy
		Plastic laminates
	Metallic based	Boron aluminum
		Alumina whiskers
	Concrete	Reinforced concrete
		Asphalt concrete
	Cermets	Tungsten carbide
		Chromium alumina
	Other	Reinforced glass
Others	Electronic materials	Semiconductors
	Lubricants	Oil
	Fuels	Coal
	Protective coatings	Anodized
	Biomaterials	Carbon implants

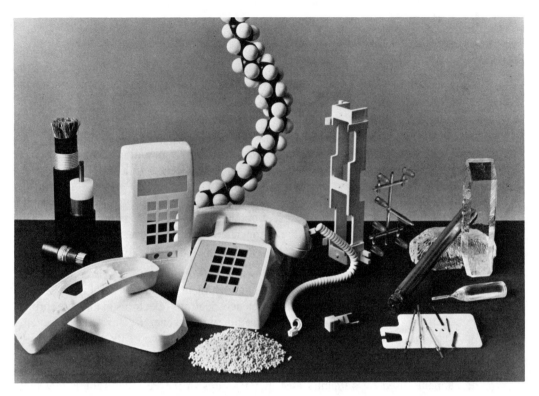

Figure 2-21 Family of materials in communication. One of major factors influencing performance of telecommunications equipment is performance of materials used in its construction. Plastics (polymers) for outdoor, under sea, and underground cables; magnetic alloys (metals); synthetic single crystals (ceramics); and fiber reinforced rubber (composites) are four principal areas in which materials research and development discoveries greatly influence telecommunications. (Bell Laboratories.)

2.3.1 Other Systems of Grouping

The grouping system used in this text follows closely grouping systems common to engineering. Slight differences exist in other grouping systems. For example, one system may divide materials into metals, polymers, ceramics, glass, wood, and concrete. Some systems divide all materials into three groups: organics, metals, and ceramics. Then they distinguish between the natural and organic materials. Others group materials as crystalline or amorphous. Regardless of the system, some inconsistencies develop due to the nearly limitless variety of materials. Still, the advantages of grouping outweigh the limitations.

This section presents the system of grouping for this text and also goes into a brief study of the general nature of each group and subgroup. Detailed study of each group follows in subsequent modules. These terms are italicized when first used (Table 2-4).

2.3.2 Metallics

Metallics or metallic materials include metal alloys. A metal in a strict definition only refers to an element such as iron, gold, aluminum, and lead. *Alloys* consist of metal elements combined with other elements. Steel is an iron alloy made through combining iron, carbon, and some other elements.

2.3.2.1 Metallic defined. The definition used for a metal will differ depending on the field of study. Chemists may use a different definition of metals than physicists. Two definitions of metals as used in engineering follow:

1. *Metals* are elements that can be defined by their properties, such as hardness, toughness, malleability, electrical conductivity, and thermal expansion. They possess relatively high strength and toughness, luster high electrical and heat conductance, and most metals resist high temperatures and prolonged stress.

2. *Metals* are large aggregations (collections of millions of crystals composed of different types of atoms held together by metallic bonds; see Figure 2-15). Metals usually have less than four valence electrons (electrons in the outer orbits of their atoms), as opposed to nonmetals, which generally have four to seven. The metal atom is generally much larger than the atom of the nonmetal.

2.3.2.2 Types of metallics. While metals comprise about three-fourths of the elements used by humanity (see metal groups in the Periodic Table), few find service in their pure form. Several reasons account for not using pure metals. Among these reasons are that pure metals may be too hard or too soft, or they may be too costly because of their scarcity; but the key factor normally is that the desired property sought in engineering requires a blending of metals and elements. Thus the combination form (alloys) finds greatest usage. Therefore, metals and metallics become interchangeable terms. Metallics are broken into subgroups of ferrous and nonferrous metals.

2.3.2.2.1 Ferrous. Ferrous is the Latin word meaning iron. Ferrous metals include iron and alloys of at least 50% iron, such as cast iron, wrought iron, steel, and stainless steel. Each of these alloys is highly dependent on possessing the key element carbon. Table 2-4 shows examples of ferrous metals.

Steel is our most widely used alloy. Sheet steel forms car bodies, desk bodies, cabinets for refrigerators, stoves, and washing machines; it serves as doors, "tin" cans, shelving, and thousands of other uses. Heavier steel, such as plate, I beams, angle iron, pipe, and bar, form the structural frames of buildings, bridges, ships, automobiles, roadways, and many other structures.

2.3.2.2.2 Nonferrous. Metal elements other than iron that consist of more than 50% of noniron elements are nonferrous alloys. The nonferrous subgroup includes both common lightweight metals, such as titanium and beryllium, and common

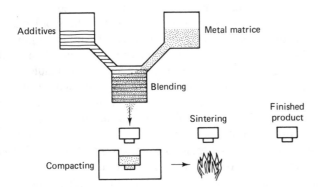

Additives

Metal matrice

Blending

Sintering

Finished product

Compacting

Figure 2-22 Powdered metal.

heavier metals such as copper, lead, tin, zinc, and alloys of brass and bronze. Among the heavier metals is a group of white metals including tin, lead, and cadmium; they have lower melting points around 230° to 330°C. Among the high-temperature (refractory) nonferrous metals are chromium, nickel, tantalum, and tungsten. Tungsten has the highest melting point of all metals, 3400° C. The combination of nonferrous alloys is practically without end.

2.3.2.2.3 Powdered Metals.
Alloying of metals involves melting the main ingredients so that upon cooling the metal alloy is generally a nonporous solid. Powder metal is often used because it is undesirable or impractical to join elements through alloying or to produce parts by casting or other forming processes. Powdered metal is called *sintered* metal. As seen in Figure 2-22, this process consists of producing small particles of metal, squeezing them together, and *sintering* (applying heat below the melting point of main alloy). The squeezing pressure with added heat bonds the metal powder into a strong (often porous) solid. Powdered metals can be ferrous, nonferrous, or a combination of both ferrous and nonferrous elements with nonmetallic elements.

2.3.3 Polymerics

Polymeric materials are basically materials that contain many parts. *Poly* means many and *mer* stands for monomer or unit. A polymer is a chainlike molecule made up of smaller molecular units (monomers). The monomer, made up of atoms, bond together covalently to form a polymer that usually have a carbon backbone.

Figure 2-12 shows a simple sketch of an ethylene monomer. The atoms of carbon (C) and hydrogen (H) form covalent bonds. In Figure 2-12, many of the ethylene monomers have joined into a polyethylene polymer. Thousands of polyethylene polymers join together to form polyethylene plastic. The same process of polymerization is responsible for the formation of man-made rubbers and plastics, natural wood and rubber, animal bone, skin, and the tissues of humans, animals, and insects.

2.3.3.1 Plastics. The term *plastic* is used to define man-made polymers containing carbon atoms covalently bonded with other elements. The word plastic also means moldable or workable such as dough or wet clay. Plastic materials are either liquid or moldable during the processing stage, after which they turn to a solid. After processing, some plastics cannot be returned to the plastic or moldable state; they are *thermosetting plastics* or *thermosets*. *Thermo* means heat, and *set* means permanent. Common thermosetting plastics include epoxy, phenolic, and polyurethane. Other plastics can be repeatedly reheated to return to the plastic state; they are *thermoplastics*. Examples of thermoplastic plastics are acrylics (e.g., Plexiglass[R] or Lucite[R]), nylon, and polyethylene. While today most plastics are produced from oil, they can also be made from other organic (carbon) materials such as coal or agricultural crops, including wood and soybean. As our limited supply of oil is depleted, the major sources of polymerics will change.

2.3.3.2 Wood. Of all the materials used in industry, *wood* is the most familiar and most used. Wood is a natural polymer. In the same manner that polymers of ethylene are joined to form polyethylene, *glucose* monomers polymerize in wood to form *cellulose* polymers ($C_6H_{10}O_5$). Glucose is a sugar made up of carbon (C), hydrogen (H), and oxygen (O). Cellulose polymers join in layers with the gluelike substance *lignin*, which is another polymer.

2.3.3.3 Elastomers. Prior to World War II, most rubbers *(elastomers)* were natural rubbers. Today synthetic or man-made rubbers far exceed the use of natural rubbers. An elastomer is defined as any polymeric material that can be stretched at room temperature to at least twice its original length and return to its original length after the stretching force has been removed. Some stretch to over ten times original length. Elastomers are able to store energy so they can return to their original length and/or shape repeatedly.

Elastomers have a molecular, amorphous structure similar to other polymeric materials. This *amorphous* or shapeless structure consists of long coiled-up chains of giant molecules (polymers) that are entangled with each other. Adjacent polymers are not strongly bonded together, and, as seen in Figure 2-23a, when a tensile force (pulling force) is applied, these coils straighten out (bond straightening) and snap back like springs to their original coiled condition upon removal of the force. When this same type of force is strong enough, not only do the polymers increase in length as a result of the bonds lengthening between individual atoms, but they also lengthen (bond lengthening, Figure 2-23b) due to the unwinding coils resulting in a temporary structure that approaches crystallinity or a more orderly structure. Remember that with crystallinity comes strength.

To further increase the strength of the elastomers, the process of *vulcanization* is used to form the necessary crosslinks (strong bonds) between adjacent polymers to be discussed later in the section on thermosetting plastics. Vulcanization is a chemical process that produces covalent bonding between adjacent polymers with the help of a small amount of sulfur. Crosslinks tie the polymers together to produce a tough, strong, and hard rubber for many uses in industry, such as in automobile tires.

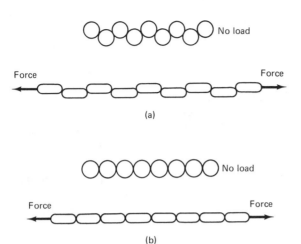

Figure 2-23 Elastomer structure. (a) Bond straightening. (b) Bond lengthening.

2.3.3.4 Other natural polymers. A most amazing natural polymer is human skin, which has no equal substitute. Animal skin or hide in the form of fur and leather has limited industrial use because synthetic materials have been developed that offer greater advantages to the designer than the natural polymers. Medical science continues to study such natural polymers as bones, nails, and tissues of humans and animals in order to synthesize these materials for replacement when they are damaged due to injury or illness. Bioengineering and biomechanics are newer fields that integrate engineering and medicine to solve material problems in the treatment of humans.

2.3.4 Ceramics

Ceramics are crystalline compounds of metallic and nonmetallic elements. *Glass* is grouped with ceramics because it has similar properties, but most glass is *amorphous*. Included in ceramics are porcelain such as pottery, abrasives such as emery used on sandpaper, refractories (materials with good resistance to heat) such as tantalum carbide with a melting temperature around 3870°C, and structural clay such as brick. Ceramics, including glass, are hard, brittle (no slip), stiff, and have high melting points. Ceramics primarily have ionic bonds, but covalent bonding is also present. Their structure usually consists of two main ingredients. One major ingredient contains the atoms that form the crystalline structure. The second ingredient forms a glassy substance, which acts as cement to bond together the crystalline structure. Silica is a basic unit in many ceramics. The internal structure of silica has a pyramid (tetrahedron) unit as diagrammed in Figure 2-24a, which shows both the graphical diagram and chemical equation. These silicate tetrahedrals join into chains with the graphical diagram shown in Figure 2-24a. In Figures 2-24a and b, note how the larger oxygen (O) atoms surround the small silicon (Si) atom. The silicon atom occupies the space opening (interstice) between the oxygen atoms and shares four valence electrons with the four oxygen atoms. Chains such as shown in Figure 2-24b are extremely long

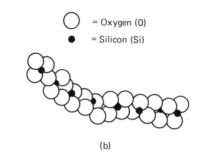

○ = Oxygen (0)

● = Silicon (Si)

(SiO₄ ion)

(a)

(b)

Figure 2-24 Ceramic structure. (a) Silicate tetrahedral. (b) Single chain of silicate.

and join together in three dimensions. The chains are held together by ionic bonds, whereas individual silica tetrahedrals bond together covalently. Silica is combined with such metals as aluminum, magnesium, and other elements to form a wide variety of ceramic materials.

2.3.5 Composites

By strict definition, a *composite* is a material containing two or more integrated materials (constituents) with each material keeping its own identity. Normally, combining of the materials serves to rectify weaknesses possessed by each constituent when it exists alone. By this strict definition, many natural materials exist as composites; for example, wood is a combination of cellulose and lignin, but wood and natural materials are not classified as composites.

Some of the most familiar composites include fiber glass, plywood, and laminated coins of silver and copper (quarters and dimes). While most of the groups of the family of materials could be classed as composites because of the way they are placed in service (such as painted steel or case-hardened steel), the composite classification commonly refers to materials developed to meet the demands in building, electronic, aerospace, and auto industries. With an ever-increasing use of composites, they are truly the material of today and the near future, because composites can be designed to be stronger, lighter, stiffer, and more heat resistant than natural materials or to possess properties required by technology that are not available in a single material. Composites allow the designer to select the right combination of materials to perform safely at the lowest cost.

The subgroups of composites shown in Table 2-4 include wood based, plastic based, metallic based, concrete, cermets, and others. It is also possible to classify composites by their structure. Composite structures include layers, fibers, particles, and any combination of the three. Layered composites, as seen in Figure 2-25, consist of *laminations* like a sandwich. The laminations are usually bonded together by adhesives, but other forces could be used, such as those provided by welding.

Fibers and particles are made into composites by suspending them in a *matrix* or by the use of *cohesive* forces. The matrix is the material component, such as plastic, epoxy cement, rubber, or metal, that surrounds the fibers or particles. *Cohesive* forces involve the molecular attraction of one constituent to the other.

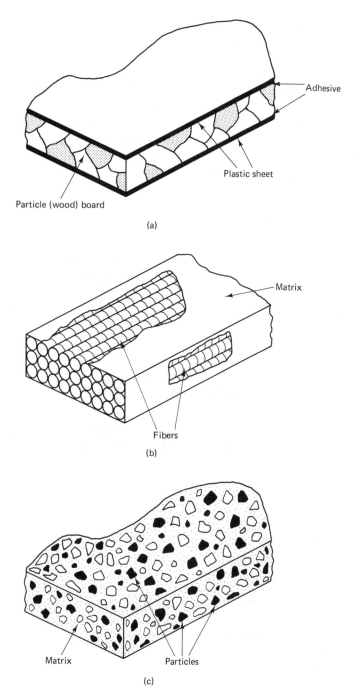

Figure 2-25 Typical composites. (a) Layered composite. (b) Fiber composite. (c) Particle composite.

Examples of layer composites (Figure 2-25a) include plywood, laminated boards such as particle board (wood chips) covered with plastic sheets, thermocouples (different metals welded together), safety glass (plastic sandwiched between sheets of glass), cardboard, and Alclad^R (copper alloy core that gains corrosion resistance from outside covers of aluminum). Particle composites include concrete (gravel, rock, and sand in Portland cement matrix), powdered metal (metal powders or ceramics bonded by cohesion), and particle board (wood chips in resin matrix), as seen in Figure 2-25c. Fiber composites include fiber glass (glass fibers in polyester plastic matrix), hardboard (vegetable fibers in adhesive matrix), boron fibers in aluminum matrix, reinforced glass (wire mesh in glass matrix), and graphite epoxy (graphite fibers in an epoxy plastic matrix), as seen in Figure 2-25b.

2.3.6 Other Materials

This group is used to include materials that do not fit well into the previously discussed groups. Comparing the materials in the other group with the rest of the family of materials, Table 2-4 reveals most come from the metallics, polymers, ceramics, and composite groups, but their applications justify a separate study. Some will be discussed in Chapter 9.

APPLICATIONS AND ALTERNATIVES

The demand for rubber continues to increase even as the major source (petroleum) of synthetic rubber is rapidly running out and natural rubber sources are strained. Automobile tires require vast amounts of rubber, with 200 million used tires being scrapped each year. So two problems exist with rubber tires: (1) they require vast amounts of rubber, but oil is a depleting resource, and (2) disposing of worn-out tires is difficult. Natural rubber still dominates the tire field, but petroleum-based synthetic rubber is also used.

Several solutions emerge to the rubber and tire problem. It is possible to increase the amounts of natural rubber through cultivation of the guayule (gu-you-la) plant, a plant that grows in desert areas unfit for most other agriculture. The guayule shrub can supplement the hevea rubber tree, which is now the major source of natural rubber. Disposing of the vast amount of discarded tires has caused problems. The used tires have been made into playground toys, thongs for feet, artificial reefs at sea to attract fish, door mats, and bumpers on boats. None of these techniques makes much of a dent in the 200 million used tires.

Worn-out tires can help ease the energy crises and meet other needs. Through recycling it is possible to obtain gaseous hydrocarbons and oil.

Experimentation with burning used tires shows that the average tire burned in special furnaces yields energy equivalent to 2.5 gallons of fuel oil. This energy is converted into steam for heating or producing electricity.

2.4 SELF-ASSESSMENT

To test your understanding of this module, answer the following questions;

2-1 How many elements have been discovered up to the present time?

2-2 Why do we need models of atoms?

2-3 Explain what an isotope is.

2-4 Using a periodic table, determine the number of protons in the element rubidium.

2-5 How many electrons does an atom of aluminum have in its outermost energy level?

2-6 Verify that the size of an atom of lithium is larger than an atom of fluorine.

2-7 The formula $2n^2$ is an easy way to determine the maximum possible number of electrons in any energy level of an atom. Using the formula, calculate the maximum number of electrons in the P energy level.

2-8 The key word in describing ionic bonding is transfer. What is the key word that can be used to describe covalent bonding?

2-9 Give examples of ionic, covalent, and metallic bonding.

2-10 Express the average interatomic distance in units of nanometers and inches.

2-11 In which family group belong the materials with large chainlike molecules that consist of smaller molecules or monomers? Are they generally amorphous or crystalline?

2-12 When a metallic contains at least 50% iron what is its classification?

2-13 A metallic consisting of metal elements and other elements has what name? Give three examples.

2-14 Define composites. List three composite structures.

2-15 In the ceramic grouping, most materials are crystalline, but which one has an amorphous structure?

2-16 List three examples of lightweight nonferrous metals or alloys, three examples of heavier nonferrous metals or alloys, and one example of a refractory metal.

2-17 What is the subgroup of metallics in which metal particles are joined by pressure and sintering heat?

2-18 What subgroup covers man-made polymers containing carbon atoms covalently bonded together with other elements? List the two divisions of this polymeric material and give an example of each.

2-19 Name three natural polymeric materials.

2-20 How do elastomers differ from other polymeric materials? How can the strength of rubber be improved?

2-21 How does the energy shortage threaten the supply of plastics and rubber?

2-22 What is the prime purpose of selecting a composite material over material from the other family groups?

2-23 List three materials that were grouped in the "other" family group of materials.

2-24 In the Pause and Ponder section you were asked to imagine how you would select materials for designing a fuel efficient car. (a) Can you recall uses of plastics in cars? (b) How many such applications can you list? (3) What other substitutes would you make?

2.5 ACTIVITIES

2-1 This module emphasizes the variety of materials in use today. To aid in your understanding of the application of various materials and assist in your becoming a wise "materials consumer," you should:

    (a) Observe the materials used in everyday products such as tables and chairs, buildings, roads, street lights, home appliances, shop tools, laboratory instruments, bicycles, automobiles, electronic equipment, clothing, and any other products you encounter.

    (b) Ask yourself, your friends, your instructor, or knowledgeable people: Why did the designer select that particular material or combination of materials for this product? Would another material be less costly, lighter, resist corrosion, save energy, recycle rather than pollute, and so on?

    (c) Gather a variety of materials and compare their features, or construct a mobile of various materials with labels on each material. Do a neat job and you will have a nice art object.

2-2 Gather the following: soup can, seamless beer can, penny, piece of solder, lead sinker, screw, and washer. Use a magnet on each to see which attracts the magnet.

    (a) What do the metals that attract the magnet possess?

    (b) The beer cans and the soup cans are also referred to as tin cans. Are they made of tin? Explain.

2-3 Refer to such periodicals as *Popular Science, High Technology, Machine Design, Plastic World, Materials Engineering,* and *Smithsonian,* or other science and technology magazines, to read about new uses of the family of materials.

2-4 Select appropriate materials to construct a container to hold a single uncooked egg. To test the success of your container, drop it from 10 meters. If you want to make a contest of various designs by other students, use the weight of the containers to determine the best design. The design of lightest weight that survives the drop with no harm to the egg would be best. If no eggs survive, the instructor would choose the most creative design employing the widest variety of materials.

2.6 REFERENCES AND RELATED READING

ASHBY, M. F., and D. R. H. JONES. *Engineering Materials: An Introduction to Their Properties and Applications.* Elmsford, N.Y.: Pergamon, 1980.

ASIMOV, ISAAC. *Building Blocks of the Universe.* New York: Abelard, 1957.

AVNER, SIDNEY H. *Introduction to Physical Metallurgy.* New York: McGraw-Hill, 1964.

BAISER, ARTHUR, and KONRAD KRAUSKOPF. *Introduction to Physics and Chemistry.* New York: McGraw-Hill, 1964.

BELLMEYER, FRED W. *Synthetic Polymers.* Garden City, N.Y.: Doubleday, 1972.

BETTS, JOHN E. *Physics for Technology.* Reston, Va.: Reston, 1964.

BROWN, THEODORE L., and H. EUGENE LeMAY, JR. *Chemistry: The Central Science, 2nd ed.* Englewood Cliffs, N.J.: Prentice-Hall, 1981.

CHRISTIANSEN, G. S., and PAUL H. GARRETT. *Structure and Change: An Introduction to the Science of Matter.* San Francisco: W. H. Freeman, 1960.

CLAUSER, HENRY R. *Industrial and Engineering Materials.* New York: McGraw-Hill, 1975.

COMPANION, A. L. *Chemical Bonding.* New York: McGraw-Hill, 1964.

DeGARMO, E. PAUL. *Materials and Processes of Manufacturing,* 3d ed. New York: Macmillan, 1975.

DRAGO, R. S. *Prerequisites for College Chemistry.* New York: Harcourt, Brace, Jovanovich, 1966.

EWEN, DALE, and LeRAY HEATON. *Physics for Technology Education.* Englewood Cliffs, N.J.: Prentice-Hall, 1981.

GAMOW, GEORGE A. *Mr. Tomkins Explores the Atom.* New York: Macmillan, 1945.

GIANCOLI, DOUGLAS C. *Physics.* Englewood Cliffs, N.J.: Prentice-Hall, 1980.

HANKS, RICHARD W. *Materials Engineering Science: An Introduction.* New York: Harcourt, Brace, Jovanovich, 1970.

SHAPIRO, GILBERT. *Physics Without Math: A Descriptive Introduction.* Englewood Cliffs, N.J.: Prentice-Hall, 1979.

SMITH, CHARLES O. *Science of Engineering Materials.* Englewood Cliffs, N.J.: Prentice-Hall, 1977.

VANVLACK, L. H. *Materials for Engineers: Concepts and Applications.* Reading, Mass.: Addison-Wesley, 1982.

Periodicals

Adhesive Age

Ceramics Monthly

Chemical and Engineering News

Chemical Engineering

High Technology

Iron Age and Metal Work International

Journal for Materials Science and Engineering

Machine Design

Materials Engineering

NASA Technology Briefs

Plastic Technology

Plastic World

Popular Science

Science 8X

STAR (Scientific and Technical Aerospace Reports)

Steel Facts

3

STRUCTURE OF SOLID MATERIALS

3.1 INTRODUCTION

The kinetic theory of matter describes atoms in a solid as being at their lowest energy state. Their motions have slowed considerably. These atoms of solids are so close to their neighboring atoms that they can no longer act as independent particles. They vibrate about three equilibrium positions, depending on the degree of remaining kinetic energy (thermal energy) they may possess. They can also move in conjunction with their neighboring atoms. But they can and do move. In the discussion about bonding in Module 2, we learned that two or more atoms can chemically join each other depending on their electronic structures. We also learned that atoms, ions, and molecules are of different sizes.

The internal structure of solids determines to a great extent the properties of the solid or how a particular solid material will perform or behave in a given application. The study of the internal structure is pursued in various ways using x-rays, as well as electron, proton, or neutron beams to disclose crystal size, crystal structures, crystal imperfections, bonding types, spacing of adjacent atoms and adjacent planes, and the different atoms in a solid. A brief knowledge of some of the more important tools in the field of spectroscopy and microscopy that permit a still deeper probe into the internal structure of materials is vital to a greater understanding of the effects of structure on the behavior of natural and man-made materials. Figure 3-1 shows photomicrographs of varying magnification produced with aid of various instruments used in materials science.

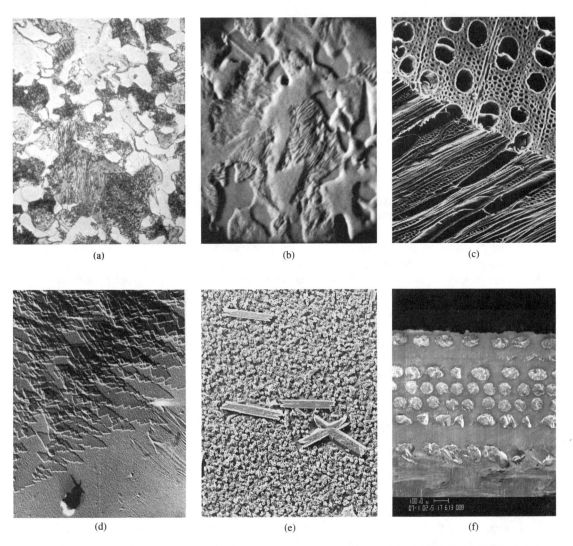

Figure 3-1 Photomicrographs of structural features of some representative materials. (a) Steel (SAE/AISI 1045) 1000×. (NASA.) (b) Steel (SAE/AISI 1045) SEM 2550×. (NASA.) (c) Wood (Maple) 415×. (U.S. Forest Products Lab.) (d) Plastic (polyethylene) dendrite crystals. (Dr. Philip H. Geil—U. of Illinois.) (e) Glass ceramic. Crystals growing in amorphous glass. 17,900×. (Corning Glass Works.) (f) Fiber metal composite. (NASA.)

3.1.1 Spectroscopy

White light when refracted (light rays undergo a change in direction when passing from one medium to another) is broken up into a number of different colors and forms a spectrum. Each spectral color corresponds to a particular wavelength. Every chemical element (as well as its atoms) produces a characteristic spectrum, which

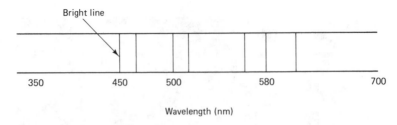

Bright line

350   450   500   580   700

Wavelength (nm)

Figure 3-2 Atomic spectrum of sodium in the visible region. (Adapted from D.C. Giancoli, *Physics*, Prentice-Hall, Inc., Englewood Cliffs, N.J., 1980.)

may be detected and measured by a spectroscope. Spectroscopy, using such an instrument, studies substances through an analysis of their spectra. In other words spectroscopy permits the precise measurement of the wavelengths of atoms' radiations, which are discrete and distinct for each atom. When atoms are exposed to high energy from an outside source, they become excited and their electrons move to higher energy levels. Becoming unstable, the electrons tend to return to their lower energy levels. The interchanges of energy between radiation and matter take place in discrete units called *quanta*. Under excitation an atomic substance will reach a state of dynamic equilibrium in which some atoms are releasing energy at the same rate that it is being absorbed by others. An atom loses energy when its excited electrons transit to a lower energy level, giving off a photon of light. The wavelength of this photon is related to the magnitude of the change of energy. Such transitions, when observed by the spectroscope, produce atomic spectra (Figure 3-2) with discrete and distinguishing wavelengths corresponding to the various electron transitions between energy levels. Such studies permit scientists to increase their knowledge of atoms, electronic structures, and various bonding mechanisms.

3.1.2 X-Ray Diffraction

One of the most useful tools in the study of crystal structures of solids is *x-ray diffraction*. An x-ray diffractometer is shown in Figure 3-3. X-rays have wavelengths about equal to the diameters of atoms (10^{-10} meters) or about the same length as the spacing between atoms (or ions) in solids. When these x-rays are directed at a solid with a crystalline structure, the waves are reflected. The equation, known as Bragg's law, $n\lambda = 2d \sin \theta$, relates the wavelength, λ, to the distance between planes of atoms, d, and the glancing angle between the incident beam and the plane of atoms, θ. Figure 3-4a shows a sketch of this relationship. If the incident beam strikes the planes of atoms at some arbitrary angle, the reflected beam may be nonexistent, because the reflected rays from the atomic planes will be out of phase and produce destructive interference (Figure 3-4b) and cancellation of the reflected beam. It has been determined that at a particular angle θ the reflected beam will be in phase and produce constructive interference (Figure 3-4b). Mathematically, this can be expressed as $\lambda_1 = n\lambda_2$, where $n = 1, 2, 3 \ldots$; the distance traveled from the different

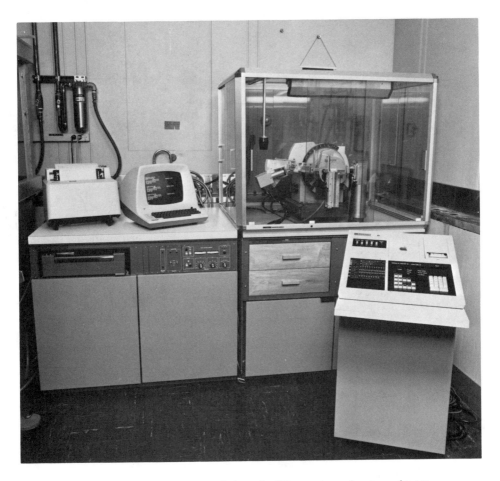

Figure 3-3 A microcomputer-controlled powder diffractometer goniometer used to analyze the composition of various substances. (General Electric.)

parallel planes of atoms represents an integral number (n) of wavelengths. Referring to Figure 3-4a, this statement may be written as

$$ACA' = BEB' - n\lambda, \qquad \text{where } n = 1, 2, 3 \ldots$$

Knowing the wavelength of the incident beam of x-rays, the glancing angle can be measured experimentally, and solving Bragg's law for $d = n\lambda/2 \sin \theta$, the interplanar distance can be calculated. Figure 3-5 shows a photographic plate upon which the constructive waves produce a series of dots, indicating that the x-rays are scattered from crystals at only certain angles. All the various angles at which diffraction occurs are determined by measurements on the photographic film. By studying the directions of diffracted x-ray beams, as well as the intensities produced by this powerful tool, much can be learned about the crystal structure of solids.

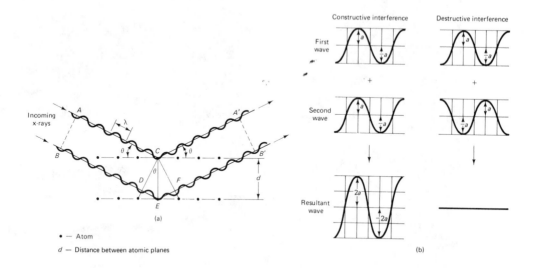

Figure 3-4 (a) Scattering of X-rays by atoms in parallel planes. (b) Constructive and destructive interference of waves. (Theodore L. Brown and H. Eugene LeMay, Jr. *Chemistry—The Central Science,* 2nd ed., Prentice-Hall, Inc., Englewood Cliffs, N.J., 1981.)

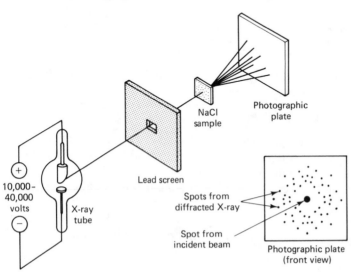

Figure 3-5 The X-ray diffraction pattern for NaCl and the experimental method by which it is obtained. (T.L. Brown and H.E. LeMay, Jr. *Chemistry—The Central Science,* 2nd ed., Prentice-Hall, Inc., Englewood Cliffs, N.J., 1981.)

3.1.3 Electron Microscopy

The maximum magnification produced by optical microscopes is about $2000 \times$. This limitation is imposed by the wavelengths of visible light, which limit the resolution of minute details in the specimen being observed. Resolving power is a term used to describe the ability of a lens to reveal detail in an image. Optical microscopes do not have the ability to discern details smaller than 10^{-6} meter (m) image. To observe objects in the 10^{-10}m range, high-energy electrons with wavelengths around 4×10^{-11} m are used. These electrons are similar to visible light in that they are both forms of electromagnetic radiation. By using magnetic lenses to focus the electrons (glass lenses could not be used), as shown in Figure 3-6a, the resolving power can be greatly increased to give magnifications over $200,000 \times$. It must be remembered that electrons are opaque to metallic specimens (i.e., the electrons are strongly absorbed by the metallic atoms). In a transmission electron microscope (TEM), shown in Figure 3-6b, the electron beam passes through an extremely thin film of the solid being studied. A scanning electron microscope (SEM) is used to study the surface of

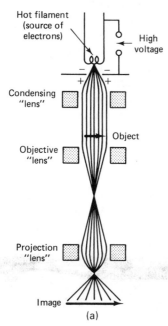

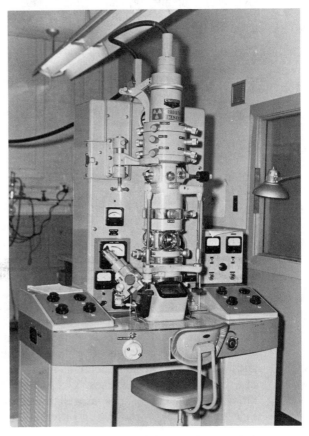

Figure 3-6 (a) Electron microscope. The squares represent magnetic field coils for the "magnetic lenses." (Douglas C. Giancoli, *Physics—Principles with Applications*. © 1980, Reprinted with permission of Prentice-Hall, Inc., Englewood Cliffs, N.J.) (b) Photograph of a transmision electron microscope (TEM.) (NASA.)

(b)

solid materials. In the replica method, the specimen is a thin plastic film on which the fine details of the solid's surface have been recorded. Field-ion microscopy with magnifications up to 5×10^6 is an example of another technique being used and developed to probe even further into the microstructure of solids.

An example of the use of electron microscopy can be found in the research and development of synthetic crystalline materials produced by molecular beam epitaxy (MBE). These new materials not found in nature have extremely thin layers of atoms deposited alternately on a semiconducting substrate or base. By controlling the chemical composition and thickness of the layers, various crystalline structures are formed with varying electrical, mechanical, and optical properties. The MBE method, as described in Chapter 9, would not be possible without the prior development of electron microscopes with small wavelengths (0.2 nm) and large magnifications, which permitted scientists to verify that their efforts were indeed producing a crystal with extremely thin atomic layers of alternating composition. Atomic layers as thin as one-ten millionth of a centimeter were determined to exist in these new man-made materials.

Knowing that (1) changes in the internal structure of solids are accompanied by a change in energy, and (2) the release of this energy in the form of heat (*exothermic*) or the absorption of this heat (*endothermic*) can be detected and measured leads to another technique that increases our knowledge about the internal structure of solids. *Differential thermal analysis* (DTA) techniques produce temperature patterns that can be interpreted to obtain information about the various structural changes that solids undergo as a result of the application of different external forces. Finally, even the vibration of atoms in a molecule can be detected using infrared spectroscopy. Such techniques, for example, are used to identify unknown polymeric materials.

3.1.4 Metallurgical Microscopy

Solid materials observed under an optical metallurgical microscope are opaque to light, much as dense solids are to electrons. Therefore, the surface features of a specimen must be illuminated by reflected light, some of which is returned through the microscope to be magnified by the objective and eyepiece lenses for direct viewing by the human eye. A rough sketch of such a scope is shown in Figure 3-7 with the transit of two light rays through the scope, one of which is not reflected back through the microscope. Such microscopes, fitted with photographic capability called metallographs, allow us to directly view or take pictures of the internal structure of solids within the limits of resolution imposed by the use of visible light. In contrast, the diffraction techniques discussed previously, which use x-rays, electrons, or neutrons, produce diffraction patterns that indirectly provide information about the internal structure of solid materials. The metallographic specimen to be viewed is polished and etched with an acid. The etching reacts with the grain boundary areas at a different rate than the grains themselves. When incident light rays strike these areas with their different crystal orientations, as well as their difference in rate of etching, the reflected light travels back to the eye in different amounts (see Figure 3-8). If most of the

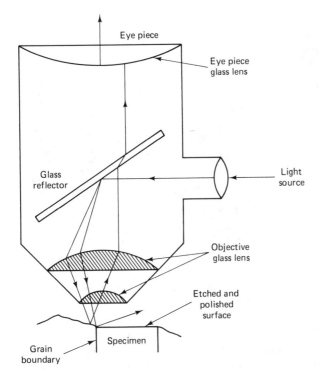

Figure 3-7 Schematic of optical system of a metallographic microscope detecting a grain boundary in a crystal.

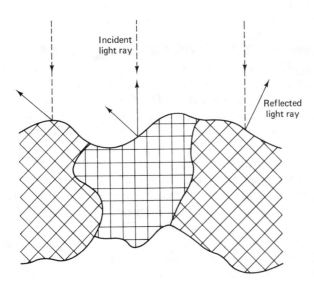

Figure 3-8 Sketch showing identification of individual crystals by reflected light rays.

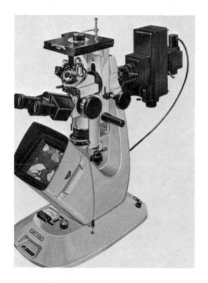

Figure 3-9 Photograph of a metallograph. (Buehler Ltd.)

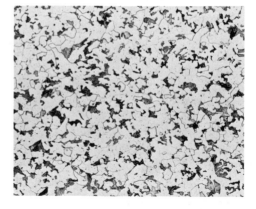

Figure 3-10 A photomicrograph of low carbon steel. (Buehler Ltd.)

incident light is reflected back to the eye, the grain will appear bright; other crystals reflecting less light will appear darker. A photomicrograph of a sample taken through a metallograph (Figure 3-9) showing the details of the microstructure including the grain boundaries, is shown in Figure 3-10.

3.2 CRYSTALLINE STRUCTURE OF SOLIDS

Definitions of solids most always contain the phrase "definite volume and shape," together with the fact that solids maintain their shape to a varying extent when subjected to external mechanical forces. Solids occur in two forms: crystalline and amorphous. A true crystalline solid possesses an ordered three-dimensional geometric arrangement that repeats itself. A metal would be representative of a crystalline solid (though not perfectly crystalline). An *amorphous solid,* on the contrary, contains no repetitious pattern of atom locations to any extent. The classic example of an amorphous solid, glass, is sometimes referred to as a supercooled liquid because of the random nature of its atomic arrangement. Figure 3-11 illustrates the different atomic arrangements of three states of matter.

In our treatment of *crystallography* (the study of crystalline structures), we will limit ourselves to the orderly arrangement of the atoms in their microscopic world. In so doing, we will represent atoms, ions, or molecules essentially as spheres of varying sizes occupying points at various distances from each other in space. Hence the need for an *axis system*. Such a system is shown in Figure 3-12.

The arrowhead on each axis points in the positive direction. The negative directions therefore are opposite to the arrowheads. The *x, y,* and *z* axes are in this case perpendicular to each other, with their origins coinciding at any convenient

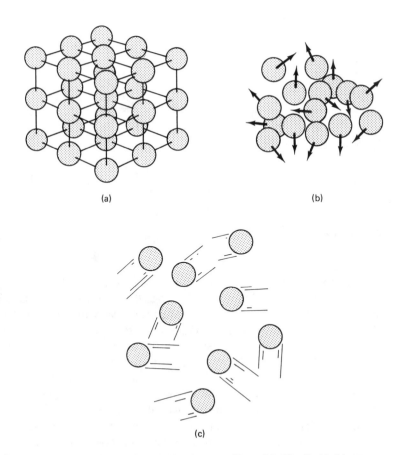

(a)   (b)

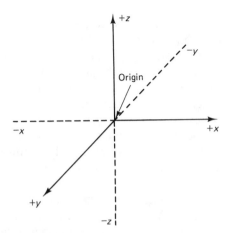

(c)

Figure 3-11 How atoms are arranged in (a) a crystalline solid, (b) a liquid, (c) a gas. (D.C. Giancoli, *Physics*, Prentice-Hall, Inc., Englewood Cliffs, N.J., 1980.)

Figure 3-12 A standard axis system.

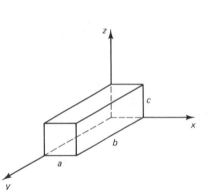

Figure 3-13 A box aligned on an axis system.

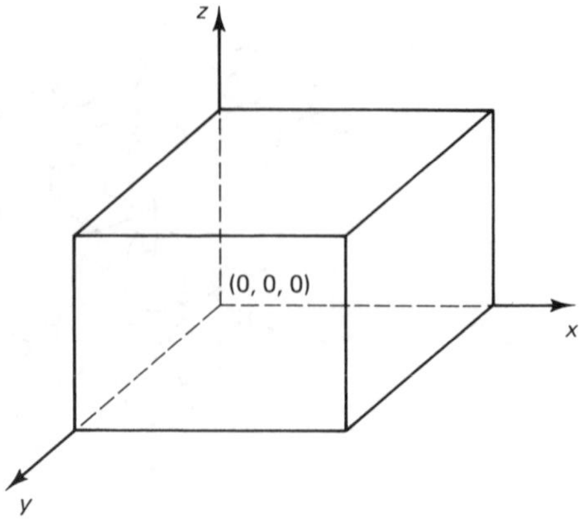

Figure 3-14 A 3-dimensional box aligned on an axis system with an origin at lower, left, rear corner.

point. A box with dimensions *a, b,* and *c,* is sketched in Figure 3-13, using this axis system. If this solid figure were a cube, these dimensions would be of the same magnitude. However, in one important crystal system they are not equal in magnitude. Another interpretation of these letters (or *lattice parameters*) is that they represent distances from the origin to points on the axes. The letter *b* would, for example, represent the perpendicular distance from the origin to a point on the *y* axis.

The three axes in Figure 3-14, which are infinite in length, form three mutually perpendicular planes that are also infinite in size. The *x-y* horizontal plane forms the base of the box. The top of the box would be contained in a horizontal plane parallel to the *x-y* plane. The *y-z* plane contains the left end of the box, and the *x-z* plane contains the rear side of the box. We should note that there are countless planes parallel to three primary planes, which may have to be referred to using some standardized technique.

3.2.1 Unit Cells

In our study of crystals, the axes system just described (to be called a simple cubic crystal system) is modified slightly for the sake of uniformity. The standardized axes system shown in Figure 3-15 retains the *x, y,* and *z* axes. The box is now called a unit cell (defined later). The angles between the principal planes are named α (Greek letter alpha), β (Greek letter beta), and γ (Greek letter gamma). For example, in Figure 3-16, α is the angle measured in degrees between the *x-y* and *y-z* planes; angle β, between *x-y* and *x-z* planes; and angle γ, between the *x-z* and *y-z* planes. The sides of the box (unit cell) labeled *a, b,* and *c* are the lattice parameters in the *x, y,* and *z* directions, respectively. These distances are also known as *intercepts*. To adequately describe a particular axes or crystal system, all six of the preceding dimensions are needed.

The term *unit cell* is used to describe the basic building block or basic geometric arrangement of atoms in a crystal. You can compare a unit cell to a single brick in a brick wall. Knowing that atoms are located at each corner of the single brick, it is

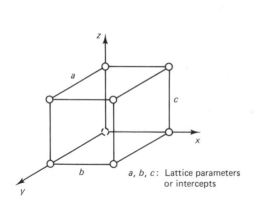

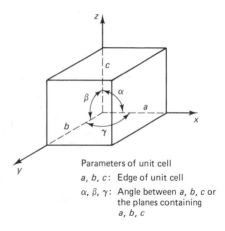

Figure 3-15 A simple crystal lattice unit cell.

Figure 3-16 Angles between planes and intercepts in a unit cell.

easy to picture the atomic structure of a crystal with such a unit cell repeating itself in three-dimensional space. The unit cell could be thought of as the minimum area of wallpaper containing a geometric arrangement that is repeated over and over again on a papered wall. However, in this latter example we are only using two dimensions, and we must remember that unit cells and crystals are three-dimensional atom arrangements of solid materials.

3.2.2 Space Lattices

If you repeat the unit cell in all three dimensions, you create a crystalline structure with a definite pattern. This larger pattern of atoms in a single crystal is known as a *space lattice* or *crystal lattice*. A space lattice is three sets of straight lines at angles to each other constructed to divide space into small volumes of equal size, with atoms (ions or molecules) located at the intersections of these lines or in between the various lines. We must remember that the lines and points in a space lattice are only imaginary. The lattice concept is used to show the position of atoms, molecules, or ions in relation to each other. Atoms may be represented by circles, spheres, or Ping-Pong or tennis balls located either at the intersection of these lines or in between these lines. A part of a space lattice is sketched in Figure 3-17 with small spheres representing atoms. We must also remember that the actual atoms in solids are located as close to each other as possible, thus attaining the lowest possible energy level. Two atoms closest to each other would be represented by two spheres touching each other. The closer the atoms are, the denser the solid.

For the purposes of our study, we will become familiar with 3 of the 14 possible space lattices. These are the *cubic, tetragonal,* and *hexagonal crystal systems.* Reference to the periodic table will disclose that 67 of 92 elements solidify into one of these three crystal structures. Table 3-1 identifies these structures.

The *cubic axis system* has already been described with its mutually perpendicular axes and equal lattice parameters. The cubic coordinate axis system allows us to form three basic unit cells by placing atoms, ions, or molecules at various lattice positions.

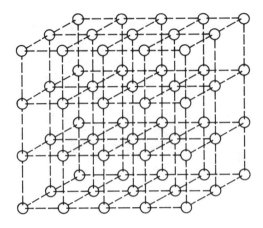

Figure 3-17 A space or crystal lattice.

The simple *cubic unit cell* (Figure 3-18) consists of eight atoms located at each corner of the cube.

The point to be made is that this representation of atoms in a solid is such as to show only location of atoms. It must be remembered that if you represented these eight atoms by hard rubber balls and arranged them in accordance with this simple cubic unit cell, all eight atoms would be touching each other.

Another unit cell (Figure 3-19) is known as the *body centered cubic* (bcc). It is similar to the simple cubic unit cell, but contains an additional atom located in the

TABLE 3-1 PERIODIC SYSTEM OF THE ELEMENTS FOR FERROUS METALLURGISTS

PERIODIC SYSTEM OF THE ELEMENTS

Adapted Primarily for Ferrous Metallurgists

Atomic size factors (in parentheses) are % smaller (−) or larger (+) than gamma (FCC) iron at 75 F. Lattice environment (Coordination No.) is taken into account; CN is 12 except 6 for interstitials H, B, C, N & O. Groups VI, VIb, VII & VIIb form ionic compounds with the metals. Atomic size is based largely on work of W. Hume-Rothery and associates and L. Pauling (Some values, such as those for H & O, are approximate). Alloying valences are those of Pauling.

H-1 ▲(−58) ⊗ XX

0	I	II	III		IV	V	VI	VII
He-2 FCC (Others)	Li-3 ⊗(+23) BCC* HCP†	Be-4 ●(−11) HCP* BCC	B-5 ◬(−29) ⊗ XX		C-6 ▲(−34) ⊗ XX	N-7 ▲(−36) ⊗ XX	O-8 ▲(−33) ⊗ XX	F-9
Ne-10 FCC	Na-11 ⊗(+50) BCC* HCP	Mg-12 ⊗(+27) HCP	Al-13 ◐(+14) FCC		Si-14 ●(+7) XX	P-15 ●(+2) XX	S-16 ●(+1) XX	Cl-17 XX

0	Ia	IIa	IIIa	IVa	Va	VIa	VIIa	VIII			Ib	IIb	IIIb	IVb	Vb	VIb	VIIb
Ar-18 FCC	K-19 ⊗(+86) BCC	Ca-20 ⊗(+56) FCC* BCC	Sc-21 ⊗(+29) HCP* BCC	Ti-22 ◐(+16) HCP* BCC	V-23 ●(+6) BCC	Cr-24 ●(+1) BCC	Mn-25 XX* FCC†	Fe-26 ● (0) BCC* FCC	Co-27 ●(−1) HCP* FCC	Ni-28 ●(−1) FCC	Cu-29 ●(+1) FCC	Zn-30 ●(+6) HCP	Ga-31 ◐(+12) XX	Ge-32 ●(+9) XX	As-33 ●(+11) XX	Se-34 ●(+11) XX	Br-35 XX
Kr-36 FCC	Rb-37 ⊗(+97) BCC	Sr-38 ⊗(+42) FCC* HCP†	Y-39 ⊗(+27) HCP* BCC	Zr-40 ◐(+15) HCP* BCC	Cb-41 ●(+10) BCC	Mo-42 ●(+8) BCC	Tc-43 HCP	Ru-44 ●(+6) HCP	Rh-45 ●(+6) FCC	Pd-46 ●(+9) FCC	Ag-47 ◐(+14) FCC	Cd-48 ⊗(+25) HCP	In-49 ⊗(+23) XX	Sn-50 ⊗(+23) XX	Sb-51 ⊗(+27) XX	Te-52 ⊗(+27) XX	I-53 XX
Xe-54 FCC	Cs-55 ⊗(+112) BCC	Ba-56 ⊗(+76) BCC	La-57 ⊗(+48) HCP* FCC†	Hf-72 ◐(+26) HCP* BCC	Ta-73 ◐(+16) BCC	W-74 ●(+11) BCC	Re-75 ●(+9) HCP	Os-76 ●(+7) HCP	Ir-77 ●(+8) FCC	Pt-78 ●(+10) FCC	Au-79 ◐(+14) FCC	Hg-80 ⊗(+25) XX	Tl-81 ⊗(+36) HCP* BCC	Pb-82 ⊗(+39) FCC	Bi-83 ⊗(+35) XX	Po-84 ⊗(+40) XX	At-85
Rn-86	Fr-87	Ra-88 ⊗(+49) FCC	Ac-89														

Note 1: The rare-earth (lanthanide, 58-71) and actinide (90-103) series are omitted.
Note 2: Valence is 4 for C; 3 for N and P.
Note 3: (1) and (2) are not alloying valences.

| Alloying Valence | 1 | 2 | 3 | 4 | 5 | 6 | 6 | 6 | 6 | 6 | 5.56 | 4.56 | 3.56 | 2.56 Note 2 | 1.56 Note 2 | (2) Note 3 | (1) Note 3 |

SUBSTITUTIONAL SOLID SOLUTIONS

● FAVORABLE SIZE FACTOR: 0 TO ± 13%
◐ BORDERLINE SIZE FACTOR: ± 14 TO ± 16%
⊗ UNFAVORABLE SIZE FACTOR: > ± 16%

INTERSTITIAL SOLID SOLUTIONS

▲ FAVORABLE SIZE FACTOR: > (−40%)
◬ BORDERLINE SIZE FACTOR: (−30) TO (−40) %
◬ UNFAVORABLE SIZE FACTOR: < (−30%)

STRUCTURE

BCC – BODY CENTERED CUBIC
FCC – FACE CENTERED CUBIC
HCP – HEXAGONAL CLOSE PACKED
XX – NOT BCC, FCC OR HCP USUALLY MORE COMPLEX
* – STRUCTURE AT 75 F
† – ALSO FCC   † – ALSO BCC

TYPE OF GAMMA IRON (FCC) FIELD IF ALLOYED WITH IRON

▶ GAMMA LOOP, LIKE Cr
▮ LIMITED GAMMA LOOP, LIKE B
◀ OPEN GAMMA REGION, LIKE Ni
◢ LIMITED GAMMA REGION, LIKE C

(INCO.)

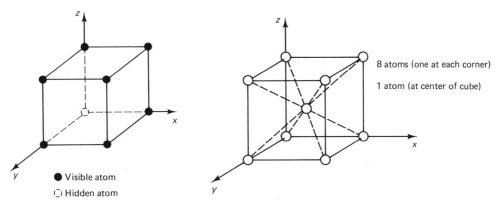

8 atoms (one at each corner)

1 atom (at center of cube)

● Visible atom
◌ Hidden atom

Figure 3-18 A simple cubic unit cell showing atoms only at corners of cube.

Figure 3-19 A body-centered cubic unit cell.

center of the cube. The third type of unit cell formed from the cubic axis system is the *face-centered cubic* (fcc). One atom at each corner and one in the center of each of the cube faces make up the complement of atoms. There is no atom at the center of the cube (Figure 3-20).

The *tetragonal crystal system* has similar unit cells to the cubic, but the sides are not equal. As an example, the *body-centered tetragonal* (bct) crystal lattice unit cell is shown in Figure 3-21. Tin forms a tetragonal unit cell. The tetragonal is similar in that the axes are all normal to each other. The difference lies in the length of the intercepts. The x and y intercepts have the same magnitude. The z intercept is larger than the x or y intercept.

The *hexagonal crystal system* (see Figure 3-22), comprised of three axes (a_1, a_2, a_3) in the x-y plane 120° apart and a fourth axis (z) at 90° to the x-y plane, contains a difference in the size of the angles between the axes, as well as a difference in the intercepts. The atoms in the lower or basal plane as well as the atoms in the top or upper plane form hexagons. The intercepts a_1, a_2, and a_3 are equal in length, and

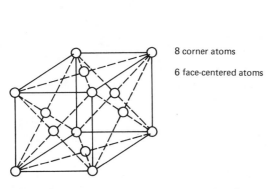

8 corner atoms

6 face-centered atoms

Figure 3-20 A face-centered cubic unit cell.

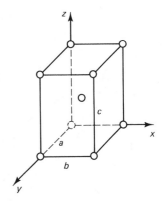

Figure 3-21 A body-centered tetragonal crystal lattice unit cell.

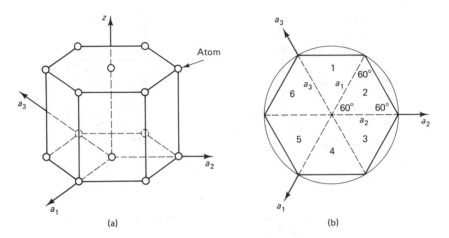

Figure 3-22 (a) A hexagonal crystal lattice unit cell. (b) Top plane of unit cell showing six equilateral triangles.

the fourth intercept c is of different length. The atoms in three dimensions trace out a right hexagonal prism.

Atoms of solid materials do not form the purely hexagonal unit cell as in Figure 3-22 because they cannot satisfy equilibrium conditions by being so far apart. In other words, they would be unstable. Consequently, they form the hexagonal unit cell called hexagonal close packed (hcp), as shown in Figure 3-23 with its three mid-plane atoms. Zinc, titanium, and magnesium form hcp unit cells.

3.2.3 Crystal, Crystallographic, Atomic Planes

The principal crystal structures differ in the location of the atoms in their unit cells. Later in this text we will learn that atoms within a solid (a steel structural member, for example) move rapidly in the direction of planes that are most densely packed

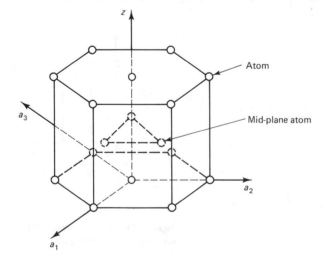

Figure 3-23 A hexagonal close packed crystal lattice unit cell.

Structure of Solid Materials   Module 3

or populated with atoms. To identify these closely packed planes of atoms within a crystal lattice or unit cell materials technicians and scientists use a set of numbers known as *Miller indices* of a given plane. The procedure for finding Miller indices of a plane may be found in materials texts such as those listed in the References. Examples of atomic planes sketched in a unit cell are shown in Figures 3-24 through 27. Note that these crystallographic planes are designated by three numerals not separated by commas and enclosed within parentheses.

Atoms within solids tend to move in coordination with each other. Planes of atoms, particularly those with the most closely packed mass of atoms, slip or move in parallel directions much like two adjacent cards in a deck of playing cards. This movement of atoms in a solid results from the application of some type of external force, such as a compressive or tensile load. The degree of movement of planes of atoms gives an indication of the ductility and strength of the particular material. Therefore, it is important to know something about planes in the various crystalline systems. We will limit our study to planes that contain the most atoms.

In the *bcc crystal system,* we note in observing the *bcc unit cell* in Figures 3-19 and 3-24 that, by use of a cutting plane through this cell, the plane containing the most atoms would be one that passed through four of eight corner atoms and the one in the center of the cube. All figures are expanded views. Actually, the atoms are much closer. This diagonal plane could be a (110) plane, as sketched in Figure 3-24. It should be noted that, although these planes contain the most atoms, the atoms are not as closely packed as they could be. A view of this diagonal plane from a position at right angles to it, as in Figure 3-25, shows that the spaces are not occupied by atoms; the diagonal has a length of $4r$ (r is the radius of each atom); and the atoms touch each other along the bcc unit cell body diagonal.

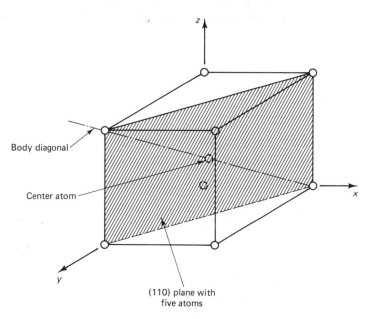

Figure 3-24   A bcc unit cell showing (110) plane.

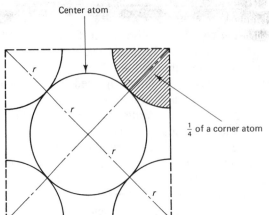

Center atom

$\frac{1}{4}$ of a corner atom

Figure 3-25 Normal view of diagonal plane (110) in bcc unit cell.

In the *fcc unit cell*, the planes with the most atoms can be represented by plane (111) as sketched in Figure 3-26 or seven other planes with the same number of atoms. Observe that the plane contains three face-centered atoms and three corner atoms. Figure 3-27 shows these atoms all in contact with each other with a minimum of unoccupied space between them. With atoms of the same size, six atoms are touching each atom in such a plane.

The *hcp unit cell* readily tells us the most closely packed planes of atoms are the basal planes and all planes parallel to them. This means that a metal with a hcp crystal lattice can only slip in one direction. In other words, this metal would be brittle. The bcc or fcc metals, having a greater number of slip planes, can slip in many directions and are thereby classified as ductile metals. It must be remembered

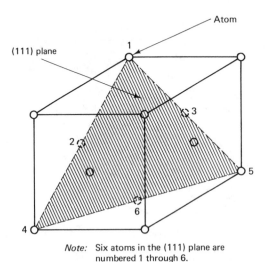

Atom

(111) plane

Note: Six atoms in the (111) plane are numbered 1 through 6. Not all atoms are sketched in the unit cell for clarity.

Figure 3-26 (111) plane in fcc unit cell.

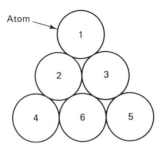

Figure 3-27 Normal view of (111) plane in fcc unit cell showing six numbered atoms.

that at this stage we are talking about the atomic structure of millions of atoms in one crystal. The vast majority of solid materials are polycrystalline (i.e., composed of millions and millions of individual crystals whose axes are oriented in random directions). This random orientation could result in a blocking effect; that is, a crystal would stop the slippage of atoms in any adjacent crystal. More details of the movement of atoms will be discussed later.

3.2.4 Coordination Number

To describe how many atoms are touching each other in a group of coordinated atoms, the term *coordination number* (CN) is used. The CN is the number of neighboring atoms that each atom has immediately surrounding it. Note in Figure 3-23 that each upper and lower basal plane of a hcp unit cell contains an atom at its center. Each atom touches six atoms in its own plane, plus three atoms above and below in adjacent planes. Consequently, the CN for these atoms would be 12. The number of nearest atoms is dependent on two factors: (1) the type of bonding, and (2) the relative size of the atoms or ions involved. In our discussion of bonding, for example, we learned that valence electrons determine the type of bonding as well as the number of bonds an atom or ion can have. Carbon (C) in Group IV has four covalent bonds and therefore a CN of 4. The Group VII elements, such as chlorine (Cl), form only one bond (CN is 1). The relative size of the atoms determines how many neighboring atoms will touch another atom. Ionic bonding involves ions of different charges, hence different sizes. The limiting factor in this case is the ratio of the size (radii) of the combined atoms of ions. The minimum ratios of atomic (ionic) radii produce various CNs.

Figure 3-28 represents the five ions occupying one of the six faces of the fcc unit cell for NaCl. The Na ion is just the right size to fit between the Cl ions at the corners of the unit cell. Thus the ions are closely packed, with each cation separated from other cations by a layer of anions. Each cation and each anion are shared equally by six oppositely charged ions. Therefore, a CN of 6 describes this geometric arrangement. The distance between the center of the cation and its neighboring Cl anions is found to be the hypotenuse of an isosceles right triangle (two legs or sides of equal length) of side R. Using this radius R and either simple right-triangle trigonometry or the Pythagorean theorem ($z^2 = x^2 + y^2$), this distance ($r + R$) can be

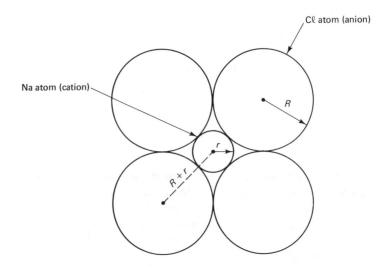

Note: Two Cℓ anions located along the axes
perpendicular to the page are not
shown for clarity.

Figure 3-28   A face plane in NaCl crystal.

easily determined to be the sum of both radii. In mathematical form, $(r + R)^2 = 2R^2$. Taking the square root of both sides and solving for the ratio of r/R, we obtain the value of 0.414. This ratio tells us that in order to have a CN of 6 we must have two ions whose ratio of radii is at least 0.414.

In the case of sodium chloride (NaCl), r/R is $0.95/1.81 = 0.520$. This being greater than 0.414, NaCl qualifies for a CN of 6. Note that as the difference between r and R decreases, higher CNs are possible. A CN of 12 is the maximum, which occurs when the atoms (ions) have the same radius and the ratio becomes 1. In other words, as the r gets smaller than R (radius of surrounding atoms), the fewer neigh-

TABLE 3-2   MINIMUM RADII
RATIOS FOR CNs

CN	$\dfrac{r}{R}$
3	$\geqslant 0.155$
4	$\geqslant 0.225$
6	$\geqslant 0.414$
8 (bcc)	$\geqslant 0.732$
12 (hcp or fcc)	1.0

r, radius of smaller atom

R, radius of larger atom

$\geqslant$, greater than or equal to

TABLE 3-3 CHARACTERISTICS OF SELECTED METALS

Element (symbol)	Crystal structure (20°C)		Atomic radius 10^{-10} m	Melting point (K)
Beryllium (Be)	hcp		1.14	1562
Aluminum (Al)	fcc		1.43	933
Titanium (Ti)	hcp		1.46	1945
	bcc	877°C		
Iron (Fe)	bcc		1.24	1810
	fcc	910–1399°C	1.27	
	bcc	1399–1538°C		
Cesium (Cs)	bcc		2.62	300
Tungsten (W)	bcc		1.37	3680

boring atoms can make contact or touch the smaller atom. Table 3-2 lists the minimum radii ratios for some common CNs.

With a CN of 12 each atom has contact with twelve other atoms. Each atom in a fcc or hcp unit cell meets this description provided their radii are of similar size. A CN of 12 is the largest number that can exist.

3.2.5 Allotropy

It is of interest to list some common materials and show their crystal structure and melting point to illustrate the concept of the movement of atomic planes within a solid. Table 3-3 contains several elements that exist in more than one crystal structure depending primarily on the temperature. This phenomenon is known as *allotropy* or *polymorphism* (*poly,* many; *morph,* shape). Actually, an allotropic material, after changing to one structure, can reverse the phenomenon and return to its previous structure; a polymorphic material does not possess this reverse phenomenon. Over one-quarter of the elements are allotropic. Steel owes its existence to this property. Because of this property, steel can be produced from iron. When iron is heated to above 910°C, its structure changes from bcc to fcc, allowing for a much greater absorption of carbon atoms (2% maximum). Figure 3-29 shows one carbon atom in an fcc unit cell of iron.

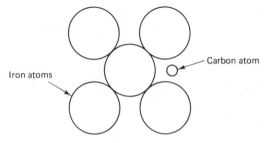

Iron atoms

Carbon atom

Atomic radius C, 0.077 nm
Atomic radius Fe, 0.127 nm

Figure 3-29 A face plane of an fcc unit cell for iron showing a carbon atom located in an interstice.

In general, we now know that any allotropic change in an element means different properties. In the case of steel, all its properties almost instantaneously change at 910°C. The best example of a change in properties with crystal structure is that of carbon (C). One polymorph of carbon is graphite, a black, greasy, low-strength material. A second polymorph is diamond, the hardest naturally occurring substance.

3.2.6 Volume Changes and Packing Factor

In discussing crystal structure changes, we mentioned that every change in atomic structure brings changes in properties of the solid. One of these changes is volume. When steel transforms from bcc to fcc structure, it decreases in volume. The explanation for this phenomenon involves the density of atoms in the various unit cells. *Density,* as we learned, is the ratio of the mass to the volume of a substance, which stays constant provided it is nonallotropic. Mass is measured in kilograms and volume in cubic meters.

The *atomic packing factor* (APF) or packing factor (PF) is the ratio of the volume of atoms present in a crystal (unit cell) to the volume of the unit cell. In calculating the volume of an atom, we assume the atom is spherical. The difference between the PF and unity (1) is known as the *void fraction,* that is, the fraction of void (unoccupied or empty) space in the unit cell. Using the *simple cubic crystal unit cell* sketched in Figure 3-30, with atoms of equal radius (R) located at each corner, the volume of the cell occupied by the eight atoms is equivalent to one atom. Each corner atom contributes one-eighth of its volume. Therefore, the volume of one atom is $\frac{4}{3}\pi R^3$. The (100) face plane of this simple cubic unit cell shows four corner atoms touching each other (see Figure 3-31). The relationship of the radius R of an atom to the lattice parameter or edge length a of the unit cell is $2R$. The volume of the unit cell is a^3. The volume of atoms is $(\pi/6)\, a^3$. Thus

$$\text{PF} = \frac{\text{volume of atoms}}{\text{volume of unit cell}} = \frac{(\pi/6)\, a^3}{a^3} = \frac{\pi}{6} \cong 0.52$$

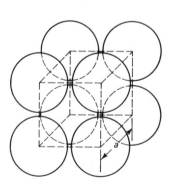

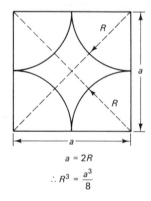

$$a = 2R$$
$$\therefore R^3 = \frac{a^3}{8}$$

Figure 3-30 A simple cubic unit cell.

Figure 3-31 (100) face plane of simple cubic unit cell.

Solving for the PF in terms of the radius R produces the same results. The void factor is therefore $1 - 0.52 = 0.48$. What the calculation tells us is that only about half (52%) of the space in the simple cubic unit cell is occupied by the atoms. This is too inefficient, so atoms of metals do not crystallize in this structure. Remember, the closer the atoms come to each other, the less energy they have and the more stable is their structure.

The *bcc unit cell,* as you recall, is quite similar to the simple cubic unit cell with the addition of one atom in the very center of the unit cell. Therefore, the bcc unit cell contains the equivalent of two atoms.

For the *fcc unit cell* there are four net atoms. Each of the eight corner atoms contributes one-eighth of an atom. Each of the six face atoms contributes one-half. The total is $1 + 3$ or 4 atoms. Notice that the atomic radius (R) cancels out in all these calculations, which tells us that the PF is not dependent on the radius of the spheres being packed if all the atoms are of the same size.

The fcc structure has the maximum PF for a pure metal. The hcp structure also has a PF of 0.74. Finally, we should note that the coordination number varies directly with the PF. As an example, the $CN_{bcc} = 8$ and $PF_{bcc} = 0.68$; the $CN_{fcc} = 12$ and $PF_{fcc} = 0.74$.

In our study of metals we will see that pure iron will change its structure upon heating from bcc to fcc at 910°C. Knowing the PF for both these structures will lead us to the conclusion that iron will contract in volume as it is heated above 910°C. This change in structure forms the basis for the production of steel, as well as the heat treatment of steel.

3.3 CRYSTAL IMPERFECTIONS

So far we have described and discussed arrangements of atoms that are perfect in every way. Now it is essential to deviate from this ideal situation because we know that in nature the perfect anything is the exception and the imperfect is the norm. To grow a perfect crystal requires laboratory conditions, and even with such a controlled environment we have achieved limited success. Of course, this lack of perfection in the microstructure of materials used by people is far from being all bad. If it were not for imperfections of many kinds in solid materials, these solids would not possess the properties that we desire them to have. An example would be the heat-treating process used with high-carbon steel to change the properties of the steel to suit certain conditions demanded by our present technological age. Without imperfections in the crystalline arrangement of atoms, these processes would be severely limited if not incapable of changing the structure and hence the properties of steel. The whole semiconductor industry owes much of its existence to the imperfections in bonding arrangements of the atoms' outer shell electrons. Therefore, it is now essential to delve into the several imperfections of a solid's atomic structure so that we can learn how to take advantage of such disorder in the atomic structure. It is important to point out here that we are still talking about crystalline materials and not amorphous (noncrystalline) materials. As you know, *amorphous materials* have no regular atomic structure. This is not to say that amorphous solid materials are of no use. There are a multitude of uses for these materials, such as glass, most polymers, and even some metals. Crystal imperfections fall into two categories; those involving impurity atoms and those in which there is some disorder in the atomic structure brought about by something other than impurity atoms.

3.3.1 Crystal Impurities

The word *impurity* comes from the use of the word when referring to the small percentage of copper in sterling silver that distinguishes sterling from pure silver. One talks about the impurities in copper that reduce its conductivity. In many cases these impurities are purposely added to improve a material's properties and/or reduce its cost. Such a solid, called an *alloy,* is a combination of two metals. Brass is an alloy, consisting of copper to which has been added some zinc. The addition of zinc has a great effect on the hardness, strength, ductility and conductivity of the pure copper. Our objective in the following discussion is to explain in simple terms why adding these impurities (the zinc atoms) to the copper atom produces such differences in the properties between the pure metal and the alloy.

Before proceeding, definitions of the words solution, solvent, solute, mixture, alloy, and diffusion are required. A *solution* is a homogeneous mixture of chemically distinct substances that forms a phase. A *phase* is defined briefly as a physically distinct material that has its own structure, composition, or both. Uses of phases in solids (specifically metals) are discussed in Module 5. The atoms or molecules of one substance are uniformly distributed throughout the other on a random basis. The substance that is present in the greatest proportion is the *solvent.* The other substance or substances present are the *solutes*.

A *mixture,* on the other hand, is a material that has no fixed composition and contains more than one phase. The components (substances) can be identified and separated by physical means. Thus a mixture of sugar and salt crystals is not homogeneous nor is it random. One can readily see the two distinct crystalline phases and, with some patience, segregate one crystal phase from the other. However, if you dissolve sugar and salt in water in a dilute concentration, they form a liquid solution. In this case the salt and sugar lose their individual identities by dissolving in water.

Air is an example of a solution of many gases dissolved in another gas. A similar situation occurs in solids, producing *solid solutions. Alloys,* then, are a combination of two or more metals forming either a mixture or a solid solution. Steel is a mixture of iron with a bcc structure and cementite with an fcc structure. Brass is a solid solution because its single-phase structure is all fcc; it is a solid solution of copper (solvent) to which some zinc (solute) has been purposely added. The zinc atoms are diffused into the atomic structure of the copper on a random, uniform basis.

Diffusion comes from the Latin verb meaning "to pour out" or "spread out." In material science it stands for the intermingling in solid materials of atoms (in metals), ions (in ceramics), or molecules (in polymers). This active movement of particles is fairly well understood through our experiences with gases and liquids. A bottle of perfume uncapped in one part of a room will soon disclose its presence to people in a distant corner of the room by the diffusing of perfume atoms through the air in the room. Salt or sugar dissolved in water will diffuse throughout the water and form a solution in which the salt or sugar atoms will be evenly distributed throughout the water.

Our main preoccupation in our present study of materials is with solids. Diffusion takes place in solids, too. We know that the individual atoms and molecules in a gaseous or liquid state are relatively far apart, offering little opposition to other atoms migrating through them. In solids the atoms are tightly held close together. If a metallic crystalline solid were formed with a perfect crystalline structure, atoms would find it impossible to move about. But in our study of imperfect crystal structures we will learn that point defects are the rule and not the exception. These defects are one main reason why atoms of a solid can actively move about within the atomic structure of the solid. Combine the presence of vacancies with the fact that each atom possesses sufficient energy to cause it to vibrate about its position in the lattice structure and it is fairly easy to visualize why certain atoms possessing higher average energy in the crystal structure can break their bonds and "jump" from one lattice site to one that is not occupied (a vacancy) in the lattice structure. Once the atom moves to a new site, it leaves behind another vacancy. In other words, the atom exchanges positions with a vacancy.

The phenomenon of diffusion is especially important in solid materials in understanding the manufacturing and functioning of semiconductor materials; the carburizing of steel in surface or case hardening; the production of metal alloys including steel, the primary alloy of iron; and the heat treating of aluminum alloys, called precipitation hardening. These metallurgical processes are discussed elsewhere in this text.

In polymeric materials, diffusion is aided by similar defects in the molecular structure. The diffusion of a penetrant into a solid polymer is of great importance. If the penetrant is a gas, the gas may permeate through the solid (even glass). Polymer films are designed to prevent the diffusion of gases and water vapor into foods that have been wrapped in the film. Liquid solvents may permeate a polymer and produce a softening of the polymer.

The doping (alloying with a very small amount of alloy addition) of silicon with phosphorus in the alloying of silicon to increase its electrical conductivity is another diffusion process that depends on point defects in the silicone structure. Many sintering processes, such as powder metallurgy and the large number of welding and brazing processes for the joining of two metals by local coalescence, depend on the transport of atoms by diffusion.

Because the internal energy possessed by atoms is related to temperature, an increase in temperature will increase the rate of diffusion. Of course, if there are no vacancies in the structure, very little, if any, diffusion will occur.

Other factors that affect the diffusion of atoms, molecules, or ions is the type of bonding of the matrix atoms (i.e., strong bonding takes more energy to break the bonds). An example is high-melting-point solids. Smaller permeating atoms stand a better chance of diffusion through a structure of larger atoms, as in the case of carbon atoms diffusing through a structure of iron atoms. Our study of the microstructure of solids tells us that a lattice structure that contains loosely packed atoms (less dense) will offer less resistance to diffusing atoms than one whose structure contains tightly packed atoms. Finally, diffusion depends on time. This translates usually into allowing sufficient time for diffusion to take place at some higher temperature, as in an oven or furnace.

In summary, diffusion is the process depended on by material scientists to change the microstructure of solids so as to vary the properties of the many solid-state materials in use by our society. Having completed a brief background discussion of solvents, solutes, solid solutions, alloys, and the very important phenomenon of diffusion, we will return to the subject of crystal imperfections. The first of two categories of such imperfections in a crystal structure involves, as mentioned, impurity atoms, which produce two types of solid solutions, substitutional and interstitial solid solutions.

3.3.2 Substitutional Solid Solutions

In a substitutional solid solution (Figure 3-32), the solute atoms replace some of the solvent atoms in the crystal structure of the solvent. Using brass as our solid solution, up to about 40% of the copper atoms can be replaced by zinc atoms. This is possible because the atoms of copper and zinc are much alike. Their atoms are about the same size (atomic radius of the copper atom is 1.278×10^{-10} m and of zinc it is 1.39×10^{-10} m). Their electron structures are comparable:

$$29^{Cu}: \quad 1s^2, 2s^2, 2p^6, 3s^2, 3p^6, 4s^1, 3d^{10}$$

$$30^{Zn}: \quad 1s^2, 2s^2, 2p^6, 3s^2, 3p^6, 4s^2, 3d^{10}$$

Their crystalline structures are both fcc with a CN of 12. Another good example of a solid solution is monel. Monel is a solution of copper in nickel (about 70% N and 30% Cu). The range of solubility goes from practically no nickel to almost 100% nickel. Again, these two elements have a common crystalline structure (fcc), and the atomic radius of nickel is 0.1246 nm. Therefore, monel is a substitutional solid solution. Another interesting fact about this type of solid solution is that atoms may fill only one type of site in the lattice structure of the solvent atoms. For example, in the alloy of copper and gold, the majority of copper atoms occupy the face-centered sites and the gold atoms the corner sites of the face-centered cubic unit cell. An *ordered substitutional solid solution* is formed. As a rule, two distinct elements may form a substitutional solid solution if the sizes of their atoms do not differ by more

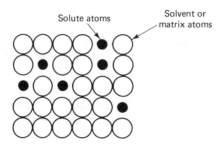

Solute atoms

Solvent or matrix atoms

Figure 3-32 Substitutional solid solution.

than 15%. Of course, there are further restrictions on the degree of solubility brought about by any differences in their crystal or electron structure.

The solute or impurity atoms, although of similar size, may be larger or smaller than the solvent atom, which will produce only a slight distortion in the lattice structure.

3.3.3 Interstitial Solid Solutions

The second type of solid solution formed by impurity atoms is the *interstitial* (Figure 3-33). If the impurity atoms take up sites in the lattice structure that are normally unfilled or unoccupied by the pure (solvent) atoms, they form an interstitial solid solution. These normally unfilled voids or vacant spaces are called *interstices*. In the fcc unit cell, we know there is a relatively large interstice in the center and smaller interstices near each corner atom. It is worthwhile to point out that steel making is made possible because of the formation of an interstitial solid solution. First, we know that iron is allotropic. At temperatures below 910°C, iron is in the bcc form.

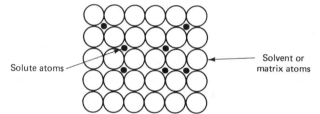

Solute atoms — Solvent or matrix atoms

Figure 3-33 Interstitial solid solution.

Above that temperature the bcc structure changes to fcc to accommodate a higher energy level of the atoms. In the fcc structures, carbon atoms can form in the interstices of the iron unit cell. At temperatures below 910°C the bcc structure contains no room for the carbon atoms to fit between the iron atoms. This fact forms the basis for many of the heat-treating procedures used to produce a multitude of steels with the many different properties required by our technological society. A detailed discussion of heat treating will follow later in this text.

3.3.4 Crystal Defects

The second category of crystal imperfections or lattice defects, a disorder of the crystal structure, is brought about by some mechanism such as thermal agitation of the crystal during its formation, the effects of gravity, or the result of high-energy radiation. Such deviations from the perfect crystal are classified for purposes of explanation as *point defects, line defects,* and *area defects.* Actually, defects occur in all combinations.

Point defects, the simplest and best understood, affect only the small volume of the crystal surrounding a single lattice site. One such point defect is a *vacancy,* that is, the absence of an atom at a lattice site in the otherwise regular crystal (see

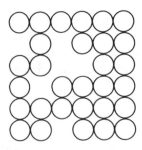

Figure 3-34 Point defects (vacancies).

Figure 3-34). As a result, the electronic bonding of the adjacent atoms is disrupted, which changes the effective radii of these atoms. This weakens the crystal. If sufficient vacancies were produced by heating of a crystalline solid, the crystal structure would lose its long-range symmetry and order, resulting ultimately in porosity or a change to a fluid. It is important to note that these local imperfections in the crystal structure produce a disequilibrium that has a great effect on the important properties of crystalline solids such as density, mechanical strength, diffusion, and electrical conductivity. Point defects do not by themselves affect strength as much as they affect diffusion—the migration of atoms.

Another point defect is called an interstitial defect or interstitialcy. This is produced by the presence of an extra atom in a void, the space between normal lattice positions. This interstitial atom may be specifically added as an alloying element or it may be an impurity atom indigenous to the solid. Other impurity atoms may deposit themselves in a lattice position reserved for atoms of the solid. In any case, these point defects produce local aberrations in the atomic arrangement in the crystal, which produce varying degrees of local disorder in the bond structure and energy distributions in the solid. It is worthwhile to mention that disorder can occur below the atomic level (i.e., at the subatomic level). The imperfections in the electronic structure of atoms exist also. This fact has been capitalized on by the semiconductor industry, which produces materials with varying electrical properties. Further treatment of point defects in the electronic structure of atoms will be included under the topic of electronic bond theory of nonmetallic crystalline solids. It should be evident now that imperfections in the atomic, ionic, and electronic structure of solids do not all have a negative effect on a solid's properties. Some imperfections can improve certain properties; others may degrade some properties. All imperfections do not affect the strength of solids.

So far we have discussed point defects in terms of metallic crystals, with the exception of the brief mention of nonmetallic crystals with the "imperfections" in their electronic structures. As we know, all metals crystallize when they solidify. So do most ceramics and some polymeric (plastic) materials. In the case of ceramic materials, the crystal structures are more complex than metals, and the bonding is ionic, covalent, or a combination of metallic and covalent. With ionic bonding the atoms, of course, become ions. In the ionic solid sodium chloride (NaCl), each Na^+ cation finds itself surrounded by six equally spaced Cl^- anions, and each Cl^- anion is surrounded by six Na^+ cations (CN = 6). Figure 3-35 is a sketch of a model of the unit cell for NaCl upon which the crystalline structure is built. Each unit cell has

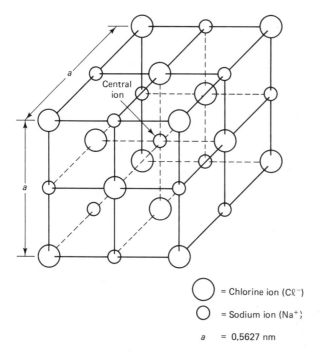

= Chlorine ion (Cℓ⁻)

= Sodium ion (Na⁺)

a = 0.5627 nm

Figure 3-35 A unit cell of NaCl.

four sodium ions and four chlorine ions associated with it. In this model the very center of the cube is occupied by a sodium ion. Each of the other 12 sodium ions is shared by three adjacent unit cells so that each contributes one-fourth of an ion to the unit cell sketched. The total number of sodium ions, therefore, is $1 + \frac{1}{4}(12) = 4$. The 14 chlorine ions distribute themselves in a similar fashion and contribute a total of 4 chlorine ions to the unit cell under discussion. Each set of ions forms a fcc structure with equal numbers of sodium and chlorine ions. The radii ratio of sodium to chlorine is 0.54, with the Na ion being almost half the size of the Cl ion. This difference in size of the ions places restrictions on other ions that could replace them. Furthermore, the replacement ions must have the same number of exterior (valence) electrons. With these restrictions in mind, point defects in ionic or ceramic crystals are the rule rather than the exception. A vacancy may consist of pairs of ions of opposite charge. An interstitialcy would consist of a displaced ion located at an interstitial site in the lattice structure.

The second type of imperfection, the *line defect,* is also known as a dislocation (see Figure 3-36). A *dislocation* is a linear array of atoms along which there is some imperfection in the bonding of the atoms. An undeformed crystal lattice is represented by Figure 3-37. Figures 3-38a and b show what appears as an extra or incomplete plane of atoms (*A-B*) that causes distortion of the crystal structure. The two-dimensional representation also has atoms behind each atom shown as open dots. The atoms in the area circled by a dashed line represent a center or core of poorly bonded atoms that extend back into the material along a line normal to the paper. This line defect is known as an *edge dislocation,* whose symbol is (⊥). A force *P* acting as shown would cause the rows of atoms to move in the direction of the force, each row moving

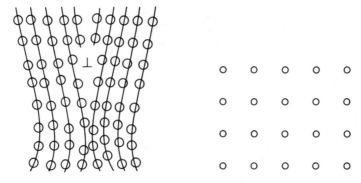

Figure 3-36 Line defects (dislocation).   **Figure 3-37** Undeformed crystal.

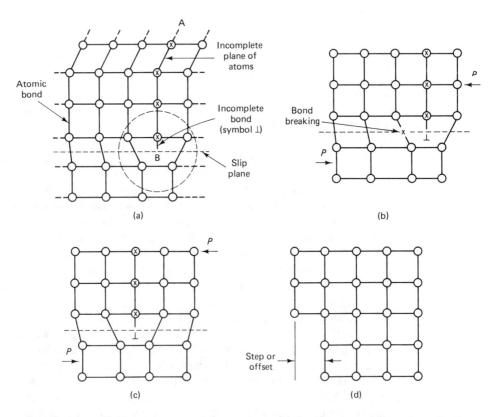

(a)

(b)

(c)

(d)

Figure 3-38 Edge dislocation movement (slip). (a) Dislocation-incomplete row of atoms above slip plane. (b) Shear force (*P*) causing dislocation to move. (c) Dislocation moved one row to the left. (d) Dislocation reaches surface of crystal producing plastic strain (deformation).

one at a time much like dominoes striking each other. As each row of atoms moves, the next follows in turn until all planes of atoms in the area have been displaced sufficiently to make all planes continuous, as shown in Figure 3-38c.

This line of local disturbance represented by the five atoms circled in Figure 3-38a, which may extend to the boundaries of the crystal, is a region of higher energy. As previously stated, this region contains the line defect or dislocation. The dislocation plane contains this line. Above this line the atoms are under a compressive stress, while those atoms below the line are experiencing a tensile stress. These bonding forces are not as strong as in a perfect crystal lattice, which permits a relatively small shear force to break the bonds, allowing the dislocation to move. The bonds reform after the dislocation passes.

The successive passage or slipping of planes of atoms has been likened to the sliding of a large, heavy rug. To move the rug requires a large force. However, if you make a wrinkle in the rug and push the wrinkle a little at a time the rug can be moved. The small movement of the rug to make the wrinkle can be thought of as the slipped portion of the rug; the other portion of the rug, the unslipped region. The wrinkle is the dislocation that separates these two regions.

The displacement of atoms (*slip*) is in a direction that is perpendicular to the dislocation line and/or plane. Where the direction of slip is parallel to the dislocation line, the line defect is called a *screw dislocation,* denoted by the symbol ↻. This is depicted in Figure 3-39. As with point defects, many line defects are actually combinations of edge and screw dislocations producing curved dislocation lines or loops that start and end within the crystal. Dislocations originate during crystallization or plastic (inelastic) deformation. More will be said about dislocations when the topic of plastic deformation is treated later in this book.

However, before leaving dislocations the following observations are made. The ideal crystal structure would contain no deformations. Experience with man-made, near-perfect crystals or *whiskers* has indicated that such whiskers contain great strength.

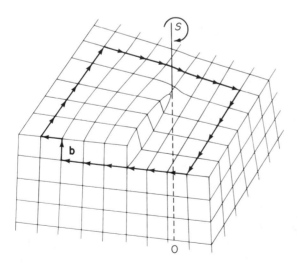

Figure 3-39 A screw dislocation. (Arthur L. Ruoff, *Materials Science* © 1973. Reprinted by permission of Prentice-Hall, Inc., Englewood Cliffs, N.J.)

The reason why a relatively little force is able to deform a crystal structure with strongly bonded atoms or ions is that only a few atomic bonds need to be broken and reformed when dislocations are present in the crystal structure. In view of our present inability to produce near-perfect crystalline solids, the problem resolves into one of determining how to control this movement of dislocations by hindering the movement (strengthening the solid) or facilitating it (temporarily weakening the solid for some purpose).

In summary, line defects have a great deal to do with the strength of a solid. An abundance of them will cause a mutual interference in their movement through a crystal preventing the planes of atoms from slipping, thereby strengthening the material. The presence of a few dislocations increases the ductility of a crystalline solid.

Area defects are the third type of imperfection and exist in the form of grain boundaries (Figure 3-40). As each crystal grows it establishes its own axes system upon which the atoms/ions orient themselves. Adjacent crystals with their differently oriented lattice structures close in on each other. The last atoms to take up position in a crystal find it more difficult to occupy normal lattice sites. Consequently, a

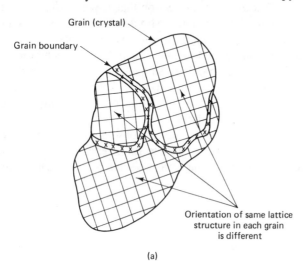

Grain (crystal)

Grain boundary

Orientation of same lattice
structure in each grain
is different

(a)

Nucleation and
dendrite formation

Grains forming
and orienting

Complete
solidification

(b)

Figure 3-40 (a) Grain boundaries. (b) Grain growth.

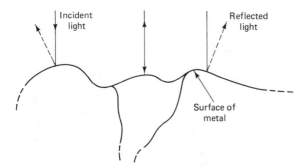

Figure 3-41 Identification of individual grains by reflected light.

transition zone is formed that is not aligned with any of the adjoining crystals. The atoms making up the grain boundary possess greater disorder, and hence greater energy than their counterparts within the crystals themselves. Furthermore, the atoms are less efficiently packed together. These factors signify that the atoms in the grain boundaries are ready to act as sources of new crystal formations (nucleation sites) once the right conditions are met. Second, they assist in the diffusion of atoms through the solid. Third, they offer resistance to the movement of dislocations and therefore modify the strength and the ability of materials to plastically deform. Fourth, they act as sinks for vacancies. A solid with a large number of individual crystals (also called grains) has more grain boundaries than a solid with a lesser number of grains. Fine-grained material will, at normal temperatures, be stronger than coarse-grained material. Figure 3-40 is a two-dimensional sketch showing grain boundaries and the different orientations of grains.

It must be remembered that a solid contains crystals each having the same lattice structure. What is different between the individual grains is the orientation of that structure within each grain. A technique for revealing the details of the grain structure of solids uses reflected light through a metallographic microscope (Figure 3-41).

3.4 SINGLE-CRYSTAL CASTINGS

Normal metal castings produce polycrystalline structures. These many crystals or grains of random orientation possess weak atomic bonds across grain boundaries since impurities gather at boundaries. Techniques to improve across-boundary strength have included mold material and alloying and producing refined *equiaxed* castings. Then casting technology worked to control the direction of grain boundaries and finally eliminate them for certain applications. *Directional solidification* (Figure 3-42) produces long grains growing continuously and parallel from one end of a casting to the other, with the longitudinal growth in the direction to withstand greatest stress. The next step involved selection of one grain and control of its solidification into a single-crystal casting (monocrystal, Figure 3-42). The monocrystal eliminates the need for boundary strengtheners, which depress melting points and limit high-temperature use, allows high treatment temperatures, and improves high-temperature corrosion resistance. The monocrystal turbine blade seen in Figure 3-42 represents the growing

Figure 3-42 Photograph of metal castings of turbine engine airfoils showing the progress in grain control from equiaxed (left) directionally solidified, to mono crystal (right). (Howmet Turbine Components Corp.)

acceptance for high temperature use of single-crystal castings. Space shuttle turbine blades were originally cast as directional solidified and then as single crystal for the main engines.

3.5 AMORPHOUS METALS AND METALLIC GLASSES

A new breed of material developed in the late 1970s by Allied Corporation's Metglas Products Department with the trademark METGLAS is an amorphous metal alloy produced by bringing molten metal alloy at about 1000°F into contact with a rapidly moving and relatively cool substrate (or chill block) in a continuous casting process. The drastic quenching operation is similar to that used in conventional solid-state thermal processing of such metals as steel to transform austenite to martensite. It differs in that the cooling is so rapid that for most metallic glasses the minimum rate is 10^5 kelvins/second (K/s). To ensure a uniform cooling rate at all points within the metal, one dimension of the liquid layer must be kept to a minimum. Thus foil, wire, or powder are the forms presently being produced. The magnitude of the cooling (about a million degrees per second) allows but 1 millisecond (ms) to solidify the metal. The end result is that atomic diffusion is prevented, which in turn precludes any nucleation and growth of crystals. A glassy state consisting of one chemically homogeneous phase with atoms packed in a random arrangement similar to that of glass or liquid metal is produced. Alloy compositions consist of transition elements such as Fe and Ni with small percentages of metalloids (B, C, Si, P). Also, mixtures of transition elements such as B and Ni are common. Initial applications of this material are in vacuum furnace brazing of engine parts for turbines in which the older, less desirable brazing transfer tape is replaced by METGLAS brazing foils.

The ductility of METGLAS foil is more than favorable with foil being folded back on itself, inducing permanent deformation without fracture. Such a property permits it to be converted to various preforms, as well as punched to exact shape to conform with various brazing joint designs.

Ferrous amorphous metals are "soft" magnetic materials, meaning they possess no preferred orientation of crystals and hence no easy direction of magnetization. In practical terms, these new materials have a vast potential for use as metal cores in electrical transformers in which billions of kilowatt hours are lost in overcoming resistance to changes in the direction of magnetization of the presently used silicon metal alloy cores. Other uses for such a family of materials are being developed daily by a multitude of industrial concerns.

3.6 SELF-ASSESSMENT

3-1 Explain the difference between a metallurgical microscope and a metallograph.

3-2 In an x-ray diffraction analysis the x-ray source used produces an x-ray with a wavelength of 1.32×10^{-10}m. The angle of maximum diffraction (θ) is found to be 10°40' with n of 1. What is the interplanar distance between the atomic planes which produced this diffraction?

3-3 Why is one type of electron microscope (SEM) only used to analyze the surface of a solid material?

3-4 Why are electron microscopes operated within a vacuum?

3-5 Glass lenses are used to focus the beam of light in an ordinary optical microscope. What type of lenses are used to focus electron beams? Explain why such lenses affect an electron beam.

3-6 Define the word spectroscopy. Is it possible to have a spectrum of something other than sunlight? If so, name some spectra other than light spectra.

3-7 What is the volume of a piece of iron having a mass of 864 g?

3-8 Under what conditions can atoms or ions form a fcc crystal structure in which each atom is in contact with twelve other atoms or ions?

3-9 What is the coordination number (CN) for the fcc structure formed by ions of sodium and chlorine producing the chemical compound NaCl?

3-10 Table salt and water both form crystals. What, if any, is the difference in the manner in which these crystals are held together as a solid material?

3-11 Using a hard ball model (rubber ball or styrofoam), construct a fcc structure to show the closest packed atom planes (potential slip planes). How many such planes are there?

3-12 A single crystal of zinc, like many other single metal crystals, is fairly ductile. There is nothing to prevent slippage of planes of atoms over one another. Explain why some of these metals, like zinc, become brittle when they are in the polycrystalline state, although they are ductile in single crystal form.

3-13 Consult a table of properties of materials and record the melting point temperatures of several common materials. Explain why there are differences in such temperatures when these solids change from a solid to a liquid.

3-14 What holds the atoms (ions) together in a compound such as NaCl?

3-15 Compare the ionic radii of negative ions with the radii of positive ions for the same elements. Tin, sulfur, or lead would be likely candidates. Explain the differences, if any.

3-16 The atomic radius of copper is 1.278×10^{-10}m. Express this radius in terms of nanometers (nm) and in inches (in.).

3-17 Whenever the density of a material is mentioned or recorded in a table the temperature is also noted. What is the explanation for this?

3-18 What is the CN for a hcp unit cell?

3-19 What is the maximum number of spheres which can surround and be in contact with a given sphere provided all of them are the same size?

3-20 Is there some relationship between the linear coefficient of thermal expansion for a solid and its melting point? List both the coefficient and the melting point temperature for a metal, ceramic, and a polymer and note the differences.

3-21 The highest atomic packing factor is obtained when atoms are of the same size. What is its magnitude?

3-22 The ideal shear strength of a perfect crystalline structure of iron has been calculated to be around 10^{10}N/m^2. This theoretical strength has never been attained. "Whiskers" of iron grown in a near-vacuum in a laboratory approach this value. With the new capabilities provided by the NASA Space Program, discuss the future ramifications of being able to grow large, flawless crystals which are undistorted by their own weight as they form in space laboratory conditions.

3-23 Observe a piece of galvanized steel. Describe and sketch the zinc coating as you observe it with the naked eye.

3-24 In the cubic axes sytem the measures of the angles α, β, and γ are how many degrees?

3-25 Sketch an fcc unit cell.

3-26 Using Table 3-1, determine the crystal structure of lead.

3-27 The rapid quenching of austenite in the heat treatment of steel produces a super-saturated and distorted crystal structure known as body-centered tetragonal (bct). Sketch the unit cell for such a structure.

3-28 What are the most likely directions for slip to occur in hcp crystals? Sketch the unit cell.

3-29 Describe the solid state in terms of kinetics and energy.

3-30 Why is lead more ductile than tin?

3-31 What steps can be taken to reduce slip in metal crystals?

3-32 Obtain a rubber band. Stretch it as far as it will go without breaking while holding it in contact with your moistened lips. Observe the temperature change. Hold the band in this stretched condition for about 30 seconds. Note any change in temperature. Release the band suddenly to its original unstretched length and touch it to your lips. Note the temperature change. Rubber is an elastic polymer when stressed. The stressed molecules align themselves and local crystallization occurs. When the stress is released the molecules return to their original arrangement. Determine if energy is absorbed or released when an amorphous material crystallizes. Record all your observations and comments.

3-33 Diffusion of atoms through a solid takes place by two main mechanisms. One is diffusion through vacancies in the atomic structure. Describe another method of diffusion.

3-34 Name two possible substances formed by alloying two or more metals.

3-35 Would chromium atoms be a likely substitute for aluminum atoms when forming a substitutional solid solution? Ruby rods are doped with chromium atoms to convert the well-known ceramic (CrO_3 in Al_2O_3) into a laser material.

3-36 Disorder in the arrangement of atoms in solids is limited to the atomic level. Cite an example which refutes this statement.

3-37 Grain boundaries in crystalline solids play an important role in the movement of dislocations through a solid. Name at least three actions that take place in the grain boundary transition zone in a crystalline material.

3-38 List some examples of solids which are allotropic.

3-39 Metals are classified as crystalline materials. Name one metal that is an amorphous solid and name at least one recent application in which the use of it is saving energy or providing greater strength and/or corrosion resistance.

3.7 REFERENCES AND RELATED READING

FULRATH, R. M., and J. A. PASK. *Ceramic Microstructures*. Huntington, N.Y.: R. E. Krieger Publishing Co., Inc., 1976.

HALL, CECIL E. *Introduction to Electron Microscopy*. Huntington, N.Y.: R. E. Krieger Publishing Co., Inc., 1983.

HIRSCH, P. B., and others. *Electron Microscopy of Thin Crystals*. Huntington, N.Y.: R. E. Krieger Publishing Co., Inc., 1977.

LOVELAND, R. P. *Photomicrography: A Comprehensive Treatise*, Vols. 1 and 2. Huntington, N.Y.: R. E. Krieger Publishing Co., Inc., 1981.

McLEAN, D. *Mechanical Properties of Metals*. Huntington, N.Y.: R. E. Krieger Publishing Co., Inc., 1977.

PHILLIPS, V. A. *Modern Metallographic Techniques and Their Applications*. Huntington, N.Y.: R. E. Krieger Publishing Co., Inc., 1971.

SCHULTZ, J. M. *Diffraction for Materials Scientists*. Englewood Cliffs, N.J.: Prentice-Hall, 1982.

4

PROPERTIES OF
MATERIALS

4.1 PAUSE AND PONDER

We have stressed the point that properties of materials are determined by their structure. Architects and engineers are limited in their creativity by the properties of the materials available to them. The demand for better computers that process more data at higher and higher rates of speed requires engineers to fully utilize the limited materials available to them.

In Figure 4-1 we see an example of creative engineering with computers. The thermal conduction modules (TCMs) were a key element in the design of the IBM 3082 processors. As very large scale integration (VLSI) circuit density increases on semiconductor chips, dissipating the heat generated from these circuits becomes more of a critical factor in the design of a circuit package. The TCM design project involved several years of computer simulation, analysis, and testing of cooling mechanisms, module materials, and structures.

The cutaway section seen in Figure 4-1 illustrates the main functional elements. The unit consists of a cooling hat, which dissipates the heat conducted from up to 133 silicon chips via spring-loaded pistons atop each chip. The TCM provides an inner chamber for the hermetically sealed unit, which employs helium gas to provide an inert ambient to enhance reliability and to provide a conduction cooling medium six times more efficient than air. A lead-plated C-ring enables hermetic sealing of the unit. The chips are mounted on a 90-millimeter-square substrate, the largest ever developed by IBM, which consists of 28 layers of ceramic sheets containing 130 meters of internal wiring. A filler port/plug enables helium

Figure 4-1 Thermal Conduction Module Cutaway. (IBM.)

gas pressurization. Some 1800 input/output (I/O) pins are brazed onto the bottom of the substrate to provide interconnections to the circuit board, the next level of packaging. A thermocouple monitors the operating temperature of the TCM unit, which is field-replaceable and can be reworked a number of times to accommodate engineering changes.

As you read through this module, try to match each property studied to a specific material and service condition from your own experiences.

A material's structure (microstructure and macrostructure) determines how it reacts to certain conditions of stress or environment. Properties, such as mechanical, chemical, or optical, reveal a material's ability to resist or react to specific conditions. This chapter will deal with the major properties involved in materials science, including the basic units of measure and methods related to the testing of properties. This knowledge is essential to the proper selection and use of materials.

The properties of a material are those characteristics that help identify and distinguish one material from another. Taken as a whole, these qualities define a material. All properties are observable and most can be measured quantitatively. Properties are classified into two main groups, physical properties and chemical properties. This classification is somewhat arbitrary. However, the major difference between the two is based on whether or not there is any change in the material during observation of the property. For example, when we observe the density, color, or hardness of a material, no change in the material is brought about. On the other hand, measuring such a property as the deflection temperature of a plastic material or the corrosion resistance of a ferrous metal does cause a change to take place in the material; the latter are therefore categorized as chemical properties.

4.2 PHYSICAL PROPERTIES

Physical properties are, in turn, arbitrarily subdivided into many categories. These subdivisions bear names such as mechanical, metallurgical, fabrication, general, magnetic, electrical, thermal, optical, and electromagnetic. Regardless of the name

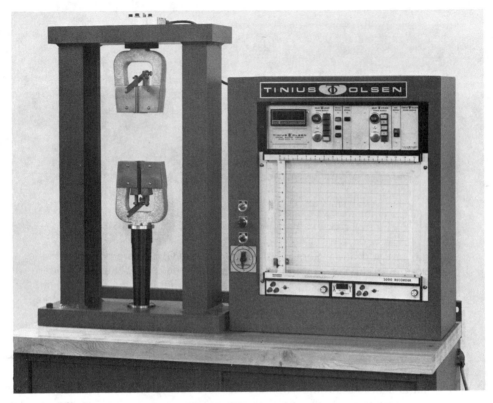

(a)

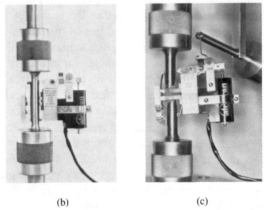

(b)   (c)

Figure 4-2 (a) Photograph of Universal Testing Machine. The grippers, with a space in between, would hold the specimen and strain indicators seen in (b) before stressed, and (c) after breaking. (Tinius Olsen Testing Machine Co.)

of the subdivision, physical properties result from the response of the material to some environmental variable such as a mechanical force, a temperature field, or an electromagnetic field (which, of course, includes visible light and, hence, optical properties). For purposes of this text, physical properties will be divided mainly into mechanical, electrical, magnetic, optical, and thermal.

4.3 MECHANICAL PROPERTIES

In selecting a material for a product such as a piston in an internal combustion engine, a designer is very interested in such properties as strength, ductility, hardness, or fatigue strength. These are some mechanical properties of a material. *Mechanical properties are* defined as a measure of a material's ability to carry or resist mechanical forces or stresses. Stress results from such forces as tension, compression, or shear, which pull, push, twist, cut, or in some way deform or change the shape of a piece of material. Many times this deformation is so minute that only delicate instruments can detect it. Figure 4-2 shows a universal testing machine used to apply loads to material specimens and, with appropriate instrumentation, to detect minute deformations of materials under load.

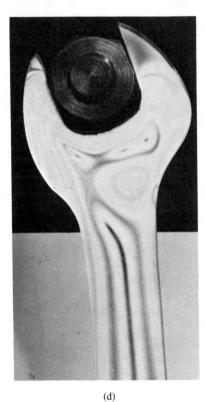

(d)

Figure 4-2 (d) Actual model of wrench for analysis of stress. (Measurement Group Inc., Raleigh, N.C.)

4.3.1 Stress

Stress is defined as the resistance offered by a material to external forces or loads. It is measured in terms of the force exerted per unit area (pounds per square inch, psi). The corresponding SI units are newtons per square meter (N/m^2) or pascals (Pa). One pascal (1 Pa) equals 1 N/m^2. Another way of defining it is to say that stress is the amount of force (F) divided by the area (A) over which it acts. Using σ (the Greek letter sigma) as the universal symbol for normal stress, we say mathematically that

$$\sigma = \frac{F}{A}$$

An assumption is made that the stress is the same on each particle of area making up the total area (A). If this is so, then the stress is uniformly distributed. When a load or force is applied to an object, we are unable to measure the stress produced by this force in the material. What we do is measure the force, identify the area over which the force acts, and measure it as well. These two quantities can then be used to calculate the stress produced in the material by the previous relationship. With the use of polarized light and photoelastic plastic, it is possible to detect concentrations of stress as seen in Figure 4-2d.

4.3.2 Strain (Unit Deformation)

Many times we assume that a body is rigid; that is, when the body is loaded with some force, the body keeps its same size and shape. This is far from correct. Regardless of how small the force, a body when subjected to a force will alter its shape. In other words, the body will change its dimensions. The change in a physical dimension is called deformation (δ Greek letter delta). Figure 4-3a illustrates a rod of original length (L_0) and original diameter (d_0) placed under a tensile load (F) and

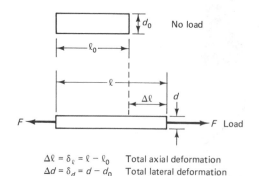

$\Delta \ell = \delta_\ell = \ell - \ell_0$ Total axial deformation
$\Delta d = \delta_d = d - d_0$ Total lateral deformation

Note: Deformations exaggerated for illustrative purposes

(a)

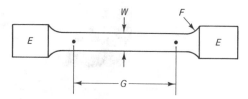

E — gripped ends, may be threaded, plain, or with hole for gripping by machine

W — reduced width to insure specimen breaks in middle — round on round specimens and flat on flat specimens

G — marked gage length to precisely measure the change in length before, during and after test

F — fillet to reduce stress concentrations

(b)

Figure 4-3   (a) Rod under a tensile load. (b) Standard tensile test specimen.

elongated (stretched). Its change in length (ΔL) equals the difference in the two dimensions ($L - L_0$) in the direction of the force. ΔL, in this example, is exaggerated for illustrative purposes. This is known as total deformation, δ. Note also that the original diameter has been reduced in size and produces a corresponding change in the lateral (at right angles to the direction of the load) direction. Note too that in one case the dimension increased and the other decreased. The change in the length is called a total axial or longitudinal deformation. The change in the lateral dimension is known as a total lateral deformation. The ratio of the total axial deformation to the original length is known as the unit axial or longitudinal strain, ε (Greek letter epsilon). The linear units are not canceled and are kept as part of the term. Therefore, in mathematical terms,

$$\varepsilon_{\text{long.}} = \frac{\Delta L \text{ mm}}{L_0 \text{ mm}} = \frac{\delta \text{ mm}}{L_0 \text{ mm}} \qquad \text{(longitudinal unit deformation)}$$

$$\varepsilon_{\text{lat.}} = \frac{\Delta d \text{ mm}}{d_0 \text{ mm}} = \frac{\delta \text{ mm}}{d_0 \text{ mm}} \qquad \text{(lateral unit deformation)}$$

In summary, when a piece of material (a body) is subjected to a load, it will not only deform in the direction of the load (axial deformation), but it will deform in a lateral direction (at right angles to the direction of the load). The ratio of the lateral unit deformation or strain ($\varepsilon_{\text{lat.}}$) to the unit longitudinal deformation or strain ($\varepsilon_{\text{long.}}$), given the symbol μ (Greek letter mu), is known as Poisson's ratio:

$$\mu = \frac{\varepsilon_{\text{lat.}}}{\varepsilon_{\text{long.}}}$$

The unit longitudinal deformation is larger than the unit lateral deformation and therefore Poisson's ratio is less than 1. For steel, it is about 0.3.

In our discussion of deformation we demonstrated it by using a tensile load. If a body was loaded in compression, the length of the body would have decreased and its width increased. One major reason for conducting tension and compression tests using standardized equipment and specimens (see Figure 4-3) is to determine the data needed to plot stress–strain diagrams for the material under investigation.

4.3.3 Stress–Strain Diagrams

The stress–strain diagram is used to determine how a certain material will react under load. Figure 4-4 is a stress–strain diagram for a low-carbon (mild) steel. Strain values (mm/mm) are plotted along the horizontal axis (abscissa), and along the vertical axis (ordinate) are plotted the stress values (MPa, megapascal). The straight-line portion of the diagram up to almost the yield point is known as the *elastic region* of the diagram. Within this range of stresses, the material will return to its original dimensions once the load, hence the stress, has been removed. In the elastic region (0 to 70 MPa), each increase in stress will produce a proportionate increase in the strain. This statement is known as Hooke's law ($\sigma = E\varepsilon$), with E the constant of proportionality.

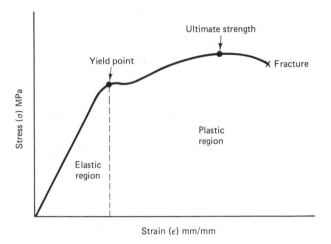

Note: Typical curve for mild steel   **Figure 4-4** Stress-strain diagram.

Beyond the yield point the material will continue to deform, but with less stress than before, because the material has begun to yield. In this region, known as the *plastic region,* plastic deformation takes place, which means that when the load is removed the material will not return to its original dimensions. It now has a *permanent set.* Note also that it takes less load to break the metal specimen than it does to reach the ultimate strength. This is because the material having yielded, the original cross-sectional area of the specimen has been reduced in size such that less material is available to resist the load.

The *yield point* represents the dividing line or transition from the elastic to the plastic region of the curve. When the stress reaches the yield point, there occurs a large increase in strain with no increase in stress. The *modulus of elasticity, elastic modulus, tensile modulus, Young's modulus, modulus of elasticity in tension,* or *coefficient of elasticity,* given the symbol E, is the ratio of the stress to the strain in the elastic region of the stress–strain diagram.* The *tensile modulus* is approximately equal to the *compressive modulus of elasticity* within the proportional limit (elastic limit of the diagram). Note that this ratio expresses the slope of the straight-line portion of the curve. Regardless of the name, this modulus is an indication of the stiffness of the material when subjected to a tensile load. The stiffness of a material is defined as the ratio of the load to the deformation produced. The higher the value of Young's modulus, the stiffer the material, as demonstrated in Figure 4-5.

Not all materials produce stress–strain diagrams (Figure 4-6) on which there is a clear indication of the start of yielding as the load is increased. Such a case is illustrated for cast iron. This situation should not be interpreted to mean that cast iron does not exhibit elastic properties under moderate loads. In other words, such materials are elastic with strains returning to zero when moderate loads, hence stresses, are removed. The modulus of elasticity for these materials is sometimes taken as the slope of a tangent to the stress–strain curve at the origin.

*The SI unit for E is the pascal (Pa).

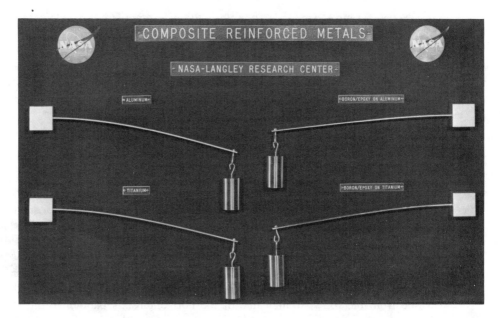

Figure 4-5 The stiffness of pure metals versus metal composites. (NASA.)

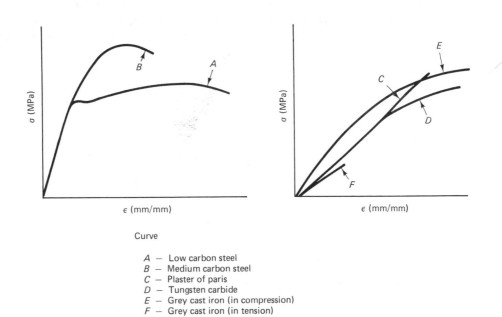

Curve

A — Low carbon steel
B — Medium carbon steel
C — Plaster of paris
D — Tungsten carbide
E — Grey cast iron (in compression)
F — Grey cast iron (in tension)

Figure 4-6 Typical stress-strain diagrams.

4.3.4 Ultimate Strength or Tensile Strength

Ultimate strength or tensile strength is the maximum stress developed in a material during a tensile test. It is a good indicator of the presence of defects in the crystal structure of a metal material, but it is not used too much in design because considerable plastic deformation has occurred in reaching this stress. Plastic deformation is not all bad. However, in many applications the amount of plastic deformation must be limited to much smaller values than that accompanying the maximum stress. The ultimate shear strength is about 75% of the ultimate tensile strength.

4.3.5 Yield Strength

Many materials do not have a yield point. Low-carbon steel is one of just a few that exhibit a point where the strain increases without an accompanying increase in stress. This poses a problem in deciding when plastic deformation begins for such materials. By agreement, a practical approximation of the elastic limit is used called the *offset yield strength*. It is the stress at which a material exhibits a specified plastic strain. For most applications, a plastic strain of 0.002 in./in. can be tolerated, and the stress that produces this strain is the *yield strength*. This is sometimes expressed as 0.2% strain. The yield strength is determined by drawing a straight line called the offset line from the 0.2% strain value of the horizontal axis parallel to the straight-line portion of the stress–strain curve. The stress at which this offset line intersects the stress–strain curve is designated as the yield strength of the material at 0.2% offset. In some cases the offset can be specified as 0.1% or even 0.5%.

  Figure 4-7 shows a typical stress–strain curve for an aluminum alloy with no pronounced yield point. The offset, offset line, and point of intersection of the offset line with the stress–strain curve are shown. When reporting yield strength, care must be taken to include the amount of offset as well as the value of the stress with the

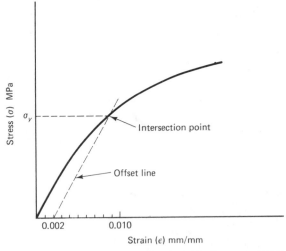

Note: For aluminum alloy 2014-T6 0.2% offset σ_y = 480 MPa.

Figure 4-7 Determining offset yield strength using stress-strain diagram.

data. In general, the yield strength of metals is much higher than that of other materials. For brittle materials, yield strength differs very little, if at all, from tensile strength. As an example, for class 40 gray iron, both strengths are 40,000 psi.

Ductile materials show a wide difference in these strengths. Figure 4-6 compares typical stress–strain curves for various materials. The yield strength in shear for a ductile material is determined to be about one-half (0.577) the yield strength in tension.

4.3.6 Resilience

The modulus of resilience (R), represented by the area under the straight-line portion (elastic region) of the stress–strain curve, is a measure of the energy per unit volume that the material can absorb without plastic deformation. If the unit of volume is one meter cubed, then the SI units of resilience are $MN \cdot m/m^3$. Figure 4-8 shows that this area is a right triangle.

4.3.7 Shear Stress

A second family of stresses is known as shear stress or shearing stress. The symbol τ (Greek letter tau) represents a shear stress. A shearing force produces a shear stress in a material, which, in turn, results in a shearing deformation. Figure 4-9 shows a shearing load acting. In Figure 4-9a, the shear force F produces an angular deformation of a block of material. This deformation is exaggerated for illustrative purposes. In Figure 4-9b, a block of material is subjected to a shear force that, if larger than the shear strength of the material, will shear a section out of the block. A hole in a metal plate can be produced by the action of a punch and hammer that delivers an impact blow, causing the metal in the plate to fail by shear.

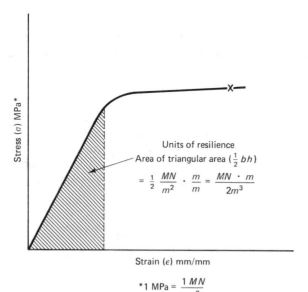

Units of resilience
Area of triangular area ($\frac{1}{2} bh$)

$$= \frac{1}{2} \frac{MN}{m^2} \cdot \frac{m}{m} = \frac{MN \cdot m}{2m^3}$$

Strain (ϵ) mm/mm

$$*1 \text{ MPa} = \frac{1 \, MN}{m^2}$$

Figure 4-8 Resilience of a material as determined by a stress-strain diagram.

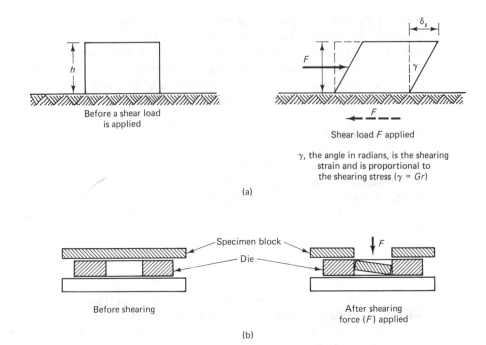

Figure 4-9 Shearing stress and strain. (a) Before shear load. (b) After shear load.

Shear strain γ (Greek letter gamma) then is the deformation (δ_s) produced by the shear force F (see Figure 4-9a) divided by the dimension h, or

$$\gamma = \frac{\delta \text{ in.}}{h \text{ in.}}$$

Note that this ratio is also an expression for the tangent (ratio of the opposite side of an angle in a right triangle to the adjacent side of that angle) of the angle labeled γ in radians.

A shear force acts parallel to the area over which it acts producing a shearing stress (τ) and a shear deformation (δ).

A stress–strain diagram can be plotted using shear stress and shear strain. In so doing it will be found that this diagram will show a definite straight-line portion (elastic region) in which the shearing stress is directly proportional to the shearing strain. Like the normal stress–strain ratio, the ratio of the two shear quantities, G, is known as the *modulus of rigidity* or *shear modulus*. In mathematical terms,

$$G = \frac{\tau}{\gamma}$$

with units of psi or pascals. Finally, we can state, as we did with normal stress and normal strain, that the two quantities can be set equal to each other (Hooke's law) and the preceding equation can be written $\tau = G\gamma$, where G, the constant of proportionality, is the shear modulus.

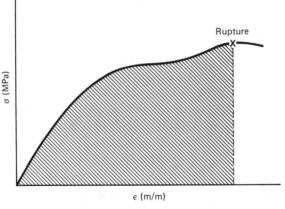

SI units for toughness — J/m^3

Figure 4-10 Modulus of toughness.

4.3.8 Toughness

The ability or capacity of a material to absorb energy during plastic deformation is known as *toughness*. The modulus of toughness (*T*) equal to the total area under the stress–strain curve up to the point of rupture (see Figure 4-10) represents the amount of work per unit volume of a material required to produce fracture under static conditions.

Impact tests are also used to give an indication of the relative toughness of a material. Notched-bar impact tests using either of the two standard notched specimens, the Charpy or the Izod, reveal the material's behavior in sustaining a shock load. These notched specimens of the material under test have either a keyhole or a V-notch cut to specifications. Figure 4-11 shows a universal impact tester for conducting

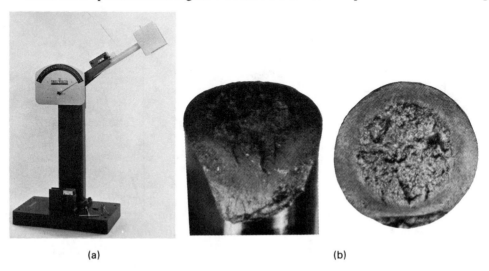

(a) (b)

Figure 4-11 (a) Impact testing machine. (Tinius Olsen Testing Machine Co.) (b) Impact failures. (John Deere Co.)

Charpy, Izod, and tension impact tests on materials. Figure 4-11a is a picture of an impact failure.

Toughness is greatly influenced by temperature. Below a certain temperature (transition temperature), a material may be brittle, whereas above this temperature the material may exhibit ductile qualities. The range of temperatures over which this change takes place is in some cases quite narrow. Therefore, it is very important to know where this transition temperature lies in relation to the possible range of temperatures over which the material is expected to perform satisfactorily.

4.3.9 Ductility

A material that can undergo large plastic deformation without fracture is called a ductile material. A brittle material, on the other hand, shows an absence of ductility. Consequently, a brittle material shows little evidence of forthcoming fracture by yielding as a ductile material would do. A brittle failure is a sudden failure. A ductile material, by yielding slightly, can relieve excess stress that would ultimately cause failure. This yielding could be accomplished without any degradation of other strength properties. Figure 4-12 shows a stress–strain curve for both a ductile and a brittle material. Note the difference in the amount of plastic deformation shown by each curve prior to fracture.

Ductility is measured in either of two ways. In the first method, the *percent elongation* is the ratio of the change in length of a specimen from zero stress to failure, compared to the original length; the quotient is then multiplied by 100%. In terms of a mathematical relationship the above factors can be written as

$$\% \text{ elongation} = \frac{L_F - L_0 \ (100\%)}{L_0}$$

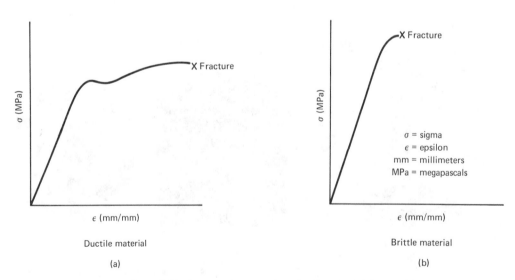

Figure 4-12   Stress-strain curves for a ductile and a brittle material.

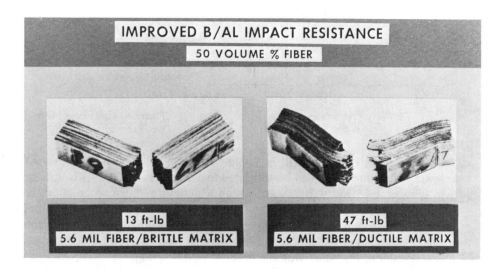

Figure 4-13 Photographs of impact specimens of fiber composites. (NASA.)

The second method, *percent reduction in area,* measures the change in the cross-sectional area of a specimen, compares it to the original cross-sectional area, and multiplies the quotient obtained by 100% or

$$\% \text{ reduction in area} = \frac{A_0 - A_F \ (100\%)}{A_0}$$

It is customary to consider a specimen that has 5% or less elongation as a brittle material. Brittle materials should not be considered as having inferior strength. Such materials lack the ability to plastically deform under load. Figure 4-13 is a series of photographs showing the effects of increased ductility on the impact resistance of a specimen of boron–aluminum composite material.

4.3.10 Malleability

Malleability, workability, and *formability* are some terms related to ductility that describe, in a general way, the ability of materials to withstand plastic deformation without the occurrence of negative consequences (rupture, cracking, etc.) as a result of undergoing various mechanical processing techniques. Terms such as *weldability, brazability,* and *machinability,* although more properly classified under chemical properties, are mentioned here as further examples of terms used to generally describe the reaction of materials to various manufacturing and/or fabricating processes in industry.

4.3.11 Fatigue (Endurance) Strength

A common service condition involves many repetitions of applied stress or reversals of stress. We know that when a horizontal beam with a rectangular cross section is subjected to a downward acting, transverse load (see Figure 4-14), the material in

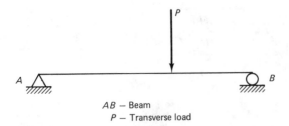

AB — Beam
P — Transverse load

Figure 4-14 Horizontal beam with a concentrated, transverse load.

the top half of the beam will be compressed (the stress s_c is compression) and the lower half of the beam will be subjected to a tensile stress (s_t). This is demonstrated in Figures 4-15 and 4-16. A shaft with a pulley or gear can be compared to the horizontal beam referred to above. The load or force, in this case, is a pulley or gear force, which causes the shaft to deflect (bend), particularly if sufficient bearing supports are not provided. Now if a transverse load is applied and removed in some cyclic fashion, the material in the beam or shaft would go from a condition of stress to a condition of zero stress. This type of cyclic loading is called *repetitious*. If the stress in the material changes due to the loading from compression to tensile, or vice versa, this is known as *stress reversal*. This latter condition is easier to visualize with a point on the surface of a rotating shaft turning at 1000 revolutions per minute whose bending stress changes from compression to tension 1000 times a minute. In practice, many failures have occurred under such conditions when the stresses de-

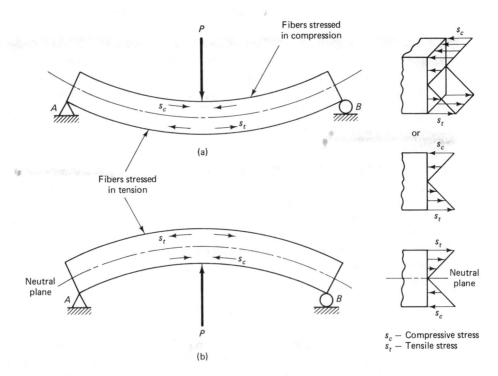

Figure 4-15 Beam deflected by a cyclic transverse load.

Properties of Materials   Module 4

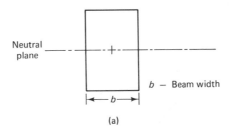

Neutral plane

b — Beam width

(a)

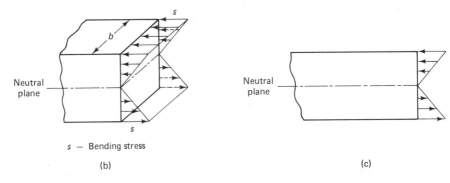

s — Bending stress

(b)

Neutral plane

(c)

Figure 4-16 (a) End view of beam showing cross section. (b) Front view (3-dimensional) of a portion of beam showing distribution of tensile and compressive bending stresses acting on a transverse plane. (c) Front view (2-dimensional) showing same stress distribution as in (b).

veloped were well below the ultimate stress and frequently below the yield strength. These failures are called *fatigue* (or *endurance*) *failures* (see Figure 4-17).

A fatigue failure starts as a tiny crack whose origin is many times traced to an inspection stamp, tool mark, or other defect on the surface. The crack produces a stress concentration that assists in the growth of the crack until eventually the area of material remaining to withstand the stress is insufficient, which results in a sudden

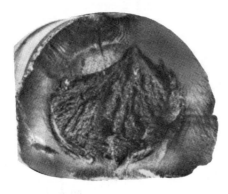

Figure 4-17 Combined fatigue failure due to combined bending and torsional loads on an axle. The outer surface was hardened. A crack began at top surface. Rough surface in center indicates last area to fail instantly when axle broke. (John Deere Co.)

Sec. 4.3 Mechanical Properties

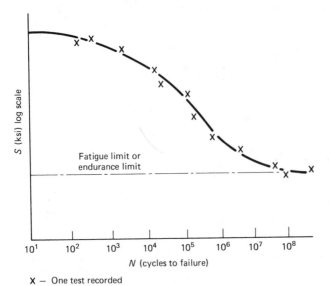

Figure 4-18 *s-N* diagram for a typical fiberglass composite material.

X — One test recorded

failure. Much empirical research is done using fatigue testing machines to determine the strength of materials under fatigue loading. Machines, developed by the National Aeronautics and Space Administration, use computers that simulate aircraft or spacecraft flights to test the material until it fails under varying conditions of load and ambient temperatures. The results of these many tests are recorded on semilog or log–log paper to produce *S* (stress)–*N* (cycles) diagrams. Typical *S–N* diagrams are shown in Figure 4-18 for a fiber-glass composite, in Figure 4-19 for a typical low-carbon steel, and in Figure 4-20 for a nonferrous metal. Figure 4-21 is a photograph of a typical fatigue failure showing two very distinct surfaces.

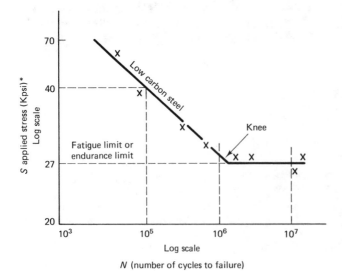

Figure 4-19 *s-N* diagram for a typical low carbon steel.

*Kpsi = Kips per square inch = 1000 pounds per square inch

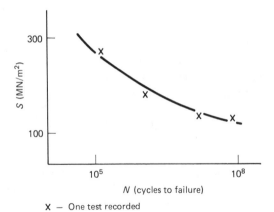

X — One test recorded

Figure 4-20 *s-N* diagram for a nonferrous metal.

Figure 4-21 Fatigue failure due to progressive fatigue cracking, with cracking progressing across most of the section before the final overload fractures the remaining metal. (John Deere Co.)

In the case of steel, Figure 4-19 shows an abrupt break or *knee* at which point the curve tends to approach a horizontal line. In other words, with any stress below the fatigue or endurance limit, this particular steel can be cycled indefinitely without fracturing. Aluminum and its alloys fracture at relatively low stresses after many cycles. Furthermore, they exhibit no fatigue limit which means that there is no stress below which they will not fracture. Therefore, in speaking of fatigue of nonferrous materials, it is necessary to express both the stress and the number of cycles in describing the life of the material. Figure 4-22 is a plot for crack lengths developed in fatigue tests that simulate aircraft flights.

The fatigue or endurance limit indicates that below this stress (S_E) a specimen of that particular material can withstand an indefinitely large number of cycles (N) without fracturing. Fatigue strength (S_F) is any ordinate on the S–N diagram. To make it meaningful, the corresponding N must be included with it. In Figure 4-19 the fatigue strength of 40 Ksi corresponds with an N of 10^5 cycles. The fatigue limit as determined empirically is generally below the yield strength. Most design stresses are lower than the fatigue or yield strength of a material primarily because of the adverse effects of surface conditions on the strength of materials. It is conservatively

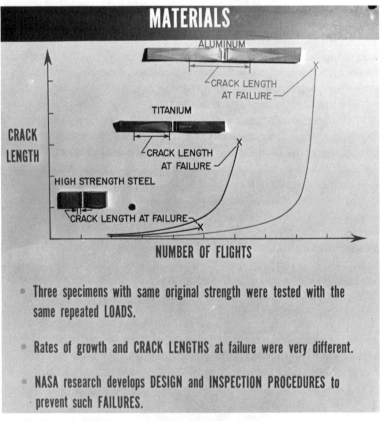

Figure 4-22 Crack propagation and failure. Three specimens with same original strength tested with the same repeated loads. Rates of growth and crack lengths at failure were quite different. (NASA.)

estimated that well over 50% of the failures occurring under service conditions are fatigue failures. This explains in part the necessity for providing steel with some form of corrosion resistance in the form of a paint or zinc coating. If a metal part is to be subjected to a corrosive environment in service, special efforts must be made to provide corrosion resistance. For without it the part is subject to fatigue failure at levels of stress even lower than the fatigue limit. Corrosion is a source of tiny cracks in the surface, which grow with time and produce failures. Some failures are catastrophic giving little, if no, warning. Shot peening and nitriding are other processes used to strengthen the outer layers of a metallic part. In view of the expense involved in running fatigue tests, particularly when attempting to duplicate service conditions in a laboratory, a fatigue or endurance ratio is sometimes used. This ratio, varying between 0.25 and 0.45 depending on the material, compares tensile strength to fatigue strength.

4.3.12 Creep (Creep Strength)

Creep is a slow process of plastic deformation that takes place when a material is subjected to a constant condition of loading (stress) below its normal yield strength. Creep occurs at any temperature. However, at low temperatures, slip (movement of dislocations) is stopped by impurity atoms and grain boundaries. Polycrystalline fine-grained materials, having more grain boundaries than coarse-grained materials, offer greater resistance to creep. At high temperatures the diffusion of atoms and vacancies permits the dislocations to move around impurity atoms and beyond grain boundaries, which results in much higher creep rates. The word *creep* implies, then, that a material plastically deforms or flows very slowly under load as a function of time. After a certain amount of time has elapsed under constant load, the *creep strain* (plastic deformation) will increase and some materials will rupture. This rupture or fracture is known as *creep rupture*. Aluminum alloys begin to creep at around 100°C. Thus, aluminum engine blocks caused serious problems when used with steel components. Although some materials will creep at low or room temperatures, this type of plastic deformation is usually associated with high temperatures. Polymeric materials creep at room temperature, but this low-temperature creep is called *cold flow*. Both steam and gas turbines used for propulsion of ships or the generation of electricity operate, by necessity, at high temperatures over a long span of years. The extremely close fits between turbine blades and their casings prevent the escape of steam or gas past the blades. If these turbine blades were allowed to change their dimensions at any time during their expected service life, during which time they are under large centrifugal loads (rotating at extremely high speeds) at extremely high temperatures for long periods of time, possibly weeks, the failure would be catastrophic. Consequently, the material from which these turbine blades are fabricated must possess, among other necessary properties, high creep resistance. Continuing research over the years has produced such materials for medium- and high-temperature service as the many titanium alloys and the "super-alloys" composed of iron, nickel, and/or cobalt.

Tensile creep tests develop data at a constant load (stress) and at a constant temperature. The amount of creep strain when plotted against time, as in Figure 4-

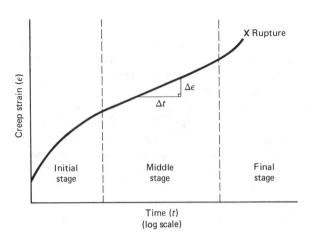

Figure 4-23 Creep curve.

23, indicates that the creep curve formed can be divided into several stages, with the last stage ending with a rupture of the material. The *creep rate* is determined, at any point, by the slope of the curve (drawing a tangent to the curve at the point). The creep rate, the slope of the middle portion of the curve, is nearly constant and represents the minimum rate. Once the creep enters the last stage, rupture soon follows. Creep rate is usually expressed in percent creep strain per hour. A typical rate might be 10^{-4} percent per hour.

Engineering materials must perform satisfactorily throughout their service lives. In many cases, service life may extend well beyond 20 years. A determination must be made as to the maximum allowable deformation that can be tolerated during the expected service life of a material before the material is chosen for the particular application.

Research in polycrystalline materials with varying amounts of grain boundary areas has shown that at high temperatures coarse-grained materials possess more creep resistance than do fine-grained crystalline materials. At low temperatures, fine-grained materials offer more creep resistance. These statements, including the phenomenon of creep itself, can be explained in terms of the movement and stoppage of mobile dislocations (linear arrangements of atoms making up imperfections in the atomic structure of solid materials) throughout the crystal structure. Creep, then, like other properties of materials, is dependent on the structure of materials.

4.3.13 Torsional Strength

Torsion is the word that describes the process of twisting. A body such as a circular rod (e.g., a shaft for transmitting power), as shown in Figure 4-24, is under torsion as a result of a force acting to turn one end around the longitudinal axis of the rod while the other end is fixed or twisted in the opposite direction. The material resists this twisting action by generating a similar twist internally. The product of the force P and the radius r perpendicular to the line of action of the force is called a *torque*. The external torque produces both a stress and a deformation of the rod. The stress

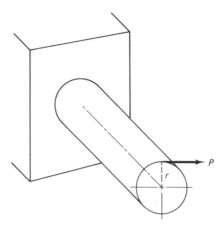

Figure 4-24 Circular rod under torsion.

is classed as a shear stress, which causes the atoms to twist past each other. The shear deformation is measured in terms of the angle γ (Greek letter gamma) and the angle θ (Greek letter theta), as indicated in Figure 4-25. As a result of the torque T, the point C on the surface of the rod moves to point C'. This deformation can be measured by angle ϕ on the surface of the rod or by angle θ shown on the cross-sectional area of the rod. The angle θ is known as the angle of twist. Both angles are exaggerated for illustrative purposes.

A torsion test machine measures the torque applied to a specimen of material along with the corresponding angle of twist (θ). The results can be plotted as a *torque–twist diagram* (see Figure 4-26), which resembles an ordinary stress–strain diagram as obtained by the usual tensile test procedure. The torsional stress is the

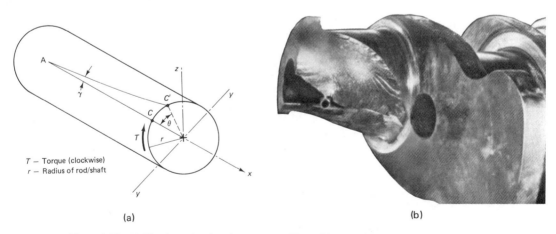

(a)   (b)

Figure 4-25 (a) Circular rod undergoing a torque with resulting deformation. (b) Torsional fatigue. Torsional (twisting) loads produce spiral types of failure. Note the curved line from (A) to (C') is part of a helix.

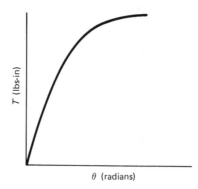

θ (radians)   **Figure 4-26** Torque-twist diagram.

shear stress (τ) produced in the material by the applied torque and is calculated using the equation

$$\tau_{max} = \frac{T\,r}{J}$$

where r is the radius of the cross-sectional area and J is the polar moment of inertia. The maximum torsional stress occurs when r is a maximum, as indicated in the preceding equation and illustrated in Figure 4-27. The maximum torsional stress occurring at the outer surfaces of the circular rod is called the *torsional yield strength*. It relates to that point on the torque–twist diagram where the curve begins to depart from a straight line. Torsional yield strength roughly corresponds to the yield strength in shear. The *ultimate torsional strength* or *modulus of rupture* expresses a measure of the ability of material to withstand a twisting load. It is roughly equivalent to the ultimate shear strength. The *torsional modulus of elasticity* as determined by the torque–twist diagram known as the *modulus of rigidity* is approximately equal to the shear modulus or the modulus of elasticity in shear (G).

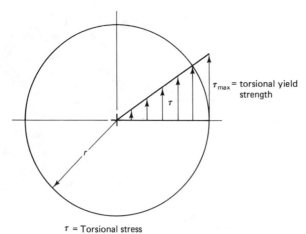

τ = Torsional stress
τ_{max} = Maximum torsional stress
    outer surface of circular
    rod

Figure 4-27 Distribution of torsional stress in circular rod.

4.3.14 Flexural or Bending Strength

A beam is a structural member that bends or flexes when subjected to forces perpendicular to its longitudinal axis. This axis in beam strength determination is called the beam's neutral axis (NA), as no normal stresses are assumed to exist along its length. The neutral axis is the edge view of a neutral plane. Figure 4-28a is a sketch of a simply supported beam. The transverse load or force P bends the beam (causes it to deform, i.e., deflect), thus producing both shear and normal stresses in the beam. The load in this example would cause the beam to deflect as in Figure 4-28b, resulting in normal stresses in compression near the top surface and normal stresses (tensile) at the bottom of the beam. These bending stresses are normal to any cross-sectional area through the beam, which makes their direction parallel to the neutral axis of the beam.

Assuming that the beam material is *homogeneous* (same material nature throughout) and *isotropic* (same properties in all directions throughout the beam material), the normal stresses will be at a maximum near the top and bottom surfaces of the beam. These normal stresses (both compressive and tensile) are known as *flexural, fiber,* or *bending stresses*. The *flexure formula,*

$$\sigma = \frac{Mc}{I}$$

relates these stresses (σ) to the bending moment (M), the maximum distance (c) from the beam's neutral axis (where bending stresses are zero) to the outer surfaces of the beam, and the rectangular moment of inertia of the cross-sectional area of the beam (I). A *flexure test* is performed using a simple beam loaded as shown in Figure 4-28. The maximum bending stress and deformation (deflection) are recorded for increments of load P. These data are plotted to obtain a stress–strain diagram. The maximum bending stress developed at failure is known as the *flexural strength*. For those materials that do not crack, the maximum bending or flexural stress is called the *flexural yield strength*. A bend test. used to determine the ductility of certain materials should not be confused with this flexure test.

4.3.15 Hardness

Hardness is a measure of a material's resistance to penetration (local plastic deformation) or scratching. One of the oldest and most common hardness tests, based on measuring the degree of penetration of a material as an indication of hardness, is the Brinell. *Brinell hardness numbers* (BHN) are a measure of the size of the penetration made by a 10-mm steel or tungsten carbide sphere with different loads, depending on the material under test. The indentation size is measured using a macroscope containing an ocular scale. *Vickers hardness numbers* (VHN) employ a diamond pyramid indentor. Otherwise, the two tests are basically similar. *Rockwell hardness testers,* using a variety of indentors and loads with corresponding scales, are direct-reading instruments (i.e., the hardness is read directly from a dial). The hardness number, for example, R_c 65, indicates the reading came from the C scale using a

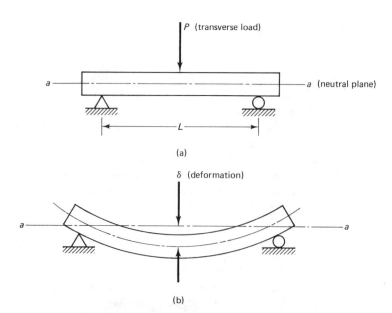

(a)

(b)

(c)

Figure 4-28 (a) Simple supported beam with a transverse load. (b) Loaded beam showing deformation (deflection). (c) Wood specimen undergoing a static bending (flexural) test showing evidence of failure due to tension. (Forest Products Laboratory, Forest Service, USDA.)

diamond indentor and a 150-kg load. It is therefore important in reporting Rockwell hardness readings to include the scale number so that the person wishing to use the information knows the type of indentor as well as the size of load used in the test.

Other tests used to report hardness include a file test (resistance to scratching), a Scleroscope test (measures the rebound of a small weight bounced off the surface of the material), and a comparison test that also uses scratch resistance. This last test compares a material's hardness to some ten known minerals arranged in order of hardness. Mostly used by mineralogists, the *Moh's scale* classes hardness of all materials between 1 (the hardness of talc) and 10 (the hardness of diamond). The scale is based on the ability of a hard material to scratch a softer material.

Materials such as very thin materials (e.g., coatings, foils, plated surfaces), very brittle materials (e.g., glass or silicon), and very small parts (e.g., gears in a wristwatch) require special care in hardness testing primarily due to their thinness and/or size. Furthermore, laboratory research in materials necessitates hardness testing on a microscale in determining the differences in hardness over the minute area of a single grain of metal or between the middle of a crystal and the grain boundary area. For such purposes a microhardness tester finds application with loads and indentations that are so small that indentations require microscopic viewing with appropriate scales for accurate measurement. Figure 4-29 shows a variety of hardness testers.

Table 4-1 provides a comparison of the approximate hardness of a variety of materials using 11 different hardness scales. These readings may be compared to the tensile strengths given in Appendix 10.6. Some correlation exists between hardness and tensile strength, but it is only approximate. For example, the tensile strength of steel (but not other materials) is about 500 times the BHN as listed in the table. As a general rule, the tensile strength of a given ductile metal can only be estimated from the hardness reading within an error of less than 10%.

4.4 CHEMICAL PROPERTIES

Chemical properties are a measure of how a material interacts with gaseous, liquid, or solid environments. Common examples include the ability of iron to resist rust when exposed to air and moisture, the resistance of wood to rotting, or the ability of rubber to withstand sunlight (ultraviolet rays) without drying and cracking. Many conditions or environments threaten materials. Polluted air is filled with elements emitted from gasoline engines, home furnaces, and industrial plants; these elements combine with materials to cause damage and shorten their service life. Some of our ancient architectural treasures, such as the Parthenon (an ancient Greek temple), and copper and stone statues are rapidly deteriorating from polluted corrosive air. Salty water from ocean spray or road salt used on icy roads promotes corrosion in automobiles. The reaction of some materials to high temperatures and fire is hazardous. Many plastics emit poisonous gases when burned. Sometimes it is desirable for materials to deteriorate or be biodegradable so the natural environment can break down the material, thus reducing solid waste.

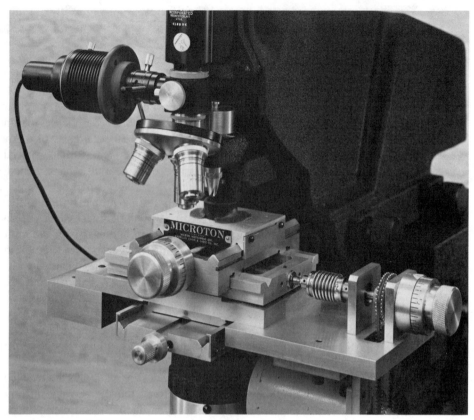

(a)

(b)

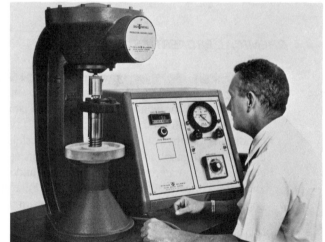

(c)

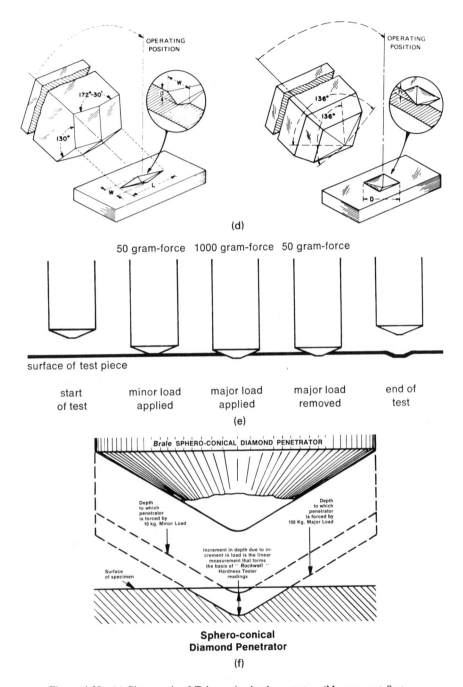

(d)

50 gram-force   1000 gram-force   50 gram-force

surface of test piece

| start | minor load | major load | major load | end of |
| of test | applied | applied | removed | test |

(e)

Brale SPHERO-CONICAL DIAMOND PENETRATOR

Depth to which penetrator is forced by 10 kg. Minor Load

Depth to which penetrator is forced by 150 Kg. Major Load

Increment in depth due to increment in load is the linear measurement that forms the basis of " Rockwell " Hardness Tester readings

Surface of specimen

Sphero-conical Diamond Penetrator

(f)

Figure 4-29 (a) Photograph of Tukon microhardness tester. (Measurement Systems Div, Page Wilson Corp.) (b) Photograph of Rockwell hardness tester. (Measurement Systems Div., Page Wilson Corp.) (c) Photograph of Air-O-Brinell metal hardness tester with digital readout of Brinell values. (Tinius Olsen Testing Machine Co.) (d) Microhardness penetrator (Vickers) indentations. (Wilson Instrument Division of ACCO.) (e) Various standard loads for the Rockwell hardness tester. (Wilson Instrument Division of ACCO.) (f) Brale sphero-conical diamond penetrator. (Wilson Instrument Division of ACCO.)

TABLE 4-1 HARDNESS SCALES COMPARISONS

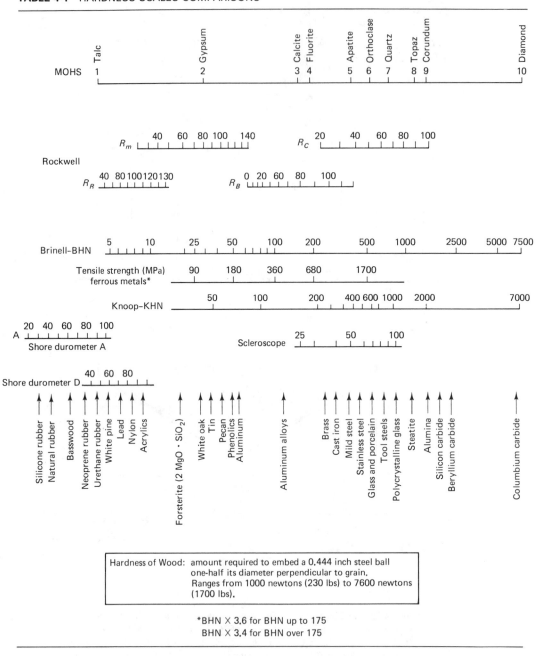

Hardness of Wood: amount required to embed a 0.444 inch steel ball
one-half its diameter perpendicular to grain.
Ranges from 1000 newtons (230 lbs) to 7600 newtons
(1700 lbs).

*BHN × 3.6 for BHN up to 175
BHN × 3.4 for BHN over 175

4.4.1 Oxidation and Corrosion

The most familiar chemical property is corrosion resistance. *Corrosion resistance* is the ability to resist oxidation. Most often, corrosion resistance is defined in terms of aqueous environments. *Oxidation* is the interaction of oxygen with elements in a material to cause structural changes due to the movement of valence electrons in the atoms of the material. An oxidized material looses electrons from atoms or ions. The opposite of oxidation is *reduction:* the gaining of electrons. For example, iron (Fe) releases electrons (*e*) during oxidation and ends up with positive ions (Fe^{2+}). The chemical reaction is shown as $Fe \rightarrow Fe^{2+} + 2e$ (see Figure 4-30a). In reduction the process is reversed as electrons are consumed: $Fe^{2+} + 2e \rightarrow Fe$. The reduction process is used to convert iron ore (iron oxide) into iron. The rusting of iron and steel results from iron's tendency to revert to the natural state so as to seek equilibrium (Figure 4-30b). The same is true of most metals because in the refined form they are more prone to oxidation.

Old pencil erasers become hard and rubber bands become stiff and crack due to oxidation. The reaction of the rubber is common for many polymers exposed to sunlight, heat, and other conditions that speed up oxidation. Figure 4-30c shows a typical thermosetting polymer with crosslinks that bond molecules together but also allow stretching. In Figure 4-30c, the crosslinks increase due to oxidation causing the rubber or plastic to lose its ability to stretch and to become hard and crystalline.

4.4.2 Outdoor Weatherability

Outdoor weatherability is a chemical property involving the ability of a material to withstand heat and ultraviolet rays from the sun, moisture, and pollutants in the air. Each of these factors affects oxidation. Rubber tires on automobiles and bicycles will develop cracks with age if they are not designed for good outdoor weatherability. A less than 1% increase in oxygen can cause severe damage to rubber.

Oxidation is normally enhanced by heat, and heat increases with oxygen. This is shown when oxygen is used to burn steel. The metal burning is done with an oxyacetylene torch (Figure 4-31a), which causes a rapid melting away of the metal as it oxidizes. Even though oxygen is not a fuel, it is highly dangerous in the presence of fire or sparks and is capable of causing most materials to burn. In the fire triangle (Figure 4-31b), removal of any element (air, fuel, or ignition) can stop the fire. Occasionally, increased temperatures reduce corrosion in stainless steels and with copper alloys.

Common ceramic materials such as silicon glass and aluminum oxide are oxidation resistant because they have already reacted to oxygen to form oxides. In the same way, an oxide film that forms on aluminum protects it.

4.4.3 Electrochemical Corrosion

Oxidation occurs in metals through an electrochemical process that is similar in operation to a car battery. In *electrochemical corrosion,* electrons flow through an electrolytic solution from one piece of the metal to another. The electrolytic solution

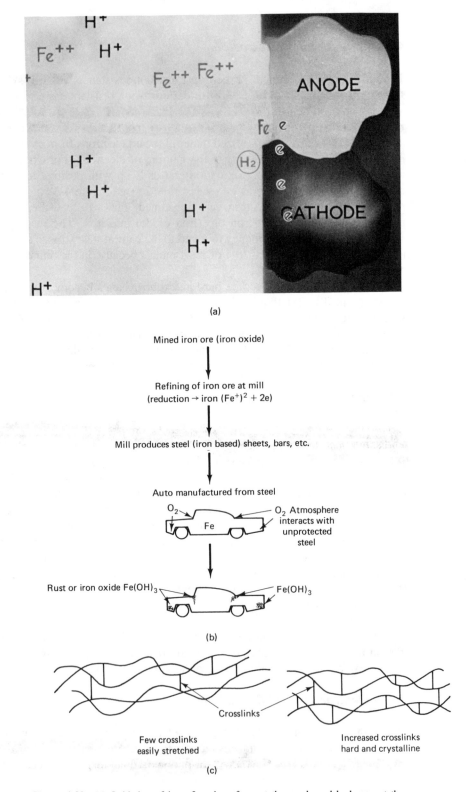

(a)

Mined iron ore (iron oxide)

Refining of iron ore at mill
(reduction → iron $(Fe^+)^2 + 2e$)

Mill produces steel (iron based) sheets, bars, etc.

Auto manufactured from steel

O_2 O_2 Atmosphere
interacts with
Fe   unprotected
steel

Rust or iron oxide $Fe(OH)_3$ $Fe(OH)_3$

(b)

Crosslinks

Few crosslinks   Increased crosslinks
easily stretched   hard and crystalline

(c)

Figure 4-30 (a) Oxidation of iron. Iron ions form at the anode and hydrogen at the cathode in local cell action. (INCO). (b) Cycle of iron and steel seeking natural equilibrium. (c) Typical thermosetting polymer. Increasing number of crosslinks produces a network structure.

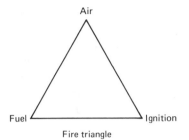

Air

Fuel / Ignition

Fire triangle

Figure 4-31 Oxidation.

(electrolyte) is a liquid such as water (H_2O) that contains ions. Recall that ions are charged atoms. Cations are positively charged ions and anions are negatively charged ions. When metals (especially dissimilar metals) are placed in the solution, one metal becomes the *anode* and the other the *cathode,* just as in a battery. Cations of one metal (anode) enter the solution to join anions. Left behind are electrons, which travel through the metal (conductor) to the cathode. Figure 4-32 illustrates electrochemical corrosion of iron that is bolted to copper. The positive iron ions carrying two positive charges (Fe^{2+}) enter the water to form bonds with the negative hydroxide ions (OH^-). Metallic bonds are being broken to form ionic bonds. Electrons (e^-) left behind in the iron travel through the bolt to the copper, where they meet positive hydrogen ions (H^+) and become neutral hydrogen atoms. The anode (Fe) is pitted as it loses atoms. This type of corrosion is also known as *galvanic action.*

Bosich (see references) lists eight types of corrosion: uniform attack, galvanic, concentration cell, pitting, dezincification, intergranular, stress, and erosion. Other authors use different labels and groupings. An explanation of each of these types of

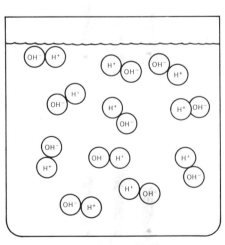

Ions in water (H_2O)

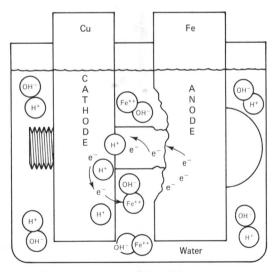

Corrosion of iron (Fe)

Figure 4-32 Electrochemical corrosion depicting ionized water and the corrosion of iron (Fe).

corrosion is beyond the scope of this book but can be gained from the books listed in the references at the end of this module.

Some authorities consider corrosion only in terms of metallics and their deterioration through chemical and electrochemical reaction to an environment. Rusting is applied only to corrosion of iron and iron alloys. Others classify metallic corrosion as not only deterioration but any interaction of metallics with an environment, whether good or bad. Our use of "corrosion" refers to a deterioration of all materials and how the material reacts with its environment.

4.4.4 Electromotive-force Series

It is possible to determine how active a metal will be by referring to the electromotive series. Table 4-2 shows the order in which dissimilar metals produce electromotive force (electron flow). The metals on top are stronger oxidizers and thus are active anodes. The bottom metals are less active and become reducing agents (cathodes). Greater galvanic action results from joining metals that have greater separation on the chart.

While magnesium and aluminum oxidize quickly, in doing so they form a hard oxide layer on the surface that prevents oxygen from reaching the metal below. Iron and steel do not oxidize as rapidly as magnesium and aluminum, but the scale on iron is soft and porous, thus allowing oxygen to penetrate farther and farther into the metal until it erodes away to nothing. A low-alloy steel known as Corten uses a small percentage (less than 0.5%) of copper to produce a compact oxide; therefore, the oxide coating on Corten also serves as a barrier and reduces corrosion to less than four times that of regular steel.

TABLE 4-2

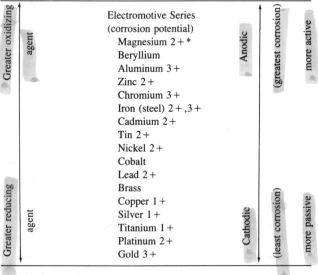

Greater oxidizing agent		Electromotive Series (corrosion potential)	Anodic	(greatest corrosion)	more active
		Magnesium 2+*			
		Beryllium			
		Aluminum 3+			
		Zinc 2+			
		Chromium 3+			
		Iron (steel) 2+,3+			
		Cadmium 2+			
		Tin 2+			
		Nickel 2+			
		Cobalt			
		Lead 2+			
Greater reducing agent		Brass			
		Copper 1+			
		Silver 1+	Cathodic	(least corrosion)	more passive
		Titanium 1+			
		Platinum 2+			
		Gold 3+			

*Oxidation numbers

Oxidation can occur on a single piece of metal due to the varied energy states of the atoms of the metal. These states result from stresses due to machining, forming, and welding, from grain boundaries, from lack of homogeneity due to alloying and casting, and from cracks and other surface irregularities. In fact, there is no way to completely stop corrosion, but it can be slowed considerably. The high-energy areas of grain boundaries become anodes, giving up atoms to the lower-energy levels. Pitting and erosion of the metal occur at the anode, while rust buildup occurs at the cathode (Figure 4-33).

The atmosphere provides excellent electrochemical mechanisms to promote corrosion. Different corrosion rates result from varied atmospheres. Oxygen and ozone, water, and pollutants such as sulfur dioxide from burnt fuel oil and coal, ammonia, hydrogen sulfide, dusts, and salts exist in varying amounts depending on

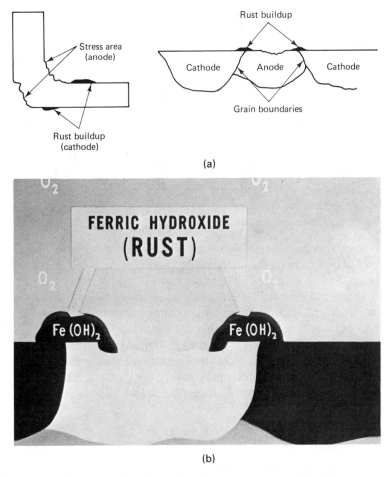

(a)

(b)

Figure 4-33 (a) Corrosion from high energy levels. (b) Conversion of ferrous hydroxide into ferric hydroxide by the action of oxygen. (The International Nickel Co.)

TABLE 4-3 POSSIBLE CORROSION RATES FOR METALS IN THREE GENERAL ATMOSPHERES

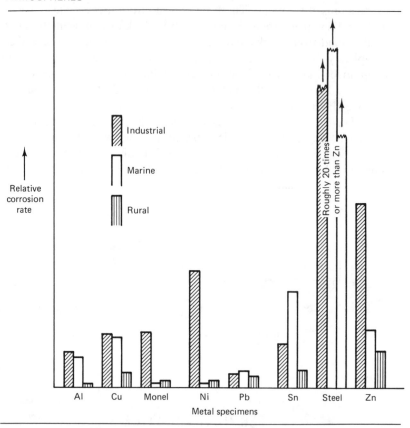

location (marine, industrial, or rural); these active chemicals produce differences in rates of corrosion for the materials exposed to them. While fluctuations in temperature do not make significant changes in corrosion rates, wind can cause dramatic shifts in airborne chemicals that enhance corrosion. Abrasive sand carried in strong winds can erode protective coatings and promote corrosion. Table 4-3 reveals the possible rates of corrosion for certain environments on metals and alloys. Industrial environments normally produce greater corrosion rates due to the many corroding chemicals. Next, the high moisture and salt in a marine environment produces more corrosion than a rural area. Rural areas in relatively dry states such as Arizona and New Mexico would produce less corrosion than rural areas in coastal states such as Virginia and North Carolina.

4.4.5 Chemical Attack

Chemical attack involves the dissolving of a material (*solute*) by a chemical (*solvent*). A common example is salt (solute) dissolving in water (solvent). Polymers present a problem in use with chemicals because of the different properties of polymers of even the same name. Therefore, it is important to follow manufacturers' recommendations when making material selections. While such ceramics as glass and concrete are very stable, even they are subject to chemical attack. Concrete will expand and crack when subjected to sulfates commonly found in the soil. Changes in temperatures and chemical concentrations also affect the reaction of a chemical to a material (see Module 7, page 338).

4.4.6 Corrosion Protection

The large variety of environments that a material can be subjected to makes corrosion protection quite complex. On the other hand, polymers have been developed that are so resistant to our natural environment that they become nearly indestructible and are solid waste pollutants. Some common methods of corrosion protection for metals include cathodic protection, protective coating, stress relieving, insulation, alloying, materials selection, and design. Figure 4-34 shows that materials are tested and then selected because of their corrosion resistance. Many rules of good design are intended to reduce the opportunity for corrosion to occur.

Cathodic protection involves the use of a sacrificial metal such as magnesium or zinc, which acts as an anode, thereby turning the metal being protected into a cathode with a reduced corrosion rate. Metals used as anodes are the more active metals, those with a greater tendency for reducing than those less active or lower on the electromotive series (Table 4-2). Common examples include the use of magnesium with steel water pipes or zincs on ships. Electric currents that pass electrons into a metal can achieve the same results.

Protective coating is the most familiar method to prevent corrosion. Paint, varnish, oil, and a variety of polymeric and ceramic coatings prevent oxygen and

Figure 4-34   Corrosion of steels in a marine atmosphere. Left: low copper steel. Center: ordinary steel. Right: nickel-copper-chromium steel. (The International Nickel Co.)

Figure 4-35 Dipping of automobile bodies. (Chrysler Corp.)

moisture from reaching the metal. Figure 4-35 shows how automobile bodies are dipped into protective coating solutions. Zinc-coated steel (galvanized steel) is an example of both protective coating and cathodic protection.

Stress relieving through heat treatment, structural changes, and design considerations provides more homogeneous metals with nearly equal energy levels. *Insulation* to stop metal-to-metal contact of dissimilar metals through the use of polymers or ceramics reduces galvanic action. *Alloying* of elements such as aluminum or chromium with steel produces oxide layers that make the metal *passive* or resistant to further oxidation. Conversely, alloying of certain dissimilar metals produces *galvanic cells* within the metal, which enhances corrosion.

Materials selection must consider the placing of materials into the proper environments to prevent oxidation, electrochemical reaction, or chemical attack. *Design* should incorporate the preceding methods, in addition to avoiding surfaces that will collect liquids and debris.

4.4.7 Water Absorption and Biological Resistance

Natural and synthetic polymers are subject to the absorption of water and biological attacks. When left untreated, dried wood will absorb moisture that serves as a good environment for the growth of fungi (small plants) and insects; they feed on cellulose and lignin and cause deterioration. Unprotected dry wood is also a host for termites and other insects. Some synthetic polymers swell through water absorption, which causes deterioration and provides a good environment for damaging microorganisms. The corrosion of metals through oxidation can result from organisms such as barnacles living on the material.

4.5 ELECTRICAL PROPERTIES

4.5.1 Resistivity

Resistivity or *volume resistivity* is the term used to describe the relationship between electric current and the applied electric field. It is a measure of the resistance to the flow of current from a microscopic level, that is, as explained in terms of the atoms, the basic building blocks of all solid materials. Resistivity, ρ (Greek letter rho), depends on the behavior and number of free or conduction electrons and not on the shape of the conductor, as is the case with resistance. This inherent property of a material, like density, will change as the structure changes as in the alloying of metals or the doping of semiconductor materials. It depends on the movement of charge carriers, electrons in metallic conductors or ions in ionic materials. In fact, it is the reciprocal of conductivity (σ) (to be discussed). Any imperfections in the crystalline structure of metallic materials, such as atoms out of their normal positions in the lattice structure, dislocations, grain boundaries, impurity atoms, or excessive vibration of atoms brought on by an increase in temperature (an increase of heat energy), will increase the collisions between electrons and the positive charged nuclei (cations), reducing their energies and preventing the transfer of that energy in the form of electron flow to some intended user.

Figures 4-36a and b show a sketch of the effects of an increase in temperature on metallic conductors and semiconductors, respectively. In the case of metallic conductors, the increased vibrational energies of the atoms as a result of an increase in energy make the passage of free electrons through the structure even more difficult. The *mean free path,* the average distance an electron can travel as a wave without hitting or deflecting off a positive ion core (atom) in the lattice structure, is decreased. Consequently, the mobility of the electrons decreases, which produces an increase in the resistivity. For semiconductor materials, resistivity decreases (conductivity increases) with an increase in temperature because more charge carriers become available to act as conduction electrons.

The doping of semiconductor materials also lowers the resistivity by increasing the number of charge carriers. Additional information on this subject can be found under the heading of semiconductors.

In the discussion of resistance, we stated that resistance varies directly with the length L and indirectly with the uniform cross-sectional area A of a conductor. To write this as a mathematical statement, we need a constant to make the units agree on both sides of the equation. Thus

$$R = \rho \frac{L}{A}$$

where ρ is the proportionality constant or resistivity in ohm-centimeters, assuming that L and A are expressed in centimeters (cm). Note that a conductor 1 cm in length with area of 1 cm^2 has a resistance R equal to the resistivity ρ.

In selecting conducting materials for the express purpose of generating heat from the flow of electricity such as Nichrome, a heat-resistant alloy of nickel and

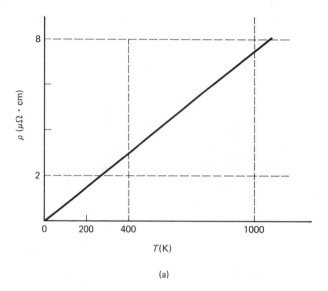

(a)

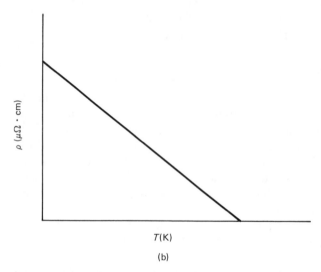

T(K)

(b)

Note: For metals curve approaches a straight line

Figure 4-36 (a) Resistivity (ρ) versus temperature (T) for a metallic conductor. (b) Resistivity versus temperature for a semiconductor material.

chromium, the material must have a carefully selected variety of properties, such as a moderate resistivity, excellent resistance to oxidation, and a capability to operate effectively at high temperatures without failure. Table 4-4 lists the electrical resistivities of some selected solids from three families of materials.

Note in our discussion that where heat is evolved in the flow of electric charge there is reduced efficiency. The incandescent light bulb is a good example. This bulb gets hot when being used. A fluorescent bulb remains cool. Consequently, the fluorescent bulb is more efficient in the use of electricity (energy). As a matter of fact, the incandescent light is only about 3% efficient; that is, 97% of the energy needed to produce light is lost mainly to heat.

TABLE 4-4 ELECTRICAL RESISTIVITIES OF SELECTED
SOLID MATERIALS

Material	Resistivity at 20°C ($\mu\Omega$-cm)*	Resistivity at 500°C ($\mu\Omega$-cm)
Silver	1.6	4.6
Copper	1.7	4.5
Aluminum	2.9	8.5
Tungsten	5.6	18.0
Steel (1040)	17	
Nichrome	105	120
Boron	10^6	—
Teflon	10^8	—
Nylon	10^{14}	—
Plate glass	10^{14}	—
Mica	10^{15}	—
Polystyrene	10^{16}	—
Diamond	10^{18}	—

*1 $\mu\Omega$ (microohm) $= 10^{-6}$ ohm

Note: Boron is a semiconductor; metals (conductors) appear at
top of the listing and insulating materials near the bottom.

Summarizing, the resistance R of an electric circuit is a function of the shape, size, and nature of a solid material. Resistivity ρ is a function of the intrinsic nature of the material itself, just like specific heat or density. Instead of thinking of resistivity, one can think in terms of conductivity, σ. Low resistivity and high conductivity are referring to a like situation in different terms. To further reinforce the concept of resistivity, remember that 1 pound of aluminum has the same resistivity as 1 gram of aluminum, whereas the resistance of 1 pound of aluminum is much different from that offered by 1 gram. Figure 4-37 illustrates an industrial use for a man-made nonconductive material.

4.5.2 Conductivity

The reciprocal of resistivity is known as *conductivity*. Using the symbol σ (Greek letter sigma) to represent conductivity, this quantity is determined for a particular material by measuring the amount of electric charge that passes through a unit cube of the material per unit of time. Consequently, σ is dependent on three factors:

1. n, the number of charge carriers in a cubic centimeter (cm^3) of material.
2. q, the charge per carrier (coulombs/carrier).
3. μ, the mobility of each carrier:

$$\mu = \frac{\text{velocity of the carriers (cm/s)}}{\text{voltage gradient (V/cm)}}$$

Figure 4-37 Pultruded fiberglass boom is superior to older metal booms in terms of mechanical, weathering, and electrical properties. The boom exhibits less than 100 microamp leakage over any 2-foot section after a 48-hour full immersion soak with a 100 KV (DC) test voltage is applied (Morrison Molded Fiber Glass Co.).

The product of these three factors with units of $1/\Omega$-cm or $(\Omega\text{-cm})^{-1}$ is the conductivity. Expressed in mathematical terms,

$$\sigma = nq\mu$$

This calculation agrees with units obtained for conductivity if the basic equation for resistivity is used ($\rho = RA/L$).

4.5.3 Dielectric Constant

The capacitance C of a conductor is measured by the amount of charge Q that must be placed on it to raise its potential by 1 volt (V). The mathematical relationship for the above is expressed as

$$C = \frac{Q}{V}$$

where C is measured in farads (coulombs/volt), Q in coulombs, and V in volts. A farad (F), the unit of electric capacitance, is defined as the capacitance of a capacitor between the plates of which there appears a potential difference of 1 volt (V) when it is charged by a quantity of electricity of 1 coulomb (C).

Two conductors separated from each other by some insulating material (including air) form a condenser. The conductors make up the plates of the condenser, and the insulation is called the *dielectric*. The condenser is charged by connecting its plates to the terminals of a battery or other source of direct current. Electrons will flow and collect on one plate until equilibrium is reached and the potential difference between the two plates equals the emf of the battery. The ratio of condenser charge Q to the potential difference V is defined as the *capacitance* of the condenser. The work required to move all the electrons (to charge the condenser) is given by the following expression:

$$U = 1/2CV^2$$

where U is the work done in joules (J), C the capacitance in farads, and V the potential difference in volts. This work is equivalent to the amount of electrical energy that is stored in the condenser.

The capacitance of a parallel plate condenser with free space as the dielectric is proportional to the area of the plates and inversely proportional to the distance between the plates. Using $C = Q/V$, the capacitance of a parallel plate condenser with air as a dielectric can be expressed as

$$C = \varepsilon_0 \frac{A}{d}$$

where A is the area of the plates, d is the thickness of the dielectric, and ε_0 (Greek letter epsilon) is the *permittivity* constant of free space whose value is 8.85×10^{-12} farads per meter (F/m).

If a dielectric such as mica is placed between the plates of the same parallel plate condenser using the same potential difference V, the charge Q on the plates will be greater and produce a corresponding increase in the capacitance C of the condenser. This result can be expressed as

$$C = \kappa_0 \frac{A}{d}$$

where κ (Greek letter kappa) is the *dielectric constant* or *relative permittivity* of the dielectric. The dielectric constant is a measure of a material's value toward making a desired capacitance when the material is placed between two conducting plates. The dielectric permits a higher potential difference to be applied than when air is used.

When a dielectric material such as a polymer, glass, or ceramic is placed between the plates of a condenser, an induced charge Q appears on the plates without any change in the voltage. This naturally increases the capacitance of the condenser. The

TABLE 4-5

Material	Dielectric constant (κ)
Al_2O_3	9.6
Mica	7
Nylon 6/6	3.5
Polypropylene	2.0

increased charge is due to the *polarization* of the entire volume of the dielectric due to the applied electric field. Polarization is the process of producing electric dipoles. A *dipole* is two equal but opposite electric charges ($+Q$) and ($-Q$) that are separated by a distance d whose product (the magnitude of one of the charges times the distance d) or Qd is known as the dipole moment or strength (p). The units of dipole moment are coulomb-meters (C-m). In dielectric materials, electrons and their atoms are not free to move much under the influence of an external electric field, but they move enough to produce polarization of the material.

Since the dielectric constant is a ratio of similar quantities, it has no units. Some typical values for κ for some nonmetallic materials at 25°C and 10^6 hertz (Hz) are listed in Table 4-5.

High values of dielectric constant are needed for capacitors, whereas low values are used as insulating materials. Much research is devoted to the miniaturization of electrical circuits, including capacitors. This means a continual search is underway to find materials that not only have high dielectric constants, but have such values at higher and higher frequencies and temperatures for use in, for example, dielectric heating applications.

4.5.4 Dielectric Strength

Dielectric strength is the voltage gradient (voltage per unit thickness, such as volts per mil) that produces electrical breakdown through the dielectric. It is an expression of the maximum voltage that a dielectric can withstand without rapture. This insulating strength is dependent not only on the usual items, such as bonding type and crystalline structure, but on moisture absorption and the nature of the applied electrical energy. In the latter the source might be a direct current (dc) or an alternating current (ac), or a combination of the two. If the magnitude of the dielectric field is sufficient, it will overcome the attraction of the electrons to their positive-charged nuclei and produce a *leakage* and eventual rupture of the material. The breakdown voltage has then been reached. With an ac source the continual reorientation of the material's atoms and electrons in addition to the deformation of the paths of the electrons results in a hysteresis loss (to be explained) that produces heat in the dielectric material. Therefore, the breakdown voltage will vary inversely with the frequency of the source voltage.

Some typical ranges of values of dielectric strength for some nonmetallic materials expressed in volts per mil (1 mil = 0.001 in.) are listed in Table 4-6.

TABLE 4-6

Material	Dielectric strength
Natural mica	1000–2000
Polypropylene	500–800
Nylon 6/6	385–470
Alumina ceramic	200–300

4.6 MAGNETIC PROPERTIES

4.6.1 Magnetic Permeability

In an electrical circuit, the conductivity σ of a conductor is a constant in that it does not depend on either the potential difference E across a unit length of a conductor of unit cross section or the current I passing through that section of the conductor. In mathematical terms, with unit value for l and A

$$\sigma = \frac{I}{E} \quad \text{(a constant)}$$

In a magnetic circuit the permeability μ of a material, a measure of the ease with which a magnetic field can be set up through a material, can be compared to the conductivity in an electrical circuit. The magnetic field represented by the symbol H with units of amperes per meter (A/m) corresponds with the voltage or potential drop. The magnetic flux density, B, expressed in units of webers per square meter (Wb/m^2) corresponds with the electric current. The ratio of B to H (B/H) is known as *permeability,* μ (Greek letter mu), with units of webers per ampere-meter (W/A-m). An important distinction between electrical and magnetic circuits is that permeability is not constant. As H changes, so does μ.

Figures 4-38a and b illustrate the permeability of a vacuum or free space in the core of an electrical coil through which an electrical current is passing. When a core of material replaces the vacuum in the coil, a change in the flux density B is detected. This change is reflected in a new value of μ. If the material is made of iron, cobalt, or nickel, an exceedingly large rise in B is noted.

4.6.2 Relative Permeability

Relative permeability, μ_r, is the ratio of two permeabilities and therefore has no units. The permeability of a material is thus compared mathematically with the permeability of free space, μ_0 (μ/μ_0). μ_0 equals $4\Pi \times 10^{-7}$ W/A-m. This ratio may be used to classify magnetic materials into three main categories:

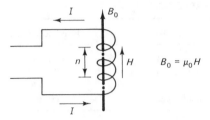

Key: n = turns per meter
      I = electric current in amperes (A)
      H = magnetic field strength (A/m)
      μ_0 = permeability of vacuum ($4\pi \times 10^{-7}$ henry/m)
      B_0 = magnetic flux density in vacuum in webers
             per meter squared (Wb/m^2)

(a)

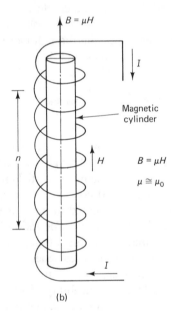

(b)

Figure 4-38 (a) Magnetic helical coil with vacuum core. (b) Magnetic coil with a solid material core.

1. Diamagnetic material if $\mu_r < \mu_0$.
2. Paramagnetic material if $\mu_r > \mu_0$.
3. Ferromagnetic material $\mu_r \gg \mu_0$.*

Actually, μ_r for diamagnetic and paramagnetic materials is nearly 1, while for ferromagnetic material it can be up to 10^6 times greater than μ_0. In the following discussion we will, on occasion, leave the world of the individual atom with its electrons and investigate the interactions of atoms with their neighboring atoms. This is necessary in order to obtain a macro view of the contributions of the individual

*The symbol $\gg$ means "much greater than."

atoms to the magnification of the bulk material. Hence we will be concerned with the microstructure of the material.

4.6.3 Diamagnetic Material

Diamagnetic materials such as pure copper or lead have zero magnetic moment because the individual atoms interact with each other and cancel out any magnetic moment for the individual atom in the absence of an external magnetic field.

4.6.4 Paramagnetic Material

Individual atoms have a magnetic moment (N–S poles), but the magnetic moments of these individual atoms do not interact with each other, which results in a random orientation that produces a zero net magnification for an aggregate of atoms. When an external magnetic field is applied, the atoms will align with the field, depending on the strength of the field and the temperature of the material. Chromium (Cr) and Manganese (Mn) are some of the metals that exhibit paramagnetism. Most materials belong to these two categories and can be considered "nonmagnetic."

4.6.5 Ferromagnetic Material

A ferromagnetic material has the ability to possess magnetic lines of force or flux without an outside magnetic field. To exhibit ferromagnetism, some of the electrons in the outer shells of the atoms ($3d$ sublevel in the case of iron, cobalt, and nickel) must not only have unbalanced spins but be suitably located so as not to be affected by the bonding or valence electrons. External magnetic fields induce the magnetism and increase the inherent magnetism already present in the material. Their response to an external magnetic field can be 10^8 times greater than for "nonmagnetic" diamagnetic or paramagnetic materials. Materials in this class are of most importance from an engineering standpoint. Iron is ferromagnetic, as are nickel, cobalt, some rare earth elements, and a variety of man-made alloys.

   This class of materials is generally subdivided into *soft* and *hard* magnetic materials. Before continuing our discussion of these materials, two additional tools need to be introduced to help us further our understanding of the behavior of ferromagnetic materials. These new aids are magnetic domains and the *B–H* or hysteresis curve.

4.6.6 Magnetic Domains

The answer to the question as to why two pieces of iron, even though their individual atoms possess magnetic moments, do not act as two magnets was not satisfactorily answered until about 70 years ago. The effect that one magnetic atom has on its neighboring atoms, known as short-range coupling, is felt only by its closest neighbors. However, the overall effect of this short-range coupling in ferromagnetic materials is the forming of macroscopically large magnetic domains. A *domain* is a

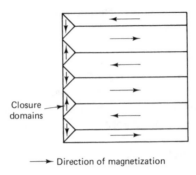

Closure domains

→ Direction of magnetization

Figure 4-39 Schematic of magnetic domains in a crystal at $B = 0$.

region of a metal or ceramic visible under a microscope in which the atom magnets (the magnetic moments resulting from the electronic structure of the individual atoms) are aligned in the same or parallel direction. This preferred direction is a function of the crystallographic structure of the atoms making up the material. In the case of BCC iron, the ⟨100⟩ direction is the easiest to magnetize, which indicates that such materials have a large degree of magnetic anisotropy (different magnetic properties). This direction of magnetization can be changed provided the necessary external energy is provided. This energy is supplied by an external magnetic field. When $B = 0$ (no external magnetic field present), each domain is balanced with a domain of opposite alignment. Therefore, no external magnetic flux is present outside the material. Figure 4-39 is a schematic of magnetic domains in a crystal.

If a magnetic field is applied, the domains whose magnetization is in the direction of the external magnetic field will grow at the expense of those domains of opposite alignment, as shown in Figure 4-40. The end result of this domain boundary migration, assuming the presence of a large enough magnetic field, is the alignment of all domains in the direction of the magnetic field. Using domains, one can readily understand why an iron nail placed in the vicinity of a magnet will become magnetized in a direction dictated by the location of the north-seeking end of the magnet used to

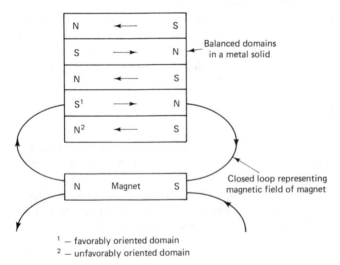

Balanced domains in a metal solid

Closed loop representing magnetic field of magnet

[1] — favorably oriented domain
[2] — unfavorably oriented domain

Figure 4-40 Balanced domains in an external magnetic field.

magnetize the nail. Proof of the existence of domains is readily available by using a colloidal solution of iron oxide particles spread over the polished surface of a piece of magnetic material. When viewed under a microscope, the particles of iron oxide will be concentrated around the domain boundaries. All ferromagnetic materials begin to lose their magnetic properties as their temperatures increase. The thermal oscillations produced tend to destroy the magnetic domains. Above a characteristic temperature known as the *Curie temperature* (T_c), ferromagnetism will no longer exist even when subjected to large external magnetic fields.

4.6.7 B–H Hysteresis Curve for Ferromagnetic Material

In our previous discussion of permeability, we used the terms magnetic flux density (B) and magnetic field (H). If we plot these two quantities, we arrive at a B–H or magnetization curve for a ferromagnetic material. Such curves are depicted in Figures 4-41 and 4-42 for soft and hard magnetic materials, respectively. Referring to Figure 4-41, an "unmagnetized" material at point a begins to receive an increasing H as the

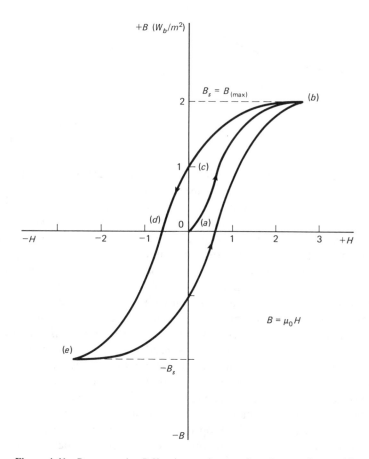

Figure 4-41 Representative *B-H* or hysteresis curve for soft magnetic materials.

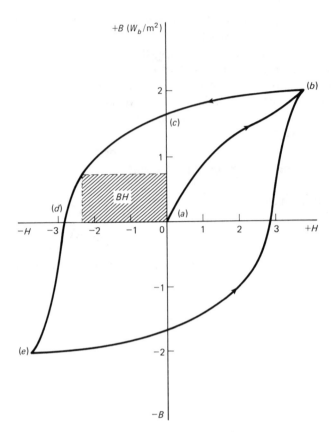

Figure 4-42 Representative *B-H* or hysteresis curve for hard magnetic materials.

external magnetic field is built up, possibly by use of an electrical coil winding as shown in Figure 4-38. Inside the material the domains are growing, as described previously, and reach point *b* where the maximum B_s (saturization) or B_{Max} is reached. At this point the material is essentially one large domain. Upon decreasing the external magnetic field (H), the domain tends to return to a balanced domain arrangement; but because of the many structural imperfections, such as crystal defects, grain boundaries, or inclusions, the balanced domain state is not reached. Such a situation can be compared to the plastic region of a stress–strain curve. Point *c* represents the remanent or remanent magnetization with $H = 0$. As the magnetic field increases in the opposite direction, that is, H increases in the negative direction (^-H), point *d* is reached, $B = 0$. The corresponding value of H at point *d* is known as the *coercive field* or *force*. This value of H is needed to demagnetize the material (balanced domains). Continuing to increase the field in the negative direction produces a similar domain growth, as described previously, but in the opposite direction (^-B_s). Reversing the direction of the field a second time produces a curve that will plot out a *hysteresis loop*, which is an indication of the energy lost in a complete cycle of magnetization. The area enclosed by the loop represents the work done per cycle in aligning the domains parallel to the direction of the changing external magnetic field.

Properties of Materials   Module 4

4.6.8 Soft Magnetic Material

Using Figure 4-41 again and treating it as a representative hysteresis curve for a soft magnetic material, note that for small values of H a relatively large value of B is produced. Second, note that relatively little remnant magnetism needs to be canceled. Third, the area enclosed by the loop is relatively small. Such a material would find use as a transformer core.

4.6.9 Hard Magnetic Material

Figure 4-42 is a typical hysteresis curve for a hard or permanent magnetic material. Comparing this curve with the similar curve for a soft magnetic material in Figure 4-41 and assuming that the scales for the axes are the same for both curves, several major differences can be observed. Hard magnetic materials gain their magnification in manufacture. Their essential property is their resistance to demagnetization. In other words, they must retain their magnetization when the external magnetic field is removed. The relatively large coercive field then is characteristic of this property, as well as the large hysteresis loop. Note the differences in the values of points c and d in both figures. The area of the largest rectangle that can be inscribed in the second quadrant of the hysteresis loop in Figure 4-42, called the *BH product,* is a measure of the power of a permanent magnet. Table 4-7 is a typical listing of magnetic properties for some nickel alloy steels.

Time and space limit discussion at this juncture of past and ongoing research and development efforts in the field of new materials (metals, ceramics, and composites) to meet the ever-increasing needs of our technological society. This treatment of magnetic materials will permit a greater understanding and appreciation of the breakthroughs in this area as they are made public in the future.

4.7 OPTICAL PROPERTIES

4.7.1 Absorption

Light, being a form of electromagnetic radiation, interacts with the electronic structure of atoms of a material. The initial interaction is one of absorption; that is, the electrons of atoms on the surface of a material will absorb the energy of the colliding photons of light and move to higher energy states. The degree of absorption depends, among other things, on the number of free electrons capable of receiving this photon energy. The electrons can then do several things. They may have sufficient energy to jump into higher energy states, which permits them to be accelerated within an electric field and thereby conduct electricity (photoconductive effect). They may collide with atoms and release their excess energy in some form of electromagnetic radiation. Or they may convert their excess energies to atoms in the form of thermal energy.

TABLE 4-7 MAGNETIC PROPERTIES OF NICKEL ALLOY STEELS[a] (INCO)

Steel[b]		Temperature		μ_i Initial permeability	μ_{max} Maximum permeability	Br Remanence, gauss	H Coercivity, oersteds	Hysteresis loss, ergs/cc/cycle
Type	Condition	C	F					
1015 (0.10 Ni)	NT	20	68	190	1590 (H = 4.7[c])	11,400	4.0	2.7 × 10⁴
		93	200	260	1680 (H = 4.2)	11,100	4.0	2.3 × 10⁴
		165	329	260	1850 (H = 3.9)	10,250	3.8	1.9 × 10⁴
		300	572	290	1900 (H = 3.3)	8,500	2.9	1.5 × 10⁴
2115 (1.04 Ni)	NT	24	75	200	1460 (H = 4.6)	10,700	4.3	2.5 × 10⁴
		107	225	190	1270 (H = 5.5)	10,020	4.6	2.4 × 10⁴
		205	401	210	1590 (H = 4.3)	9,300	3.7	2.0 × 10⁴
		304	579	230	1580 (H = 3.6)	7,320	2.9	1.7 × 10⁴
2315 (3.46 Ni)	NT	20	68	180	1370 (H = 6.4)	13,100	4.2	2.6 × 10⁴
		92	197	240	1520 (H = 6.2)	13,600	4.5	2.4 × 10⁴
		207	405	110	1360 (H = 6.0)	11,400	4.2	2.0 × 10⁴
		320	608	160	1410 (H = 6.0)	10,000	4.1	1.8 × 10⁴
2515 (4.91 Ni)	NT	25	77	110	1330 (H = 6.3)	13,700	5.9	3.7 × 10⁴
		98	208	120	1490 (H = 6.3)	12,930	5.8	3.1 × 10⁴
		205	401	180	1490 (H = 5.3)	10,890	4.7	2.2 × 10⁴
		295	563	180	1310 (H = 6.1)	11,120	5.2	2.8 × 10⁴
9 Ni (8.56 Ni)	NNT	23	73	140	700 (H = 13.5)	12,500	10.8	5.7 × 10⁴
		104	219	130	740 (H = 12.7)	12,050	10.2	5.2 × 10⁴
		193	379	130	850 (H = 9.5)	11,700	8.9	4.3 × 10⁴
		308	586	220	1000 (H = 8.6)	10,900	7.2	3.4 × 10⁴
4340 (1.88 Ni)	NT	25	77	43	330 (H = 30)	13,700	22.5	11.4 × 10⁴
		114	237	34	440 (H = 25)	12,700	13.9	7.6 × 10⁴
		167	333	40	420 (H = 28)	12,000	18.0	7.9 × 10⁴
		302	576	85	460 (H = 23)	10,250	11.0	4.9 × 10⁴
4340 (1.90 Ni)	QT	27	81	64	535 (H = 22)	13,700	18.0	8.6 × 10⁴
		87	189	64	510 (H = 23)	12,700	17.0	8.0 × 10⁴
		157	315	80	490 (H = 24)	12,300	17.5	7.8 × 10⁴
		288	550	82	500 (H = 20)	10,800	16.0	6.1 × 10⁴

[a]These data are based on the initial magnetization curve and the magnetic hysteresis loop for applied field strengths (magnetizing forces) up to about 90 oersteds.[32]

[b]See Table 1 for details of composition and heat treatment of steels. N = Normalized. NN = Double Normalized. Q = Quenched. T = Tempered.

[c]Oersteds.

4.7.2 Photoelectric Effect

Certain metallic materials will become charged positively when exposed to electromagnetic radiation (Figure 4-43). Since no electron charge can be carried by such radiation, the conclusion reached is that the radiation must interact with the electrons attached to the metallic atoms and cause the ejection of electrons from the surface of the metal. Such a phenomenon is known as the *photoelectric effect*. The ejected electrons are called *photoelectrons*. Furthermore, it is found that not all electromagnetic radiation produces this effect. Only when a certain frequency of radiation is reached (*threshold frequency*) does this effect occur. The threshold frequencies differ with different metallic elements. As an example, the metal sodium will produce this effect once the frequency of the incident radiation reaches 5.6×10^{14} Hz. The corresponding wavelength of 5.4×10^{-7} m places such radiation in the visible light portion of the electromagnetic spectrum. Zinc requires a threshold frequency of 8.0×10^{14} Hz, which is in the ultraviolet region of the spectrum. The intensity of the radiation plays no part in producing the photoelectric effect.

To adequately explain this phenomenon, Einstein in 1905 proposed that this radiation be treated not as a series of waves (wave phenomenon) but as a series of particles (light photons) that interact with single electrons. Most of the questions dealing with light radiation confine themselves to large-scale bodies such as glass lenses. The films of materials taken by x-rays and the electron microscope are some examples of radiation exhibiting wavelike properties. The pattern of electrons that projects itself onto the x-ray film and the highly magnified images of the internal structure of a material as viewed under an electron microscope are diffraction patterns of waves typical of those exhibited by the diffraction of light through a grating. When we begin to probe into the land of the atoms with its subatomic electrons, we use the particle nature of radiation to explain the results of such interactions of radiation not with a whole surface but with individual electrons.

A photon of light (a quantum of electromagnetic radiation) exchanges its energy with a single valence or conduction electron. We have learned that such electrons

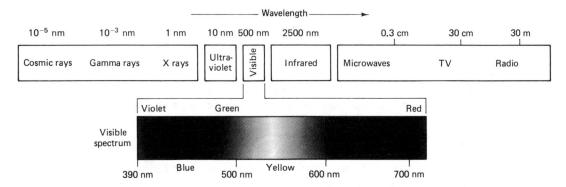

Figure 4-43 Wavelengths of electromagnetic radiation characteristic of various regions of the electromagnetic spectrum. (Theodore L. Brown and H. Eugene LeMay Jr., *Chemistry,* 2nd ed., Prentice-Hall, Inc., Englewood Cliffs, N.J., 1981.)

possess the highest kinetic energies and are the outermost electrons in an atom. When a photon of light with sufficient energy ($E = hc/\lambda$) strikes a valence electron, it causes the excited electron to leave a solid metal. Note that the photon reacts with a single electron, not with an entire atom or a solid metal electrode. The kinetic energy (KE) of the ejected electron (now called a photoelectron) can be expressed in two terms. The first term, the work function W is that part of a photon's energy (work) expended in digging out or removing the valence electron from the surface of the metal. The second term represents the remainder of the photon's energy appearing in the form of kinetic energy of the escaping electron. The photon disappears, having done its job of transferring its energy, and a photoelectron is emitted from the surface. The electrons in a metal that are eligible to receive energy from colliding photons may be in various energy levels of an atom. Those with the highest energies will require the least energy expenditure to break free from the surface; hence, they will have the greatest kinetic energy. Using electron volts (eV) as units, this energy transfer can be expressed mathematically as

$$KE = E - W$$

Suppose violet light is incident on a piece of cesium (Cs). Reference books can be consulted to find the photoelectric properties of metallic elements. Referring to such sources, we find the threshold frequency for Cs to be 4.6×10^{14} Hz; the work function W is 1.9 eV; and the wavelength (λ) of violet light is 4×10^{-7} m. To find the frequency f of violet light, we use

$$f = \frac{c}{\lambda} = \frac{3 \times 10^8 \text{ m/s}}{4 \times 10^{-7} \text{ m}} = 7.5 \times 10^{-14} \text{ Hz}$$

Note this frequency exceeds the threshold frequency for Cs. The photon's energy E is next calculated using

$$E = \hbar f$$

where $\hbar$ is Planck's constant (6.625×10^{-34} J-s) and 1 eV $= 1.60 \times 10^{-19}$ J. Thus

$$E = \frac{(6.625 \times 10^{-34} \text{ J-s})(7.5 \times 10^{-14} \text{ Hz})}{1.6 \times 10^{-19} \text{ J}} = 3.1 \text{ eV}$$

The kinetic energy (KE) of the photoelectron emitted can then be determined:

$$KE = E - W$$

$$= 3.1 - 1.9 = 1.2 \text{ eV}$$

By synchronizing the light emission from many electrons, the light produced is monochromatic (one frequency) and coherent (photons are in phase with each other). More and more materials are being found whose electromagnetic radiation can be controlled by technicians. These include solids, liquids, and gases. Semiconductor materials can produce laser action that emits photons with wavelengths in the 7 to 8×10^{-7} m range. A *laser* (light amplification by stimulated emission of radiation) is a special case of the emission phenomenon we have been discussing. The light produced is further amplified through stimulating the emission. A *maser* (microwave amplification by stimulated emission of radiation) differs from lasers mainly in the wavelength of the radiation produced.

Many uses for these well-controlled sources of energy will continue to be discovered for boring holes, welding delicate parts of the human eye, detecting and measuring pollutants in the atmosphere, and carrying human communications through space with little to no interference. Present applications of the optical behavior of materials, including glass and plastics that produce light through the interaction of electromagnetic radiation with the material's electrons, are numerous. Some common examples are worth mentioning. The coating of TV screens with zinc sulfide (ZnS), a semiconductor material upon which electrons act to produce light, is one. A second is the photo tube that "reads" the sound track on motion picture film, ultimately producing an electrical signal that can be amplified and broadcast to an audience. Solar batteries using semiconductor materials and exposure meters carried by photographers for properly setting their cameras are further examples of the many applications of such devices.

Luminescence is the reemission of photon energy at wavelengths in the visible spectrum as the result of the absorption of electromagnetic radiation from some outside source. If reemission occurs at the same time that the material is absorbing the radiation, the phenomenon is called *fluorescence*. In a fluorescent lamp, electrons emitted by the incandescent cathodes collide with electrons of the mercury atoms that fill the tube. The collisions cause the emission of radiation in the invisible ultraviolet range. This radiation, in turn, strikes the fluorescent or phosphor material coating the inner side of the tube and causes this material to emit radiation in the visible range of the spectrum.

Should the light emitted continue after the radiation producing it has been removed, the light is called *phosphorescence*. The color TV picture tube uses phosphors that phosheresce.

4.7.3 X-Rays

X-rays, discovered by Roentgen in 1895, can be produced in two ways. The first process uses a high-energy electron capable of knocking an electron from an inner energy level of an atom completely out of the atom. This removal of an electron, say from the K level, immediately creates a hole into which an L-shell electron can fall. As an L-shell electron falls back to the lower-energy K level, the sudden decrease in potential energy is emitted as a photon of electromagnetic energy characteristic of x-rays (atomic radiation). A hole is produced in the L shell, which calls for the transition of another electron from an outer level to the L shell, which produces an additional x-ray. Several x-rays are emitted as electrons cascade down to fill the lower energy level vacancies or holes, until eventually the atom captures an electron from the surrounding region and changes from an ion to a neutral atom in its lowest equilibrium condition. The wavelength of the x-ray depends on the particular energy levels involved, as do the colors in visible light emitted from a given jump. X-rays lie in the wavelength ranges between 1×10^{-12} and 1×10^{-8} m. Another process of generating x-rays (also called cathode rays) is through the sudden braking (deceleration) of high-energy electrons by impacting them on a metal target, which

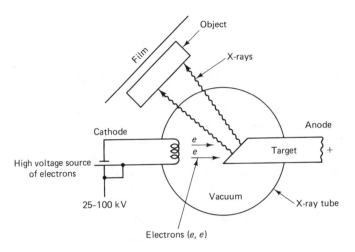

Figure 4-44 Schematic of an x-ray tube.

results in a conversion of part of the electron's kinetic energy into a quantum of electromagnetic radiation or x-ray. Such a technique is shown schematically in Figure 4-44.

4.7.4 Thermionic Emission

Another way to exchange energy with the free electrons in a solid metal is to heat the metal to such an extent that the thermal energy of the electrons is sufficient to emit them from the surface. This is known as *thermionic emission*. A typical low-power vacuum diode has a cathode consisting of a nickel tube coated with barium oxide, which is heated by an insulated filament contained within the tube.

4.7.5 Transmission

Optical devices possess many desirable properties, but the most important is their ability to transmit light. One measure of this ability is *transmittance*, defined as the percentage of an incident light ray remaining after passing through 10 millimeters (mm) of a material. It may also be expressed as a ratio of transmitted light intensity to the incident light intensity (*intensity* is the measure of the strength of a light source). Some materials will not let light pass through (*opaque*). Metals are opaque. When white light is incident on a metal surface, most of the light is reflected, and some is refracted into the surface of the metal where the photons strike the electrons, thus transferring discrete amounts of their energy to the electrons and raising them to higher energy levels. In the case of copper, the 3*d* electrons absorb the energy of the photons corresponding to blue wavelengths through collisions, and reflect the others, which give a reddish color. The energy absorbed causes the atoms concerned to increase their energy levels, which allows them to increase their vibrations, which, in turn, results in an increase in temperature, an indicator of heat.

In covalent and ionic solids the electrons are bound to the atoms by the very nature of the covalent and ionic bonds. It takes more energy to break such bonds and free the electrons with increased energies sufficient to move them to higher energy levels. Therefore, visible light passes through such solids without reacting with the atom's electrons. Thus, such solids are termed *transparent*. If impurities are present in such a solid, the photons of the visible light could react with the electrons of the impurity atoms, which are not involved in covalent or ionic bonding. In a ruby, the chromium atoms absorb the photons of blue and green light, but allow the remaining photons to pass through, which are mostly of wavelengths that correspond to red light.

Many polymer materials are transparent. Others are *translucent*. That is, these materials, because of their structure, allow some light to pass through, but most of the light is scattered due to the reflection and refraction of the light as it transmits across phase boundaries and interacts with pores in the structures. Research in new ceramic materials allows the production of single-phase, pore-free ceramics that are transparent.

The bending of light rays is due to the fact that the speed of light within a dense medium is less than in air. Many of us have had the experience of trying to pick up an object in a shallow pool of water, only to find that the object's actual location is displaced from its apparent location.

The ratio of the speed of light in a medium (v) to its speed in air (c) is called the *index of refraction of the medium (n)*:

$$n = \frac{c}{v}$$

Diamond has a refractive index of 2.4; polyethylene is 1.5; water, 1.33; and for air, $n = 1$. Another way in which the index of refraction can be defined is in terms of the angles of incidence (θ_i) and refraction (ϕ):

$$n = \frac{\sin \theta_i}{\sin \phi}$$

The refractive index n is for the medium the incident light penetrates into. Solids with dense atomic packing generally have higher indexes of refraction. In crystals the index of refraction is different for different directions, which can result in the incident rays being split into two polarized rays.

In explaining the phenomenon of *polarized* light, we use the concept of the electromagnetic wave. All electromagnetic radiation of whatever frequency consists of two fields, one electrical and the other magnetic. Figure 4-45 shows these two fields designated by the vectors **E** and **B.** Each field is separated by 90°; that is, for each **E** there is a corresponding **B** set up in a plane at right angles to each other. For simplicity, only two fields are shown. It must be remembered that any radiation is composed of many, many waves with **E** and **B** vectors representing the electronic and magnetic fields pointing in all directions. Using just the **E** field and Figure 4-46, if you position yourself as standing out on the x axis and looking back at the

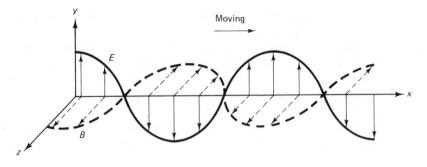

E — Electric field vector, shown in the xy plane
B — Magnetic field vector, shown in the xz plane

Figure 4-45 Electromagnetic fields in a light wave.

origin of our *x–y–z* axes system, the many **E** fields could be represented by the sketch in Figure 4-46. These **E** vectors could also be represented by their respective components in the *y* and *z* directions, thus simplifying the representation of these **E** vectors. A schematic of unpolarized light would then look like the sketch in Figure 4-47a. This view of *unpolarized* light represents **E** vectors that oscillate back and forth along the *y* and *z* axes. If the radiation consists only of **E** vectors along only one axis, it is known as *polarized*. This can be accomplished by the use of a filter that absorbs the electric field oscillations in one particular direction. A representative sketch of polarized light is shown in Figure 4-47b. The light in this example is polarized in the *y* direction.

Unpolarized and polarized light may be further understood by comparing them to vibrating a rope. If you fix a rope at one end and vibrate the other end up and down in all directions, this would be a model for ordinary or unpolarized light. If you vibrate the rope in just one direction, say the vertical direction (up and down), this would be representative of polarized light. Furthermore, if the same rope was pulled through the slots in a picket fence the rope could be vibrated only in this up and down manner. The picket fence then could serve as a model of an optical filter (a polarizer) that permits light vibrations in only one direction and absorbs (prevents)

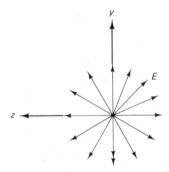

Figure 4-46 A point view of electric field vector (*E*) in an electromagnetic radiation of unpolarized light.

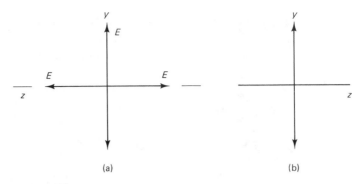

Figure 4-47   (a) Unpolarized light with E vector resolved into y and z axes. (b) Polarized light in y direction.

other vibrations from being transmitted. When light is transmitted through material, the light may become polarized due to the differences in the material structures in different directions. A material whose structure varies with the orientation of the material is not isotropic (anisotropic). As you recall, isotropy is a quality of a material that has identical properties in all directions. The term optical anisotropy refers to a material having different optical properties in different directions.

Most polymeric materials, being insulators, have no free electrons. In addition, their internal structure contains boundaries and pores that interact with light and cause it to be transmitted in different directions at different speeds, hence at different refraction indexes. Polarization of light is used in many applications, such as liquid crystal displays (LCD), polaroid sun glasses, and photoelasticity, where polarized light is used to reveal the amount of anisotropy in a transparent polymer (polystyrene) model of a loaded structure (see Figure 4-48).

4.7.6 Photoelasticity

Certain transparent materials when subjected to stresses ($\sigma = P/A$) become *birefringent* or double refracting. These substances will divide an incident ray of light into two beams that travel through the material at different speeds. In particular, these two beams are polarized at right angles to each other (see Figure 4-47).

This material property of birefringence is taken advantage of by constructing models of structures such as buildings using transparent materials having double-refraction properties. The models are subjected to forces (strains) and the internal stresses produced are analyzed by translating the stress fringes or isochromatics into stress magnitudes. An *isochromatic* appearing as a dark band (see Figure 4-48) represents a region or locus of constant difference in principal stress. A pattern of alternate bright and dark lines represents the variation in refraction produced in the model. A variation of this technique is to coat an actual structure with a photoelastic coating such as Photostress. When the part is subjected to loads, the strains are transmitted to the coating, which then becomes birefringent.

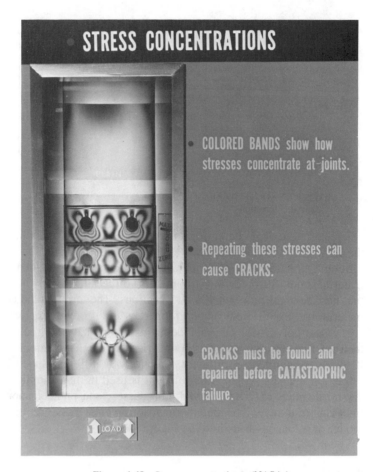

Figure 4-48 Stress concentrations. (NASA.)

4.8 THERMAL PROPERTIES

4.8.1 Heat Capacity

The efficiency of a material to absorb thermal energy is known as *heat capacity*. Heat capacity (Cp) is the number of joules needed to raise one unit mass (1 gram mole) of a substance 1 K at constant pressure. Most solids have about the same Cp at room temperature. If the heat capacity of a substance is compared to the heat capacity of water, the *specific heat of a substance (s)* is obtained. Cp for water is 1 calorie/g °C or 4.184 J/g °C, which is a substance that requires a large expenditure of energy to produce a change in temperature. Substances with high specific heats do not change their temperature appreciably. Using units of cal/g K at room temperature, specific heats (*s*) are for silver, 0.06; iron, 0.1; glass, 0.14; and polyethylene, 0.54.

Properties of Materials   Module 4

4.8.2 Thermal Expansion

Nearly all solid materials expand when heated and contract when cooled. Isotropic materials expand or contract equally in all directions. This temperature deformation results from the changes in the distances between adjacent atoms due to changes in thermal energy. The *coefficient of linear thermal expansion* α (Greek letter alpha) describes this effect in the following equation for unit deformation (ε):

$$\frac{\Delta L}{L_0} = \alpha \, \Delta t$$

where $\Delta L/L_0$ is the change in length of some dimension in inches, L_0 is the original length of that dimension, and Δt is the change in temperature. $\Delta L/L_0$ is another way of expressing unit deformation (ε, strain) brought about by temperature changes. Typical units of α are 10^{-6} (mm/mm °C). For 1020 steel, $\alpha = 11.7 \times 10^{-6}/°C$; for graphite, $\alpha = 5 \times 10^{-6}/°C$; and for nylon, $100 \times 10^{-6}/°C$.

In metals there is a close correlation between the coefficient α and the melting point (T_m). The lower the coefficient, the higher the T_m. Knowing the T_m gives an indication of the strength of the bonding forces.

A bimetallic strip exemplifies the use of the coefficient in temperature-control relays. Upon a temperature change, two metals bonded together and having different coefficients will bend, making electrical contact with a switch.

Polymers have small bonding forces (van der Waals bonds) and consequently larger coefficients than metals. In summary, the covalent and ionic bonded materials have the lowest coefficients (strongest bonding), polymers the highest coefficients (weakest bonding), with metals somewhere in between these limits.

4.8.3 Thermal Conductivity

The ability of a nonmetallic solid to transmit heat, its *thermal conductivity,* depends on *phonons* (quanta of energy). The phonons or "particles" behave somewhat like gas atoms and transfer the solid's lattice vibrations from a region of higher energy (high temperature) to a region of lower energy. The phonons encounter more collisions as they move with the speed of sound through a material as the temperature increases. This results in a lowering of the thermal conductivity with increase in temperature.

In metals the free or conduction electrons also serve as carriers of thermal energy received from the phonons. Alloys added to metals reduce the thermal conductivity, much as porosity and other crystal imperfections do in all materials. Polymers lacking free electrons and a less crystalline structure (orderly atom arrangement) have poor thermal conductivities. Such a characteristic makes polymers good thermal insulators.

Thermal conductivity (k) is the ratio of heat flow to temperature gradient. *Heat flow* is the flow of heat energy per unit area per unit time. The term *temperature gradient* describes a temperature difference per unit distance. Letting q represent the heat flow or flux and $\Delta t/\Delta x$ the temperature gradient, then thermal conductivity k can be seen to be the constant of proportionality in the equation relating these quantities, as follows:

$$q = k \frac{\Delta t}{\Delta x}$$

Therefore, a typical set of units for k at 293 K might be

$$\frac{J}{cm \cdot s \cdot K}$$

Expressing k in these units, silver has a value of 4.1, steel (average) 0.5, glass (average) 0.01, and polyethylene 0.004.

4.8.4 Thermal Resistance

Insulating materials have low thermal conductivities and can retard the transfer of heat. The thermal conductivity of most materials is temperature dependent. In other words, the value changes with a change in temperature. Refractory materials, on the other hand, have thermal conductivities that have a minimum dependence on temperature. Porous materials such as textiles, rock wool, cork, foamed plastics, and man-made insulating tiles are good insulators partly due to their ability to trap air, which itself has a low thermal conductivity. Furthermore, the entrapped air is free from circulating currents, which aid the transfer of heat. Numerous applications for insulating materials exist in today's society, particularly in the area of heat (energy) conservation. Figure 4-49 shows two vital applications that exemplify the several thermal properties mentioned in this brief discussion.

4.9 TESTING AND INSPECTION

Materials are tested to determine their basic properties. In particular, testing of materials has as its objective the performance of tests to determine numerical values for properties. As stated previously, properties of engineering materials mostly can be placed into the following major classes: physical, mechanical, chemical, thermal, electromagnetic, and optical. Testing for engineering materials is primarily concerned with mechanical properties. Another facet of inspection deals with the mathematical treatment of materials in analyzing the effects of the size and shape of material parts and the type of loading on the strength of machine parts. This approach is contained in courses of study under the titles of Mechanics of Materials, Strength of Materials, and the like. Finally, there is another specialized area of materials knowledge and skills, material inspection, linked to the previously mentioned techniques.

Inspection differs from testing in its objectives. The objectives of inspection are to examine parts or materials for the presence of *discontinuities* and *defects*. In other words, inspection concerns itself with estimating the degree to which a product conforms with design specifications. The American Society for Nondestructive Test-

Properties of Materials   Module 4

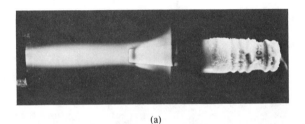

(a)

(b)

Figure 4-49 (a) Thermal resistance. High alumina (Al_2O_3) content spark plug insulators can withstand thermal shocks from below zero ($-73°C$) to white hot heat (over $1000°C$). Note frost still on top of insulator. (American Ceramics Society.) (b) Various thermal insulating materials systems protect space vehicles from high heat of reentry into the earth's atmosphere. (Corning Glassworks.)

ing (ASNT) defines a *discontinuity* as an interruption in the normal physical structure or configuration of a part, such as a crack or porosity. A discontinuity may or may not be detrimental to the usefulness of a part. A *defect* is a discontinuity whose size, shape, location, or properties adversely affect the usefulness of the part or exceed the design criteria for the part. ASNT has five major inspection systems for detecting such discontinuities. These systems are (1) radiographic testing (RT), (2) ultrasonic testing (UT) (see Figure 4-50), (3) eddy current testing (ET), (4) magnetic particle testing (MT), and (5) liquid penetrant testing (PT). In addition to these five major systems, inspection techniques using holography, microwaves, liquid crystals, infrared, and leak testing are available where circumstances dictate. As is evident from the preceding data, inspection techniques involve tests most of which are classed as nondestructive (NDT).

Two major terms used in conjunction with a discussion of testing and inspection are *quality control* (QC) and *quality assurance* (QA). The objective of QC is to *statistically* determine how much testing and inspection are required to assure products will meet design specifications and service life expectations. For example, in the production of an automobile engine, each part cannot, due to high cost, be given complete inspection to see if it meets exact dimensional tolerances, if the metal is used free of all external and internal defects, and if each part has the necessary

Figure 4-50 Ultrasonic testing setup. (NASA.)

strength. Rather, a *statistical sample* is taken of a batch of parts as they are produced. Through statistical analysis the sample is determined to be representative (probably the same) as all parts made in the process. Whole engines are also statistically chosen to be samples of the engine-making process. They are tested to failure (destructive test) as a means of providing *quality assurance* (QA) for the engine itself. QA, the goal of any QC program, is used to refer to the total set of operations and procedures used in manufacturing a product whose goals are conformance of a product to design specifications.

APPLICATIONS AND ALTERNATIVES

Materials and their properties occupy such a dominant role in society that they are often ignored or taken for granted. The packaging industry represents a prime example, especially with regard to food and products consumed by humanity. The "tin can" holds a variety of foodstuffs ranging from soup to meat to nuts. Actually, the tin can uses very little, if any, tin, an expensive metal due to its inaccessibility. Figure 4-51 shows two examples of the sheet steel used in can making. Figure 4-51a depicts the tinplate referred to as 55# 2CR 25 tin coating. The 55 is the number of

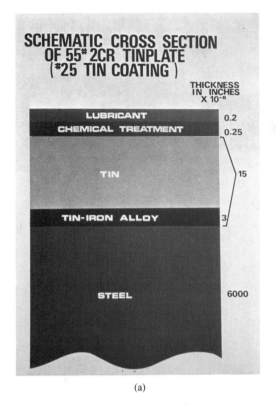

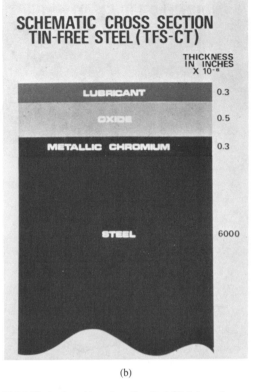

Figure 4-51 (a) Schematic cross section of 55# 2CR tinplate. (American Can Co.) (b) Schematic cross section of tin-free steel (TFS-CT). (American Can Co.)

pounds for 31,360 in.^{2} of plated sheet steel; 2CR indicates the cold rolling process used to thin out the steel; #25 tin coating means 1/4-lb tin coats the 31,360 in.^{2} with 1/8-lb on each side. Note the steel thickness of 0.006 in. with only 0.00015-in. tin plating and a 0.000003-in. tin–iron alloy resulting from the electrolytical deposition process. Lubricants of dioctyl sebacate (AOS) or acetyl tributyl citrate (ATBC) are added to minimize abrasion in the processing. Malleability of this composite sheet is important to permit the deep-drawn two-piece cans shown in Figures 4-52a and b. Tin plating provides for soldering and rust prevention. Beer and carbonated beverages present special corrosion problems. Tin-free steel (TFS) (Figure 4-51b) was introduced in 1965 to provide good resistance both before and after enameling, exceptional enamel adhesion, and resistance to under-cutting corrosion of enameled plate that is exposed to beer and carbonated beverages; TFS still maintains a lustrous surface that resists discoloration when the enamel is baked. As seen in Figure 4-51b, tin-free steel, chromium type (CT), uses a very thin film (0.000003 in.) of metallic chrome covered by a thin (0.000005 in.) layer of chromium oxide.

(a)

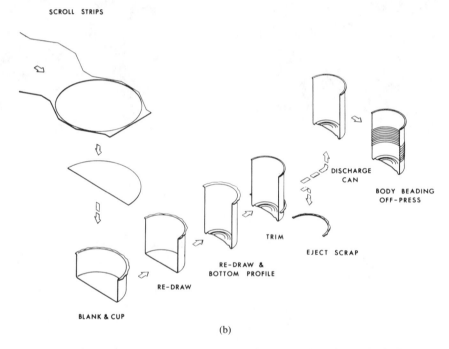

(b)

Figure 4-52   (a) Photograph of the various stages in deep drawing a tin can. Beginning with a blank of 8.3 to 8.5 inches in diameter, a tin can is drawn with final dimensions of 3-3/16 diameter, and 4-3/8 inches tall. (American Can Co.) (b) Diagram showing the stages required to produce the parts in (a). (American Can Co.)

Many interesting problems confront the materials scientist and engineers who must package food and other products. Just as beer and carbonated beverages can cause corrosion of their metal containers, the container may cause the contents to spoil or discolor if it is not compatible with the contents, so a protective coating is used on the container. Material testing to ensure the compatibility of content to container is of utmost importance to both the can maker and the producer of the contents. Spoiled contents can greatly affect the public opinion of a company; normally, the can maker would have to pay for the spoiled contents if their container caused the spoilage.

To gain practice in applying your knowledge of properties of materials, pick up a copy of one of the periodicals listed at the end of other modules. Read an article about a new product system or material development. You will encounter many references to properties.

4.10 SELF-ASSESSMENT

4-1 What qualities define a material?

4-2 Cite one property of a material that cannot be observed.

4-3 Define a chemical property.

4-4 Give two examples of a mechanical property.

4-5 Define stress and express it in appropriate SI units.

4-6 Which quantity, stress or strain, can be measured in a stress–strain tensile test?

4-7 How can you distinguish a ductile material from a brittle material with the aid of a stress–strain diagram?

4-8 Why should the test conditions under which some material property was determined be included along with the property?

4-9 Do all materials have a yield point?

4-10 Factors of safety are defined either in terms of the ultimate strength of a material or its yield strength. In other words, by the use of a suitable factor the ultimate or yield strength is reduced in size to what is known as the design stress or safe working stress. Which factor of safety would be more appropriate for a material you are dealing with that will be subjected to repetitious, suddenly applied loads?

4-11 Stiffness of a material is readily apparent in studying the material's stress–strain curve. Explain.

4-12 Product liability court cases have risen sharply in recent years resulting from poor procedures for selecting materials for particular applications. Assuming a knowledge of a material's properties is a valid step in the selection process, cite two examples where such lack of knowledge could or did lead to failure or unsatisfactory performance.

4-13 Scissors used in the home cut material by concentrating forces that ultimately produce a certain type of stress within the material. Identify this stress.

4-14 A material may be tough at room temperature. In technical terms, describe what this means. What happens to this toughness in most materials as the ambient temperature is decreased?

4-15 Make a sketch and fully dimension an Izod impact test specimen.

4-16 At what level of stress in terms of yield stress do most fatigue failures occur?

4-17 From your experience, what typical automobile part requires complete fatigue testing prior to development?

4-18 Identify the origin of most fatigue failures and explain the significance of your answer in terms of handling a metal workpiece in the machining process.

4-19 Torquing a nut on a bolt to specified values using a torque wrench is standard practice in many industries. Explain why this is so.

4-20 A structural steel beam may deflect under its transverse loading. Is this considered bad practice? Is it allowed? Cite a case where deflection of beams (wood or steel) must be limited to a minimum.

4-21 In selecting a material that will be subjected to abrasion, what property would you expect the material to possess to a very high degree?

4-22 How does the chemical industry protect its tanks and piping from corroding due to the action of highly corrosive liquids?

4-23 What would your explanation be for the development of surface cracks on the surface of four-year-old automobile tires?

4-24 Name one method large automobile manufacturers use to inhibit corrosion of the steel chassis parts of trucks and passenger vehicles.

4-25 Explain why the electrical conductivity of semiconductor materials increases as temperature increases.

4-26 Iron and steel parts in electrical machinery, by necessity, must possess high permeability and be preferably made of "soft" iron. What is the significance of this statement?

4-27 Give an example of the use of a metal with high residual magnetism.

4-28 Photoelectric cells are used in a variety of ways. Are solar batteries used in space such an application? If so, how efficient are these devices, in your opinion, in converting solar energy into electrical energy?

4-29 Describe two ways of producing x-rays.

4-30 Distinguish between translucence, transparency, and transmittance.

4-31 Photoelasticity is sometimes defined as a visual, full-field technique for measuring stresses in plastic models of parts, usually before the parts are made. In relation to the material presented in the text, does this definition really define the term? Knowing that this stress analysis of plastic models uses a polariscope, where does photoelasticity play a part in the technique?

4-32 Thermal resistivity is as important as thermal conductivity. Illustrate both properties by citing an example of each from your own experience or knowledge.

4-33 Using magnetic chucks in a machining operation, how does a knowledge of magnetic properties of metals pay big dividends?

4-34 How does quality control differ from quality assurance?

4-35 Why does a deep drawing operation as illustrated in Figure 4-52 require so many phases?

4-36 Solid objects are three dimensional. A bar with a rectangular cross section under load will deform in how many directions? If it deforms in more than one direction, are the amounts of deformation the same? Which direction would have the greatest deformation?

4-37 Given μ for a material as 0.25 and the strain in the lateral direction as 1.50×10^{-6} mm/mm. What is the axial deformation expressed in inches?

4-38 An aluminum rod 1 in. in diameter ($E = 10.4 \times 10^6$ psi) experiences an elastic tensile strain of 0.0048 in./in. Calculate the stress in the rod.

4-39 Express the stress in Problem 4-38 in SI units with an approved prefix.

4-40 A 6-ft steel bar is deformed 0.01 in. in an axial direction. What is the unit deformation?

4-41 A shaft supported by thrust bearings placed 4 ft apart will fail if the total deformation exceeds 0.00025 in. What is the maximum allowed strain?

4-42 The minimum yield stress for a material is 48,000 psi. What is the factor of safety if the allowed working stress is 24,600 psi?

4-43 Express the stress units of Problem 4-42 in megapascals, the recommended units in the SI system. Use appropriate abbreviations.

4-44 A 1-in.-diameter steel circular rod is subject to a tensile load that reduces its cross-sectional area to 0.64 in.^{2}. Express the rod's ductility using a standard unit of measure.

4-45 If a force (push or pull) of 40 lb is used to turn a wrench such that the 40-lb force is applied at right angles to the wrench at a distance of 16 in. from the center of its jaws, what torque is produced?

4-46 An angle of twist of 0.48° can be expressed as how many radians?

4-47 A standard-weight pipe with a nominal diameter of 1 in. has an outside diameter (O.D.) of 1.315 in. and a thickness of 0.133 in. Calculate its polar moment of inertia.

4-48 Determine the bending stress on the top surface of a rectangular beam whose height is $11^1/_2$ in. if its moment of inertia is 951 in.^{4} and its bending moment is 82586 lb-in.

4-49 Knowing that a common steel has a BHN of 207, determine its approximate tensile strength.

4-50 Using Table 4-4 with room temperature of 68°F, determine the resistance of a copper wire with a length of 100 m and radius of 1 mm.

4-51 Monochromatic green light of frequency 6.2×10^{14} Hz produced by a laser consists of 5×10^{15} photons, which pass a point in 1 second. How much energy (joules) does each photon possess? (Hint: $E = hf$)

4-52 The minimum energy needed to extract electrons from inside a metal is known as the work function W. W depends on the metal and its surface and is equal to the product of the threshold frequency f_0 and Planck's constant, $\hbar$. If $W = 1.9$ eV, what is the minimum frequency of the light illuminating a metal surface that will release photoelectricity?

4-53 Yellow light with a wavelength of 5×10^{-7} m traveling in a vacuum enters a glass plate ($n = 1.5$). What is the velocity of yellow light in the glass plate in meters per second (m s^{-1})?

4-54 Explain why water is chosen to cool automobile engines. What is the effect of using more than the recommended percentage of antifreeze in the automobile's cooling system?

4-55 The specific heat (specific heat capacity) of water is 1.0 kcal kg^{-1} K^{-1}. Write this value using a different set of SI units.

4-56 Even though heat and work are different forms of the same quality (energy), custom has continued to express them using different units. Name two customary units used to measure heat energy.

4-57 The coefficient of thermal linear expansion may be expressed in units such as m/m °C^{-1}. What must you do to convert these units to °F^{-1}? Would you do the same thing if you were given units of m/m K^{-1}?

4-58 Is the coefficient of linear expansion the same for iron and iron alloys (steels)? If not, how do you explain the difference and what is the implication in designing steel structures?

4-59 Both the coefficient of thermal expansion and the thermal conductivity of most materials are temperature sensitive. Explain this sentence in terms of any changes in their respective values. Can you find a material that has a coefficient of thermal expansion that is nearly zero?

4-60 For the major heading of properties develop a list of at least two properties then next to it write a service condition from your own experiences in which the property is important.

4.11 REFERENCES AND RELATED READING

AMERICAN SOCIETY OF HEATING, REFRIGERATING AND AIR CONDITIONING. *Handbook of Fundamentals*. New York: 1981.

Annual Book of ASTM Standards Part 11—Metallography; Nondestructive Testing. Philadelphia: American Society for Testing and Materials, 1981.

Annual Book of ASTM Standards Part 41—General Test Methods, Nonmetal, Statistical Methods; Space Simulation; Particle Size Measurement; Laboratory Apparatus, Durability of Nonmetallic Materials; Metric Practice; Solar Energy Conversion. Philadelphia: American Society for Testing and Materials, 1981.

BOSICH, JOSEPH F. *Corrosion Prevention for Practicing Engineers*. New York: Barnes and Nobel, 1970.

BURKE, J. J., and V. WEISS, eds. *Nondestructive Evaluation of Materials*. New York: Plenum Press, 1979.

CORDON, WILLIAM A. *Properties, Evaluation, and Control of Engineering Materials*. New York: McGraw-Hill, 1979.

DAVIS, HARMER E., GEORGE E. TROVELL, and GEORGE F. W. HANCH. *The Testing of Engineering Materials*, 4th ed. New York: McGraw-Hill, 1982.

FLINN, RICHARD A., and PAUL K. TROJAN. *Engineering Materials and Their Applications*. Boston: Houghton-Mifflin, 1975.

GLAESER, W. A., and DAVID A. RIGNEY, eds. *Source Book on Wear Control Technology*. Metals Park, Ohio: American Society for Metals, 1978.

JONES, S. W. *Materials Science—Selection of Materials*. London: Butterworth, 1970.

LACY, EDWARD A. *Fiber Optics*. Englewood Cliffs, N.J.: Prentice-Hall, 1982.

LOOK, DWIGHT C., Jr., and HARRY J. SAUER, Jr. *Thermodynamics*. Belmont, Calif.: Wadsworth, 1982.

Metals Handbook (7 vols.). *Properties and Selection of Metals* (Vol. 1). Metals Park, Ohio: American Society for Metals, 1961.

Modern Steels and Their Properties. Bethlehem, Pa.: Bethlehem Steel Corporation, 1972.

NUSSBAUM, ALLEN, and RICHARD A. PHILLIPS. *Contemporary Optics for Scientists and Engineers*. Englewood Cliffs, N.J.: Prentice-Hall, 1976.

ROFFE, STAND, and JOHN BARSOM. *Fracture and Fatigue Control in Structures: Applications of Fracture Mechanics*. Englewood Cliffs, N.J.: Prentice-Hall, 1977.

ROLLE, KURT C. *Introduction to Thermodynamics*. Columbus, Ohio: Charles E. Merrill, 1973.

SANDOR, BELA I. *Experiments in Strength of Materials*. Englewood Cliffs, N.J.: Prentice-Hall, 1980.

———*Strength of Materials*. Englewood Cliffs, N.J.: Prentice-Hall, 1978.

UHLIG, H. H. *Corrosion and Corrosion Control,* 2nd ed. New York: Wiley, 1971.

VAN VLACK, L. H. *Elements of Materials Science and Engineering*, 3rd ed. Reading, Mass.: Addison-Wesley, 1975.

WILSON, J., and J. F. B. HAWKES. *Optoelectronics: An Introduction*. Englewood Cliffs, N.J.: Prentice-Hall, 1983.

ZEBROWSKI, ERNEST, Jr. *Fundamentals of Physical Measurement*. Scituate, Mass.: Duxbury Press, 1979.

5

METALLIC MATERIALS

5.1 PAUSE AND PONDER

In Module 1, you were given the characteristics of the ideal engineering material and informed that such a material does not exist. Materials selection requires a compromise with what is ideal—the constraints of materials and processes and what will do the job at the lowest cost. Cost not only involves the price of the material itself but the expense of processing the material and maintaining it as a finished product. Many times the weight of the material determines its acceptability. For instance, the design criteria may specify a corrosive environment. Aluminum, a corrosion-resistant metal, may fit the bill; but if high creep strength is also necessary, then a stainless steel may be appropriate. Increasing corrosion resistance often reduces certain fabricating properties such as forgeability. Increasing fabricating properties such as forgeability may reduce the stainless steel's ultimate strength. The designer must develop a list of all the requirements of a part or product. Some properties are absolutely necessary, while others are desirable but may yield to material limitations.

Consider products that have been with us several decades or more, such as the automobile, electric toaster, vacuum cleaner, bicycle, typewriter, sewing machine, electric mixer, classroom desk, and movie projector. Observe the older models in your classrooms, at grandparents' or older acquaintances', and in museums. You will notice many parts of steel, iron, wood, and zinc castings prior to the 1960s. Beyond that time, many different metals and plastics became substitutes for the traditional engineering materials. The scrap car (Figure 5-1) is a good example.

Figure 5-1 Dismantled car. Broad grouping of materials: (1) light steel, (2) cast iron, (3) heavy steel, (4) chrome steel, (5) plastic, (6) glass, (7) copper, (8) battery, (9) cotton padding, (10) cardboard, (11) zinc, (12) stainless steel, (13) rubber, (14) spring steel, (15) aluminum, (16) carpeting, (17) leather, (18) fiber glass, (19) ceramics, (20) carbon, (21) mastic compound. (Bureau of Mines, U.S. Department of Interior.)

Compare the types of steel, iron, zinc castings, and plastics and their weights used in the earlier car with those in more recent cars. Between 1970 and 1980 the weight of one major U.S. automaker's average four-door car dropped from 3895 to 3228 pounds. The earlier models consisted of over 63% steel. The decrease in weight was helped by more than doubling of the use of aluminum (from 2% to 9%) and plastic (from 3% to 6%). Also look at trade journals such as *Machine Design, Plastics World, Iron Age, Metal Progress,* and *Materials Engineering* to notice how the advertisements and articles promote one material over another; for example, high-strength low-alloy steel (HSLA) is one effort by steelmakers to develop new materials for cars.

Figure 5-2 shows the results of a General Motors study of the mass savings in body panels, which compared steel to alternate materials as designed for the 1985–1987 model cars. The 151 kilograms (kg) of steel used in body panels of their average model car could be reduced to 110 kg by using a combination of 76 kg of fiber glass (glass SMC) and 34 kg of steel. Also, that 151-kg mass could go down to 98 kg using 64 kg of aluminum and 34 kg of steel. Note the mass reduction of the graphite composite (carbon SMC) and steel of 70 kg, but the cost bar shows a tremendous increase in cost using this advanced graphite composite. What causes that tremendous cost increase? Module 8 will help with the answer. General Motors did introduce another composite plastic (Enduraflex™), glass reinforced reaction injection molded (RRIM) panels, into the Fiero line of 1984 cars.

As you read this module and the trade journals, and come into contact with products old and new, keep the characteristics of the ideal material in mind and the properties of materials covered in Module 4. Ask

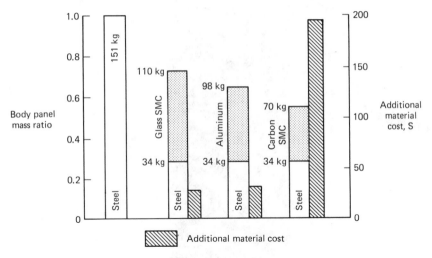

Figure 5-2 Mass savings in body panels of front-wheel drive compact using alternate materials (Courtesy of General Motors).

yourself why a plastic or another metal replaced another metal. Some of these changes are subtle, such as zinc to aluminum for motor housings, while others offer greater contrasts, such as chrome-plated zinc or steel to plastic for automobile trim. Did the change involve a new means to make the part lighter weight, an improved alloy, a more pleasing design, superior electrical or mechanical properties, or safety considerations? While you may not obtain specific answers to all these questions, your attention to the selection criteria will enhance your understanding of metals and other materials and make you a wiser materials consumer.

Humanity has depended on metals since earliest times, and today we take metals for granted. Metals and their alloys vary not only in composition, but in their properties. The study of metal properties starts with the atomic structure of metals, for it is the atomic structure that determines the various properties. In other words, to know something about strength, ductility, toughness, brittleness, or the electrical or thermal conductivity of metals is to know something about the subatomic particles (electrons), the structure of atoms, the crystalline structure, and the bonding of these atomic structures of metals.

In this module we will further examine the microworld of solid substances, specifically metals, to discover some of the fascinating reactions that take place that help explain why metals have been and will continue to be the most widely used of engineering materials (see Figure 5-3). As an example, the millions of tons of steel produced each year would not be possible without the ability to understand and control the movement of atoms of iron and carbon while in a solid state. In another situation, aluminum rivets are kept in refrigerated storage until just before being used in aircraft fabrication; refrigeration affects movement of the aluminum's atoms. How do you

Figure 5-3 The Steel Triangle, the United States Steel Corporate Center in Pittsburgh. The partially-completed building exposes such steels as USS "T-1" which is nearly three times the strength of carbon steel and bears great loads as interior columns; high strength, low-alloy steel (HSLA), Cor-Ten forms a tight adherent, dark russet, ferrous-oxide surface coating and requires no other protective coating. USS Ex-Ten, a HSLA steel serves as floor beams, and some core columns; and much carbon steel finds use in areas not requiring maintenance and high strength. (United States Steel Corporation.)

explain why high-speed, tungsten-steel cutting tools, which cut metal at speeds that make tools red-hot, do not lose their temper, but actually get harder?

5.2 NATURE OF METALS

Metals have been useful to humanity through the ages because they are "strong" when subjected to the external forces encountered under service conditions, yet become "soft" enough to yield to a machine cutting tool or to a compressive shaping force. Above a certain temperature, they melt and become a liquid that is capable of being shaped by casting.

We have realized only in recent times that the properties of all types of solid materials, including metals, arise from their atomic architecture, that is, from the manner in which their atoms arrange themselves into a crystalline order, from the number and types of imperfections found in this structure, and from the bonding forces that keep the collection or structure of atoms bound or joined together.

The "softness" quality of metals can be explained by an understanding of the atomic structure and metallic bond of the metal atoms to form a crystalline structure. We recall that the electrons in the metallic bond are free to move about their positive ions in an electron cloud or gas, which acts to glue or bond the ions together. This free movement, within limits, also allows for the movement of the atoms under the influence of an external load. This slight movement, only visible under the most powerful microscopes, is called *elastic deformation* or *elastic strain*. Once the external force such as a bending force is removed, the internal electrical forces that cause the atoms to move will decrease, allowing the atoms to return to their normal position; they leave no sign of ever being moved. If you bend a piece of spring steel such as a machinist's rule or vegetable knife, it will return to its original shape, thus experiencing elastic deformation.

If you were not careful and applied too much external force by excessively bending the knife or rule, the atoms might move too far from their original positions to be able to move back again when you released the external force. Consequently, the rule or knife would be permanently bent and no longer fit for use. This permanent deformation is known as *plastic flow, plastic slip, plastic deformation,* or *permanent set*. When auto makers stamp out a metal car body from low-carbon steel in a huge die press, they use this "softness" quality of metals. The term *cold working,* to be defined later, is applied to this stamping operation and many other metal working processes that produce plastic deformation in a metal. Cold-working operations include rolling, heading, spinning, peening, bending, pressing, extruding, drawing, and others.

The microstructure of metals can be modified in a number of ways. By now we know that this last statement can be interpreted to mean that through advances in metals technology we can affect the atomic structure of metals in a precise, controlled manner in the designing of metal alloys with the desired properties. In this module we will classify, for learning purposes, the basic methods of changing a metal's properties into the following categories: (1) work or strain hardening, (2) thermal processing under equilibrium conditions (solid solution hardening), (3) thermal processing under near-equilibrium conditions (annealing and grain refinement), and (4) thermal processing under nonequilibrium conditions. Of the four categories, only the first does not primarily involve thermal processing. Prior to discussing thermal processing techniques, one needs to understand the material on the internal structure of solids as presented in Module 3 and have a working knowledge of phase and phase diagrams, to be covered in this module.

Work hardening is a way to change a metal alloy's structure in order to alter its properties by actually performing work (*cold working*) on the metal itself. Work is a form of energy. If we can find a way to deliver work to these metal atoms, we can give them the energy necessary for the atoms to increase their movement and thus their ability to diffuse through the metal structure. Figure 5-4 shows four industrial cold-working methods, and the photomicrographs demonstrate grain structure before and after work hardening.

In our opening remarks about metals' usefulness, we mentioned that metals

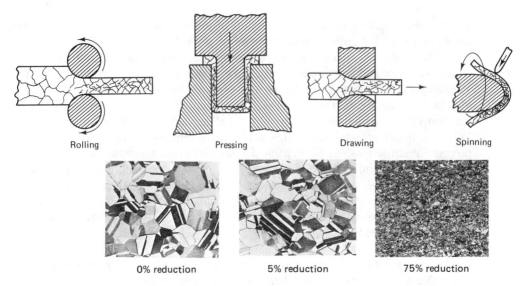

Rolling   Pressing   Drawing   Spinning

0% reduction   5% reduction   75% reduction

Figure 5-4   Some work hardening (cold working) processes.

have the ability to yield to a mechanical force that can form them into desired shapes. An external, mechanical shaping force is a force that causes a metal to exceed its elastic deformation limit and *plastically deform without fracturing*. This force and the deformation produced are the means for transferring sufficient energy to the atoms to allow them to plastically flow or move. This movement of atoms, one row at a time along planes of close-packed atoms, shifts the positions of the atoms in relation to each other. From our study of crystal structures, we know that the spacing between atoms is not only critical, but varies with the particular crystal structure. Fcc structures have the closest packed atoms and the greatest number of close-packed planes of atoms, which take less energy to allow for slip of the atoms. As more and more slip takes place, more dislocations are produced. The more the slip, the more the dislocations and the more the distortion of the lattice structure. The end result is that the deformed metal is *stronger* than the original undeformed metal and offers greater resistance to further deformation.

In summary, through cold working we have (1) reduced the metal's ductility, (2) reduced the effectiveness of the metal atoms to slip, (3) reduced movement of dislocations in the structure, (4) created distortion of the lattice structure, and (5) ended up with a stronger metal that requires a greater force or greater amount of energy to further deform it. At the same time, changes in other properties such as electrical conductivity have also occurred. This condition of noticeable increase in energy for further deformation or increase in the metal's yield strength is known as *work* or *strain hardening*.

5.3 PHASES

A *phase* is a homogeneous part or aggregation of material that differs from another part due to difference in structure, composition, or both. Some solid materials have the capability of changing their crystal structure with varying conditions of pressure and temperature. These varying conditions cause these materials to change phase. Water can exist as a gas, liquid, or solid. Note that each of these general phases contains the same *components* (i.e., the same basic chemical elements or chemical compounds). The components in the case of water are the elements hydrogen and oxygen. Water in the solid phase can have different phases because it forms different crystal structures under different conditions of temperature and pressure. Iron, being allotropic like many metallic elements, can have different crystal structures, and hence different phases (see Figure 5-5). At room temperature iron has a bcc structure. Heated to above 910°C (1670°F) iron's structure transforms to fcc. These allotropic or polymorphic forms are also referred to by Greek letters: α (alpha), γ (gamma), and δ (delta).

It is worthwhile to pause briefly to recall that *phases,* being physically distinct, with their own characteristic crystal structure must, of necessity, possess different properties. No elements found in nature are 100% pure. During the processing of metals, additional impurities are unintentionally introduced partly due to the high costs.

Now back to the question, still unanswered, of the effects of adding other substances to a pure element. The results produced can be many and varied. One result of this addition of impurities might be the formation of three different solid phases. Second, solid solutions may result. A third possibility is one or more compounds could form. Finally, one or more of the preceding results could coexist, depending on the prevailing conditions of pressure, temperature, and degree of concentration of the components of the system. Some examples follow that illustrate the diversity of the results. A fcc structure of solvent Cu atoms will substitute Ni atoms in a solid solution of copper and nickel, known as a substitutional solid solution. The substitution of Zn atoms for Cu atoms in a fcc structure forms brass, a disordered or random substitutional solid solution (see Figure 5-6a). If the ordering were 100%, a compound would be formed. The alloy Cu–Au (Figure 5-6b) has Cu atoms occupying the face-centered sites and Au atoms the corner sites of the fcc unit cell.

Ordered solid solutions form generally at lower temperatures or they come into existence when a disordered solid solution becomes unstable at some lower temperature. The fundamental lattice structure during this particular transformation may or

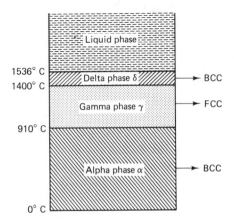

Figure 5-5 Allotropic forms of iron (3 phases: BCC, FCC, BCC).

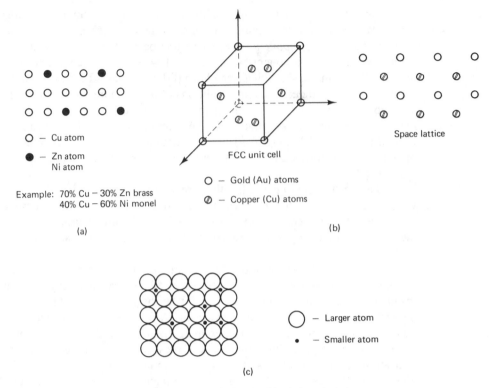

Figure 5-6 (a) Disordered (random) substitutional solid solution. (b) Ordered crystalline compound (Cu₃Au). (c) Interstitial solid solution.

may not change. It is important again to point out that this modification in the arrangement of the atoms brings on an alteration in physical properties. In some cases the ordered arrangement is harder and has greater electrical conductivity than the disordered arrangement.

Another type of crystal structure is formed when some atoms take up positions in between the regular solvent atoms, that is, in the vacant spaces between lattice points. These solute atoms and the solid solutions formed are *interstitial*. The *carbon atoms in ferrite* occupy the interstices (spaces) in this bcc allotropic form of iron. For solute atoms to form interstitial solid solutions (Figure 5-6c), they must be small enough to fit in the interstices between the normal lattice points.

It is time to ask if it is possible to determine ahead of time whether an element will form a separate phase or a solid solution when added to another element. In other words, can we predict the degree of solubility occurring in certain systems? Certainly, this lies at the heart of metals technology in developing alloys (solid solutions) that possess ductility for subsequent forming operations or two-phase materials with greater strength and hardness than the pure substances alone. A set of general rules known as the W. Hume–Rothery solubility rules provides this guidance, which like all such statements is not to be considered foolproof. In brief, these rules state that *to form a solid solution* (Refer to Table 3-1, page 62):

1. The difference in atomic diameter between the solvent atom and the solute atom should be less than 15%.
2. Elements that do not readily form compounds (near neighbors on the periodic table such as Fe and Co) have a tendency to form solutions in one another.
3. Elements that have the same lattice structure tend to form a complete range of solid solutions.

A pause to discuss these rules in regard to other previous definitions of solid solutions is worthwhile at this time. Two metals are soluble to some extent in each other. If the size difference in their atoms is less than 8% and the other conditions are satisfactory, there is almost complete solid solubility. In other words, the two metals are soluble in each other in all possible proportions. Examples of this are Monel (Ni–Cu) and brass (Cu–Zn) (Figures 5-7a and 5-7b). Copper's atomic radius is 1.28×10^{-10} m and nickel's is about 1.24×10^{-10} m (Table 10.16). Both have a fcc structure. Note their position in the periodic table. They are elements 29 and 28, respectively. The fcc Cu-Ni alloys can range from near 0% Ni to almost 100% Ni.

There are some 25 alloys used in industry that are based on this mutual solubility of copper and nickel in all proportions. These nickel alloys, known as *Monel alloys*, contain about 30% copper and have combinations of high strength and good corrosion resistance. Some high-strength Monels, in addition to being nonmagnetic, are equivalent to heat-treated steels having tensile strengths approaching 200,000 psi. In contrast to pure Cu, which has relatively low strength, the alloys of Cu and Ni permit much higher strength due to the interaction between the different atoms in the solid solution producing increased resistance to slip.

Another common alloy, *brass,* is a solid solution of (Cu) and (Zn) (Figure 5-7b). Zinc's atomic radius is about 1.33×10^{-10} m. It is element 30 in the periodic table. These atoms differ in size by less than 15%. In this system, Zn atoms can replace Cu atoms up to a maximum of 50% and produce disordered substitutional

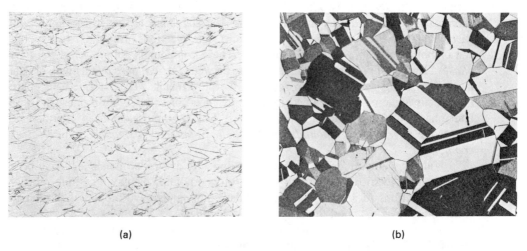

(a)   (b)

Figure 5-7 Solid solutions. (a) Monel (NI 67.5% Cu 30.18%). (b) Brass (Zn 70% Cu 30%). (Buehler Ltd.)

solid solutions. We can add tin (Sn) to Cu; Sn, the fiftieth element of the periodic table, has a radius of about 1.40×10^{-10} m. This indicates that we are approaching the 15% limit. In fact, only about a maximum of 10% of the Cu atoms can be substituted by Sn atoms to produce the single-phase alloys known as *bronzes*.

For an interstitial solid solution to form, the added atoms must be sufficiently small to fit into the interstices between the solvent atoms. From our study of atoms and the periodic table, we know that the atom's radius is measured from the center of the atom to the outermost electron. We also know that this radius decreases as we move from left to right in a period of the table (Table 3-1, page 62). Taking these facts into consideration, the five elements with radii less than 1×10^{-10} m are hydrogen (H), carbon (C), boron (B), nitrogen (N), and helium (He). A classic example is the interstitial solid solution formed by C in Fe in the production of steel. Pure Fe has a bcc structure at room temperature. This phase is called α (alpha) iron. In this form the Fe is relatively weak and incapable of being shaped. Above 910°C, iron, being allotropic, changes to a fcc structure known as γ (gamma) iron. From our study of unit cells of the various lattice structures, we know that the interstices between atoms in a bcc structure are quite small. In the fcc unit cell a relatively large interstice exists in the center. Added C to α Fe allows only 0.025% of the C atoms to take up positions in iron bcc structure. If instead the α Fe is heated to above 910°C and then the C atoms are added, the γ Fe will accommodate up to 2% of the C atoms. This interstitial solid solution is given the name of *austenite*. The formation of steel and cast iron will be treated extensively when phase equilibrium diagrams and heat treating are discussed later.

In studying the formation of compounds in liquid or gaseous states, it may be fair to conclude that most chemical compounds are nonmetallic. Second, they involve the exchange or sharing of electrons. Third, one of the elements involved must have a positive valence and the other a negative valence (oxidation states). Fourth, the algebraic sum of these oxidation states must be zero in the compound formed. Experience indicates that two metals show no inclination to join together in some chemical way to produce compounds. This last statement is true in the liquid and gaseous states. We are now discussing solid state reactions. In solids, two metals can combine to form *metallic compounds*. These compounds are *intermediate alloy phases*. Their compositions are intermediate between the two pure metals with crystal lattices differing generally from those of the pure metals. The intermediate phase may occur alone or accompanied by the pure metal or solid solution phases. These later phases then are called terminal phases. The three most common intermediate phases are *interstitial compounds, intermetallic* or *valency compounds*, and *electron compounds*.

Valency compounds are formed by chemically dissimilar metals and have poor ductility and electrical conductivity. Electron compounds display a definite ratio of valence electrons to atoms. These particular compounds have properties similar to solid solutions (i.e., good ductility and low hardness).

Interstitial compounds involve the same five elements with the relatively small atoms (H, C, B, N, He) mentioned in the discussion of interstitial solid solutions. Many of these compounds may form when an excess of solute atoms is present that exceeds the saturation point for an interstitial solid solution. The smaller solute atoms join with the matrix atoms to produce interstitial metallic compounds. These intermediate phases transform directly from the liquid phase to the solid compound phase at a fixed temperature and a fixed composition. Like most compounds, these interstitial compounds, such as Fe_3C, TiC, or W_2C, are hard and possess great strength but poor ductility.

Many of these compounds have great technological importance in the strengthening of solid materials. They will be eventually called on, as an example, in a heat-treating process to precipitate out at some lower temperature to impede the slippage of planes of atoms over one another, thus producing an increase in strength and hardness of steel.

The liquid (or molten) state of a metal may consist of a single pure metal. It may comprise a solution of two or more metals that are completely soluble in each other. A third possibility is that the liquid state represents two *immiscible* metals. Finally, it could contain two insoluble liquid solutions. Note that the word insoluble is synonymous with immiscible. Substances that are not soluble (insoluble) are immiscible. Very few metals are insoluble in the liquid state. Aluminum and lead come close to being immiscible both in the liquid and solid state. In the solid state they solidify into two separate layers with a clear line of contact showing no appreciable degree of diffusion. Most metals are soluble to some degree in their liquid phases. When these metals are cooled under equilibrium conditions from their liquid phases, they solidify and produce various solid phases. Assuming they are completely soluble in the liquid phase, upon cooling they may be:

1. Completely miscible in the solid state.
2. Insoluble in the solid state.
3. Partly soluble in the solid state.

Other products may be formed, such as intermediate phases. Also, if the liquid phases are not completely soluble, other results may be produced in the solid state. To understand why and how these various results occur requires a good knowledge of the construction and use of *phase diagrams*.

5.4 PHASE DIAGRAMS

Phase or equilibrium diagrams serve as maps for finding one's way through the many solid-state reactions that occur in crystalline metals. Equilibrium conditions may be defined as slow heating and/or cooling of the metals to permit any phase change to occur. To describe completely the conditions of equilibrium for a particular system, three externally controlled variables of temperature, pressure, and composition must be specified. Normally, phase diagrams record the data when the pressure is held constant under normal atmospheric conditions. Consequently, phase diagrams essentially are graphical representations of a metallic system under varying conditions of temperature and composition. To determine the data necessary to plot these many phase diagrams, specialized equipment and techniques are used. Cooling curves for

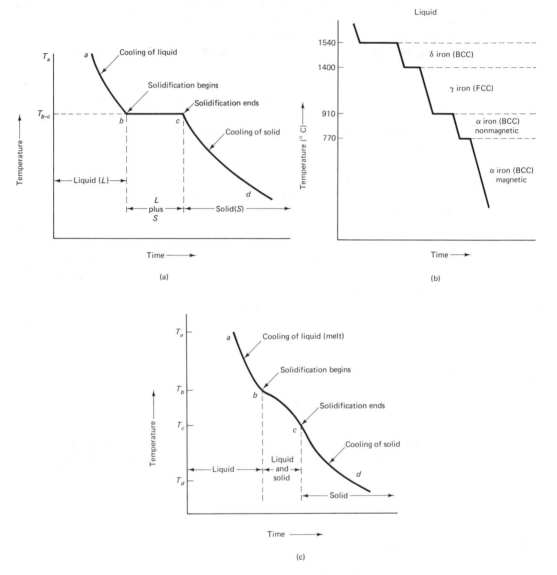

Figure 5-8 (a) Cooling curve for pure metal. (1) Heat pure metal to point T_a (liquid phase or melt); (2) cooling of liquid metal $a - b$; (3) point b pure metal starts to *precipitate* out of solution; (4) point c *precipitation* complete; pure metal completely solid. Curve from b to c straight horizontal line showing constant temperature $(T_b - T_c)$ since thermal energy absorbed in change from liquid to solid; (5) more cooling of solid pure metal from c to d and temperature begins to fall again. Change in volume follows same pattern. (b) Cooling curve for pure iron. (c) Cooling curve for a metal alloy. (1) Two metals heated to point a (liquid phase or melt with both metals soluble in each other); (2) cooling of alloy $(A + B)$ in liquid phase; (3) point b solidification begins; (4) point c solidification complete (both metals in solid form). Sloped $b - c$ due to changing from liquid to solid over range of temperature T_b to T_c since metals A and B have different melting/cooling temperatures; (5) further cooling from c to d of solid state metal alloy.

a pure metal, a pure iron, a metal alloy, and a nonmetal are sketched in the following pages to illustrate the phase changes that occur, if any. Note in the case of pure iron that three of the four phase changes occur while the iron is in the solid state. The α iron stage existing at the higher temperatures is nonmagnetic.

Cooling Curves. A cooling curve is a graphical plot of the structure of a material (usually a pure metal or metal alloy) over the entire temperature range through which it cools under equilibrium conditions.

Figure 5-8a, b, and c shows the phases that the particular metal or alloy is in at any particular temperature and time. In the case of a pure metal or alloy, three phases are indicated: (1) liquid phase (melt); (2) a combination liquid–solid phase; and (3) a final solid phase near or at room temperature (Figure 5-8c).

For the temperature to remain constant, there must be a balance between the heat being withdrawn from the metal and the heat being supplied. In this case, the metal is giving up heat (latent heat of fusion), which balances the heat being removed. The result is a net change of zero as measured by a thermometer. As a metal cools from a higher to a lower temperature, its electrons require less thermal energy as a result of the relatively more ordered, more dense, and more bonded positions occupied by the atoms. As the electrons surrender their excess thermal energy (in the form of heat), they move closer to their nuclei and slow down their motions to match the energy level of their new positions (*equilibrium* position). Cooling curves for non-metals such as glass or plastics (Figure 5-9) show no clearly defined melting points.

Samples of an alloy are taken at various temperatures and quickly cooled to capture the structure at that particular temperature. Metallurgical microscopes using visible light study the grain structure, including the size, shape, and distribution of grains with a maximum magnification of about 2000×. X-rays, gamma rays, and electron beams probe the crystal structure with great resolution and magnification. Electron microscopy permits the detailed chemical analysis of selected areas within the solid material, as well as the character of an individual dislocation. The data collected are used to construct various representations, particularly of the structure of the alloy under varying conditions of temperature, pressure, and composition. Most, if not all, of these graphical representations indicate the structure of a system

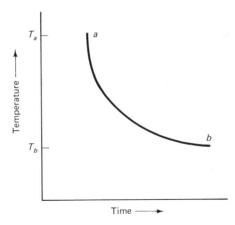

Figure 5-9 Cooling curve for a non-metal. No phase transformation. Nonmetallic heated to temperature T_a and allowed to cool to room temperature T_b. With no phase changes, continuous plot of temperature and time produces a smooth curve.

under *equilibrium conditions*. Equilibrium from a material's standpoint has been defined as a condition of minimum energy. This condition might be compared to a metal spring that is neither stretched nor compressed. Once the spring is stretched or compressed, a force is set up inside the spring that tends to return the spring to its free length or position. The spring has stored up the work done on it in the form of potential energy. Anywhere there is disorder in the crystalline structure of a solid (grain boundaries, imperfection, etc.) is a place of high energy similar to the example of the spring under the action of some force. The magnitude of the energy difference is proportional to the magnitude of the disorder. Therefore, sufficient time must be allowed for atoms to diffuse in order to establish equilibrium conditions at any particular pressure, temperature, and composition. In practice, equilibrium conditions are normally departed from either because of the length of time involved (and attendant cost) or purposely to produce a solid phase that is in a *nonequilibrium* condition.

In our previous discussion we have learned that atoms possess higher energies at elevated temperatures than at some lower temperature. This higher energy permits the atoms to diffuse quicker. At lower temperatures, this kinetic energy is lessened. One can see that this reduced atomic mobility coupled with the disequilibrium energy of a crystal structure at positions of high-energy content presents a broad spectrum of possibilities for the diffusion (precipitation) of atoms producing a phase transformation in the solid state of a system.

The phase diagram for water makes a good departure point for an understanding of phase diagrams for systems comprising more than one component. In the case of water, the component is the compound H_2O. The system under study would then contain a definite amount of H_2O. The one-component diagram for a definite amount of water shown in Figure 5-10 is a graphical plot of pressure along the vertical axis versus temperature along the horizontal axis. These variables are expressed in SI units.

This pressure–temperature fixed composition diagram actually contains three curves. The fusion curve is the straight line that separates the solid from the liquid phase. The curved line separating the liquid from the vapor phases, called the vaporization curve, is a plot of the various pressures and temperature combinations at

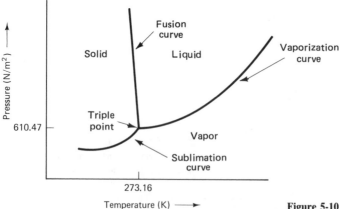

Figure 5-10 Phase diagram for water.

which the liquid and vapor phases coexist in equilibrium with each other. The same can be said of the sublimation curve separating the solid from the vapor phases. Each curve forms a boundary between two different phases. Special note should be made of the point where all three curves intersect. This point, known as the *triple point*, is where all three phases coexist. For water the triple point occurs at a temperature of 273.16K (~0°C) and a pressure of 610.47 N/m².

In our study of alloys, not only do we have more than one component, but we want to study the solid-state reactions while varying the amounts of these components. To accomplish this, several approaches are used. Fortunately, in practice, most metallurgical processes are conducted at normal atmospheric conditions, which simplifies matters considerably.

If the cooling curves for many combinations of two metals (metal alloy) were plotted, as Figure 5-11, a phase or equilibrium diagram would be formed. Points *b* in the cooling curves at which solidification began can be connected to form a smooth line, or *liquidus*. Points *d* at which solidification is complete for each alloy are also connected by a smooth line, a *solidus*. There are various types of phase or equilibrium diagrams of metal alloys depending on whether or not the two components are soluble, insoluble, or partially soluble in each other in the liquid or solid phase.

The *liquidus* is the name of the curve that represents the temperatures at which solidification begins. The *solidus* is the curve which passes through all points at which solidification is completed. In the binary stage, the liquid plus solid, a two-phase slushy region, the process of solidification takes place. It is evident that the points where the liquidus and solidus curves meet represent compositions of 100% A or 100% B (i.e., pure metals). In between are an indefinite number of varying compositions of the two pure metals A and B. Metal A could represent Cu and metal B,

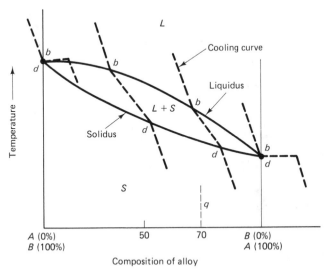

Figure 5-11 Set of hypothetical cooling curves for an alloy system phase diagram. If cooling curves for several combinations of two metals (metal alloy) are plotted as shown here, a phase or equilibrium diagram forms. Points *b*, in cooling curves where solidification begins, are connected to form smooth line (*liquidus* line). Points *d* where solidification is completed for each alloy, are also connected by a smooth line (*solidus* line).

Ni. The process of solidification begins with the formation of seed crystals or embryos from the liquid. Seed crystals, embryos, or nuclei (don't confuse with the nucleus of an atom) subsequently grow into full crystals. In order for these nuclei to form, supercooling or undercooling of the liquid phase is necessary; that is, a temperature decrease below the equilibrium temperature is needed to make the phase transformation proceed at a measurable rate. This supercooling is an indication of the removal of the latent heat of solidification. Again, we can call on our example of the spring that, once deformed, stores up energy within itself. Supercooling is comparable to the amount the spring is deformed. For a nucleus to form, energy is needed to create the nucleus by overcoming surface tension effects. Nucleation can be aided by the presence of impurity atoms, which are almost always present. As each of these nuclei grows larger, it forms dendritic or treelike crystals with a random distribution of crystallographic planes relative to one another (see Figure 5-12).

These crystals are called *grains* and are surrounded by regions of high energy known as *grain boundaries*. If many nucleation sites are formed, a fine grain size in the solid state is the result. To obtain this fine grain, a very minute amount of some element is purposely added, which acts as impurities upon which the nuclei can form. In other words, the average grain size is proportional to the degree of supercooling.

Figure 5-13 is an example of a typical phase diagram for two metallic elements that are completely soluble in both liquid and solid phases. The left-hand vertical

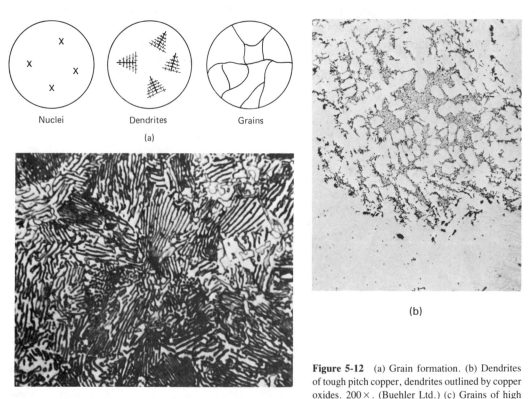

Nuclei   Dendrites   Grains

(a)

(b)

(c)

Figure 5-12   (a) Grain formation. (b) Dendrites of tough pitch copper, dendrites outlined by copper oxides. 200×. (Buehler Ltd.) (c) Grains of high carbon steel. (American Iron and Steel Institute.)

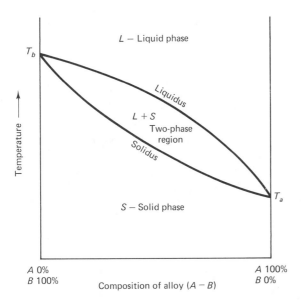

Figure 5-13 Phase diagram. Metals *A* and *B* completely soluble in both liquid and solid phases.

axis represents a pure metal B with a melting point of T_b. Metal A is represented by the right-hand vertical line with a melting point temperature less than metal B. The upper curved line of the cigar-shaped $L + S$ region is the liquidus and the lower line is the solidus. The pure metals and any composition of the two are in the liquid (melt) phase upon heating above the liquidus. Upon cooling to the solidus, they solidify. When an alloy is at a temperature above the solidus but below the liquidus, it exists as part liquid and part solid in a two-phase region labeled $L + S$.

We may wish to know what the *actual composition of this two-phase region* is for a particular alloy at a specific temperature. Using an alloy with a composition of 60% A and 40% B (60A–40B), we will plot this composition on Figure 5-14 to first determine the chemical composition of this alloy at any given temperature. Vertical line *X-Y* represents our alloy. At point 1 the alloy is at temperature (T_1) in the liquid solution phase. Upon slow cooling, point 2 is reached at a temperature of T_2. Our alloy is now entering the two-phase region. Part of our alloy is liquid and part of it has formed a solid solution. To determine the composition of these two different phases at any point, such as 3, a horizontal temperature line called a *tie line* is drawn through point 3. This tie line will cross both the liquidus and solidus curves at points labeled L_3 and S_3, respectively. From both these points, drop vertical lines to the abscissa (horizontal axis) to determine the composition of these two phases at the temperature T_3. The solidus phase formed so far has a composition of 20% A–80% B determined by proceeding from point 3 along the tie line to the solidus point S_3, then vertically downward to the horizontal axis at the point labeled 20% B. Similarly, the composition of the liquid phase is found to be 80% A–20% B. This time we proceed along the tie line to the liquidus and vertically downward to the horizontal axis. Note also that the tie line drawn across the two-phase region ties in the solid region to the left of the two-phase region with the liquid phase immediately adjacent to the right. As the alloy continues to cool to temperature T_4, observe using the same

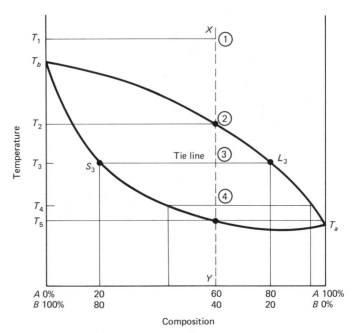

Figure 5-14 Phase diagram tie line.

procedure, that both phases continue to increase in the percentage of metal A. Upon reaching the solidus at temperature T_5, the last liquid, very rich in metal A, solidifies at the grain boundaries. Through diffusion of the atoms all the solid solution will be at an overall composition of 60A–40B.

This is also an excellent procedure to follow to identify the phases in a two-phase region. In Figure 5-15, a portion of a phase diagram is shown with the phases labeled γ and L. Gamma (γ), a Greek letter, is a common symbol used to identify a distinct phase. Using the previous procedure, you should be able to readily label the two-phase region as $\gamma + L$. The γ phase, being below the liquidus, is a solid phase.

Before leaving our discussion on learning how to determine the composition of an alloy at any temperature, you should test yourself at this point by determining the compositions of our 60A–40B alloy (Figure 5-14) at temperatures T_2 and T_4. If you followed the procedure outlined, the alloy at temperature T_2 is in the early stages of solidifying with the solid phase composed of almost 90% B–10% A and the liquid phase 42% A–58% B. At temperature T_4 the last of the liquid phase is beginning to solidify. The liquid is about 98% A–2% B and the solid phase is almost 60% A–40% B.

The next problem to be tackled with phase diagrams is to determine the *relative amounts of liquid and solid existing at a specific temperature in a two-phase region.* So far we know that at temperature T_3 two phases exist: the solid phase has a composition of 20A–80B, and the liquid phase has a composition of 80A–20B. How much of the liquid has solidified upon being slowly cooled to the temperature T_3?

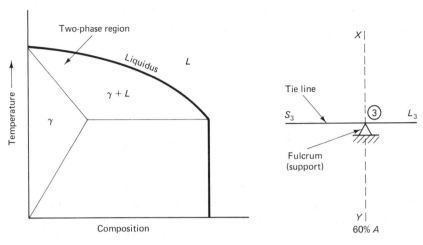

Figure 5-15 Two-phase region of a typical phase diagram.

Figure 5-16 Tie line from Figure 5-14.

To determine this percentage, we use what is known as the *lever rule*. The lever rule gets its name from treating the tie line as though it was a level being supported (fulcrum) at the point where the vertical line (*x-y*) representing our alloy crosses the tie line. Figure 5-16 is a sketch of the tie line removed from Figure 5-14.

The complete tie line from S_3 to L_3 represents the total weight (100%) of the two phases present at temperature T_3. The vertical line (*x-y*) representing our alloy cuts the tie line into two parts or lever arms; one part forms S_3 to point 3 and the second, 3 to L_3. The length of the first lever arm, S_3 to point 3, represents the amount of liquid phase present; the right lever arm, point 3 to L_3, represents the amount of solid phase present. Note that the left lever arm or left part contains S_3 on the solidus line, yet it represents the amount of liquid phase. The right lever arm contains a point L_3 on the liquidus, but this arm represents the amount of solid phase present. Thus, the lever rule states that these two lengths are inversely proportional to the amount of the phase present in the two-phase region. Observe that point 3 is closer to the liquid phase, and therefore the quantity of the solid phase is greater than the quantity of liquid at that temperature. The lengths of these two parts or lever arms of the tie line are measured in units of composition. Now let us apply this lever rule to our alloy 60A–40B. Using Figure 5-17, a repeat of Figure 5-16, the tie line's length in terms of composition is 60 (80% − 20% = 60), leaving out the percentage sign.

The length of S_3 to 3 (left lever arm) is 40, leaving 20 for the length of 3 to L_3. Next, ask yourself what percentage of the whole tie line is the right lever arm,

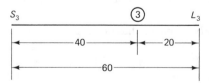

Figure 5-17 Tie line from Figure 5-16.

3 to L_3. To find the answer, you would take the length of 3 to L_3 (20) and divide it by the total length of the tie line S_3 to L_3, and multiply the result by 100%:

$$\frac{20}{60} \times 100\% = 33.3\% \quad \text{(solid formed)}$$

The left part (left lever arm), using the same technique, represents 66.6%:

$$\frac{40}{60} \times 100\% = 66.6\% \quad \text{(liquid formed)}$$

As a final step, we must remember the "inversely proportional" words used in the lever rule. This means 33.3%, which was computed using the lever arm closest to the liquidus, is the percentage of solid solution in the two-phase region at the temperature T_3. Inversely, the 66.6% represents the amount of liquid present. In summary, at temperature T_3 our alloy 60A–40B consisted of two phases. The solid phase of composition 20A–80B comprised 33.3% of all material present, and the liquid solution made up of 80A–20B represented the remainder of the two phases of 66.6% of the total material present.

Let's take a final look at Figure 5-14 to note another metallurgical process that is going on. This process is *diffusion*. For our purposes, note that the solid formed at T_2 consists of a composition of 10A–90B. Assuming slow equilibrium cooling, our alloy will ultimately completely solidify with a composition of 60A–40B. What that tells us is that sufficient time must be given for the atoms to migrate through the solid phase to satisfy the basic principles of equilibrium; that is, given sufficient time atoms will diffuse away from places of high concentration such that a uniform distribution of atoms in the crystal structure is once again achieved. Research into the diffusion process discloses that the rate of diffusion slows as the difference in concentration of different atoms is reduced. Second, diffusion in a solid slows with reduced temperature. Therefore, in practice it is extremely difficult to achieve equilibrium cooling. Time is money. Also, slow cooling produces large grain size, which is usually undesirable. Faster cooling rates are the rule, not the exception, which prevent complete diffusion. The end result is that the initial crystals formed are of one composition. As the crystal grows in a dendrite fashion by atoms attaching themselves to the original crystal, the composition of the total crystal changes. The original higher melting point central portion is surrounded by lower melting point solid solutions. This condition is known as *coring* or *dendritic segregation*. Segregation is always present to some extent when metals are melted and subsequently solidified in the making of steel (steel ingots) and in the casting of metal parts. A metallurgical process that attacks the problem of cored structures will be discussed later.

To continue our study of phase diagrams, the next step is an understanding of the *eutectic reaction*. The word eutectic is taken from the Greek and means "to melt well." Figure 5-18 labels the intersection of the liquidus and solidus as the eutectic point (E). The temperature that corresponds to it is labeled T_E, which is the lowest

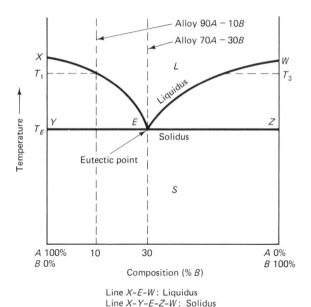

Figure 5-18 Metals A and B completely soluble in liquid phase and completely insoluble in the solid phase (hypothetical).

Line X–E–W: Liquidus
Line X–Y–E–Z–W: Solidus

temperature at which a liquid solution will remain completely liquid. The alloy 70A–30B has the lowest melting point of any alloy in the system AB. Not only does it have the lowest melting temperature, but this eutectic alloy solidifies completely at this temperature rather than over a range of temperatures. Alloy 90A–10B begins to solidify at temperature T_1 and completes its solidification at temperature T_E. The two-phase region to the left of the eutectic point is composed of metal A and liquid. The two-phase region to the right of the eutectic point is made up of metal B and liquid. All these alloys, once they reach the solidus, will solidify as separate phases. Thus, the region below the solidus consists of a mixture (not a solution) of two solid phases. This eutectic reaction for the eutectic alloy, in our example 70% A–30% B, can be written in equation form as follows:

$$L \rightleftharpoons S_1 + S_2$$

The double arrow indicates reversibility. Cooling the liquid phase produces two solid phases (S_1, S_2). Heating the two solid phases produces a liquid (melt) (L). In the case of the lowest melting composition of the system AB, the eutectic composition 70A–30B, its two solid phases will be completely distinguishable from each other when viewed under a metallurgical microscope. The parallel wavy line pattern formed as viewed by the microscope is known as a lamellar type (Figure 5-19). It can be compared to the stripes of a zebra.

Referring to Figure 5-20, alloys to the left of the eutectic mixture are known as *hypoeutectic* mixtures. Those mixtures to the right of the eutectic are known as *hypereutectic*. Those alloys closer to the eutectic composition will contain more eutectic mixture in the solidified alloy. All phase diagrams drawn depicting the

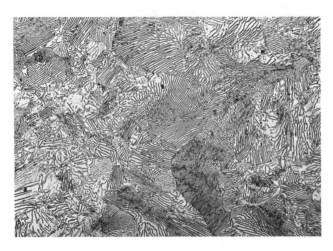

Figure 5-19 Two Phase Lamellae Composition. (Buehler Ltd.)

complete insolubility of two metals in the solid state have been labeled as hypothetical. Most metals are soluble to some degree in each other. An alloy that comes close to being completely insoluble is aluminum silicon (AlSi), Figure 5-21. Silicon is only slightly soluble in aluminum. Therefore, we could substitute Si for metal A and Al for metal B and make considerable use from these diagrams in understanding this alloy system (Figure 5-20).

Our interest now is turned to the more realistic situation in which the metals are to some degree soluble in each other in the solid state. This situation is depicted in Figure 5-22. Notice that this diagram is basically the same as the hypothetical phase diagram, Figure 5-18, with the addition of points Y and Z and the lines extending

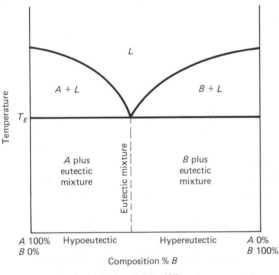

Metal A — Silicon (Si)
Metal B — Aluminum (AL)

Figure 5-20 Hypoeutectic and hypereutectic mixtures.

Figure 5-21 Al-Si alloy as cast ×250, unetched. (Buehler Ltd.)

from these points. Lines *Y-S* and *Z-T* are *solubility curves,* also known as *solvus lines.* Before learning more about these lines, note that there are three areas of single-phase solutions, three regions of two-phase solid solutions, and one eutectic mixture.

The solid alpha (α) phase is composed mostly of metal A with some metal B. The solid beta (β) phase is rich in metal B along with some metal A. The α phase differs from pure metal A by the presence of some metal B. Using a tie line, the two-phase region between α phase and the liquid (*L*) phase is labeled α + *L*. The main solid phase, α + β, is determined in the same manner. Figure 5-22 shows the case of partial solubility of one substance in another. As the temperature increases, the more solid (composition B) can be dissolved in the liquid. Note the solvus lines

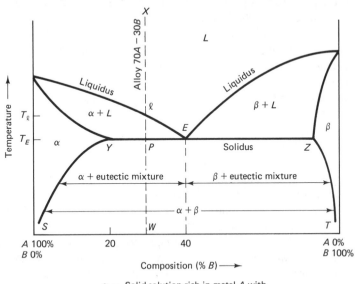

Figure 5-22 Metals *A* and *B* completely miscible in liquid phase and partially miscible in the solid phase.

end at points Y and Z in Figure 5-22. Point Y is the limit of solubility of metal B and solid solution. At the lowest temperature the degree of solubility is a minimum (98% A–2% B); that is, only 2% B is soluble in A but as the temperature increases this degree of solubility of B in A increases to a maximum limit (80% A–20% B).

Now it is time to follow a particular alloy (70% A–30% B) and briefly explain what phase transformations take place as it is cooled from somewhere in the liquid phase. At point l (Figure 5-22) on vertical line X-W and temperature T_1, the liquidus line is reached. Solid solution α is beginning to form. As the alloy cools, this solid solution becomes richer in metal B until the solidus is reached at point P and temperature T_E. The last liquid to solidify has a composition of 60% A–40% B. The solid phase has a composition, assuming that diffusion has kept pace with nucleation and growth of crystals, of 80% A–20% B. Applying the lever rule, we could determine the relative amounts of these two phases present. Note that the last liquid to solidify has the eutectic composition and therefore begins to solidify and form a eutectic mixture consisting of alternate layers of crystals of α and crystals of β phases. At this point we are just below temperature T_E. Here is where the solvus lines come into the action. If you were to draw in the tie line for this temperature, it would be pretty close to the line Y-Z. As the temperature continues to lower, the tie lines that you might draw get longer and longer. Add to this observation the fact that the solvus line S-Y shows the maximum solubility of metal B in metal A at various temperatures, and this solubility decreases as the temperature is reduced to room temperature. Metal B is in solution in the α solid phase. Therefore, the excess β phase must precipitate out of solution. At room temperature our alloy will consist of α phase with some excess β phase precipitated within it, plus the eutectic mixture. This eutectic mixture is also made up of both α and β phases.

One last solid state reaction will be discussed. The *eutectoid* reaction takes place within the solid state. The suffix "oid" means "resembling or like." Therefore, the eutectoid reaction is like the eutectic reaction. As you recall, the eutectic reaction involves the transformation of a liquid phase into two solid phases ($L \rightleftharpoons S_1 + S_2$). The eutectoid reaction then is the transformation of a solid phase (S_1) into two new solid phases (S_2 and S_3). An appropriate equation describing this reaction is:

$$S_1 \rightleftharpoons S_2 + S_3$$

Figure 5-23 shows the many similarities of this type of reaction to the eutectic reaction (see Figure 5-22). In industry the eutectoid reaction is most important as it forms the basis for the heat treating of many metallic materials. Point E is the eutectic point. Point H is the eutectoid point located well below in the solid region of the diagram. As shown, the gamma solid solution at point H transforms into the eutectoid mixture made up of two solid phases. The prefixes hypo- and hyper- are attached to particular compositions of metals on either side of the eutectoid mixture in a similar fashion to those compositions related to the eutectic composition. This very crude diagram will be refined and studied in more detail when it is presented elsewhere in this book in relation to the subject of heat or thermal treatment of metals. At that time the diagram will be called the iron–iron carbide phase diagram—the heat treater's main road map to the processing of steel and cast iron.

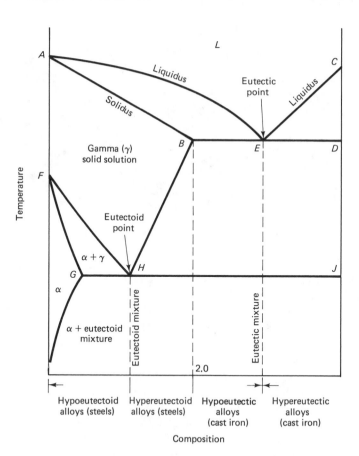

Figure 5-23 Eutectoid reaction.

Multiphase Metal Alloys. So far our discussion has involved solute and solvent atoms that have the same type of crystal structure and thus form relatively pure *homogeneous* mixtures or solutions. Now we advance to alloys that are classified as mixtures of more than one phase that are different from each other in their composition or structure. Under the microscope, the boundaries between phases can be seen in some metal alloys (Figure 5-24). Solder is an example of this type of multiphase metal alloy, with the metals tin and lead present in two separate phases. Steel is a more complex multiphase mixture of different phases, some of which are solid solutions. There are many more *multiphase metal alloys* than single-phase solid solution alloys, primarily due to the greater flexibility of their properties with more variables to control to produce differing properties. In other words, the properties of multiphase metal alloys such as steel depend not only on the phases present, but on the structure, amount, shape, distribution, and orientation of these phases (Figure 5-24). The ability to control these phases allows us to produce desired properties in metals. For example, low-carbon ductile steel is used on auto bodies, and harder and more brittle steel is used for making metal files.

Phase or Structural Transformations. Just how can we bring about these transformations to produce a metal alloy that has certain properties that we desire? First, let's consider the atomic world once more. We know that a few metal atoms

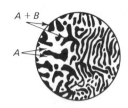

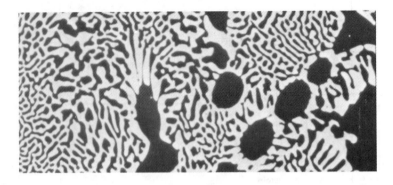

Figure 5-24 Multi-phase metal alloy. Two different compositions (*A* and *A* + *B*) in the solid state. 65 Al-Cu alloy as cast ×250, Fe (NO₃)₃ · 9H₂O. (Buehler Ltd.)

can change their structural arrangement as a result of a change in temperature. Remember the name of this phenomenon? *Allotropy*.

We know that there is a definite limit of solubility of one solid material in another and that this limit also depends on temperature. If this limit is reached, we see atoms coming out of solution to form a multiphase metal alloy. Remember the name of this phenomenon? *Precipitation*.

We know that atoms can move about in a solid just as cigarette smoke moves in a room, but certainly not as fast. Remember this phenomenon? *Diffusion*. And the rate at which the atoms move varies with the temperature. The atoms move at faster rates because they have more energy. What energy? *Thermal energy*. Where does this energy come from? From an external source of heat. The more heat, the more the energy, and the more the movement. By now you should have a pretty good idea about the answer to the next question. In looking over the various ways that we can change atomic structure in a metal alloy, what is the one common quality or characteristic in all of them? Yes, it is thermal energy.

5.5 THERMAL PROCESSING: EQUILIBRIUM CONDITIONS

By controlling the thermal energy, that is, by controlling the supply or removal of heat, we can make metal atoms move. This art and science of controlling thermal energy for the purpose of altering the properties of metals and metal alloys is known as *thermal processing* or *heat treating*.

As with most subjects in materials science, thermal processing concerns a metal's atomic and crystal structure. So far in our study of materials technology, we have accomplished many learning objectives dealing with the "inner workings and hidden mechanisms" in the world of a metal's atoms. Crystal structures, bonding, diffusion, crystal imperfections, dislocations, allotropy, solid solutions, phases, multiphase solid mixtures, solute, solvent, and saturated solid solutions are but some of the many terms that now have meaning for us.

When we studied phases and solid solutions, we used cooling curves for pure metals and metal alloys. These curves are graphic pictures showing the relationships between the variables of time and temperature. Many of these single cooling curves are plotted to give a graphical picture of the possible phases a metal alloy could be in at any particular temperature and at any particular composition. Such phase diagrams are also known as equilibrium diagrams. *Equilibrium* means that the time variable is not controlled but allowed to run its course. In other words, equilibrium implies that phase changes shown on equilibrium diagrams are produced under conditions of slow cooling with no restraints on time. This allows the atoms that are diffusing through a solid material the time to seek and find the equilibrium position of lowest energy level. Phase diagrams show us the following:

1. What phases are present for a particular alloy composition and temperature.
2. The extent of solubility between two metals.
3. The maximum solubilities of each metal in the other.
4. The alloy composition with the lowest melting point (eutectic).
5. The melting points of the pure metals making up the alloy.

To learn how thermal processing can produce phase changes, we will use iron and its alloy steel as examples of ferrous metals. We will not discuss other industrial materials that are heat treated such as glass, but leave this important area of study for independent research.

Steel, being an alloy of iron and carbon, is also allotropic, existing in several forms. Referring to the steel equilibrium diagram (Figure 5-25), which is a graphical record of the various phase transformations of steel in the solid state, and using the eutectoid composition, we see that above 727°C steel is in a solid state with a structure called *austenite,* a single-phase fcc solid solution. Figure 5.25a is a simpler version of the complete diagram shown in 5-25b. Austenite is characterized by its ability not only to be deformed but to absorb carbon. The fcc unit cell of iron atoms contains interstices that are large enough for the small carbon atoms to occupy, producing an interstitial solid solution. Many *hot-working* operations in industry take place with steel heated to such a temperature as to produce this austenitic phase. Furthermore, the austenite region is the starting point for many of the thermal processes about to be discussed.

At temperatures below 912°C, pure iron changes to a stable phase called alpha ferrite or *ferrite*. This ferrite phase can accommodate up to a maximum of 0.02% carbon (by weight), which produces a solid solution with a bcc structure. The bcc structure contains interstices in its unit cell that are too small to accommodate the carbon atom. Another component in this diagram is *cementite,* an intermetallic compound with the chemical formula Fe_3C. Also called iron carbide, this brittle substance can retain up to 6.67% carbon. It forms a network around the pearlite structure in hypereutectoid steels. On the simplified steel equilibrium diagram (Figure 5-27), cementite is represented by the vertical line at the extreme right end of the diagram. When steel with the eutectoid composition forms at 927°C, it produces a lamellar two-phase mixture of ferrite and cementite called *pearlite* (see Figure 5-26). Using tie lines across the two-phase regions of Figure 5-27, a hypoeutectoid steel can be described as a mixture of ferrite and pearlite (Figure 5-28); eutectoid steel as having

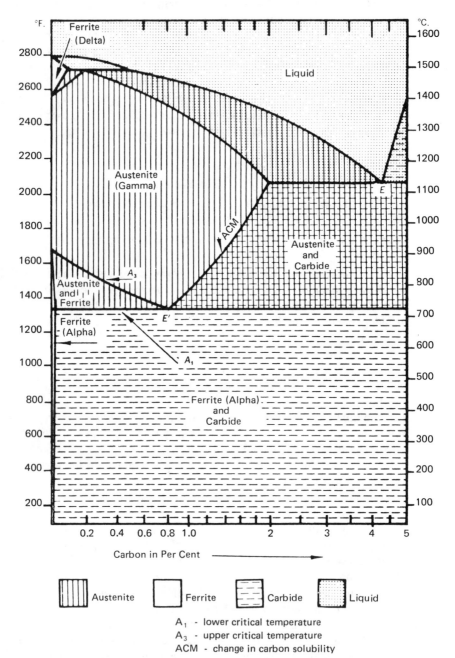

Figure 5-25 (a) Simplified iron-carbon equilibrium diagram. (United States Steel.) (b) Iron-carbon equilibrium diagram. Dash lines show true equilibrium of iron and graphite. Solid lines show a metastable phase diagram of iron and iron carbide (Fe_3C). The metastable diagram is used in the same manner as a true equilibrium diagram. The distinction between the two is negligible above 2100°C. (American Society of Metals.)

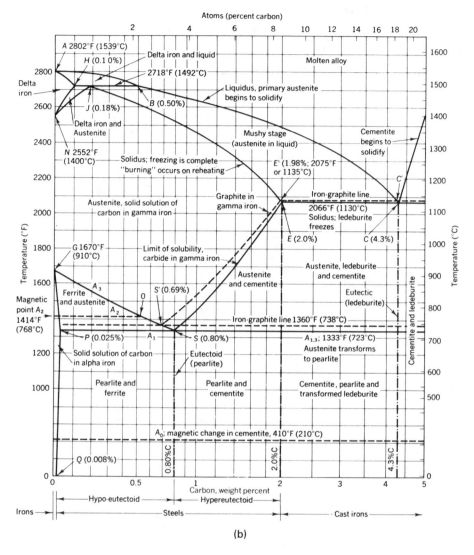

Figure 5-25 (continued).

a pearlitic structure; and hypereutectoid steel as a mixture of pearlite and cementite. Such structures can be discerned under the microscope. At this point it should be evident that there are countless alloys of carbon and iron (carbon steels) with varying amounts of carbon producing a corresponding variety of steels with different properties to serve the demands of industry. Later in this text other elements will be discussed that also play a part in producing numerous specialty steels (alloy steels) that find ever-increasing uses.

It is most important to understand that steel is only possible because iron is allotropic and carbon atoms are small enough to fit in between the iron atoms in a

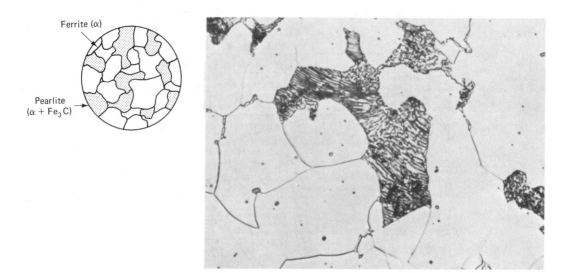

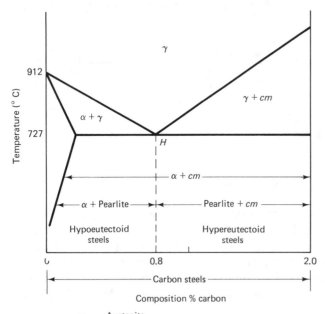

Figure 5-26 Pearlite. Lamellar two-phase mixture of ferrite and cementite. The white areas in this hypoeutectoid steel are grains of proeutectoid alpha ferrite. The dark areas are pearlite colonies consisting of alternate lamellae of alpha ferrite and Fe_3C (cementite). The resolution of the two phases in the pearlite region depends on magnification and orientation and varies from area to area. (Buehler Ltd.)

γ — Austenite
α — Alpha ferrite or ferrite (pure iron)
cm— Cementite, Fe_3C with 6.67% C (iron carbide)
H — Eutectoid point

Figure 5-27 Simplified steel equilibrium diagram.

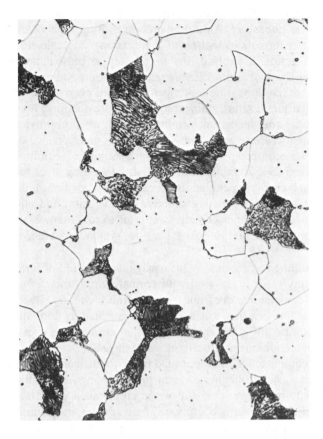

Figure 5-28 Hypereutectoid steel slowly cooled, $1000 \times$. White areas are cementite and dark areas are pearlite. (United States Steel Corporation.)

fcc austenite structure (interstitial solid solution). The many different properties of carbon steel are produced as a result of changes in the amount of carbon and the difference in the abilities of ferrite and austenite to dissolve carbon. Austenite can dissolve almost 100 times more carbon than ferrite. But austenite does not exist at room temperature under equilibrium conditions, whereas ferrite does. It is worthwhile to note on the iron–iron carbide phase diagram (Figure 5-25) the locations of wrought iron (almost pure iron), steels, and cast irons (greater than 2% carbon). Most important of all is the realization, first, that everything we have discussed about the allotropic forms of iron, the diffusion of carbon atoms, phase transformations, austenite, ferrite, cementite, and pearlite takes place in the *solid state*. Not once did we get anywhere close to the melting point of iron, 1535°C. Second, all phase transformations took place *under equilibrium conditions*.

5.6 THERMAL PROCESSING: NEAR EQUILIBRIUM CONDITIONS

Several thermal processing techniques used with ferrous metals approach equilibrium conditions as represented by the iron–iron carbide equilibrium or phase diagram. By definition, such processes bring about a change in a material's properties. Therefore, they must affect the material's atomic structure. The first thermal process to be discussed is generally known as *annealing*. Annealing processes bring about changes mostly by producing phase transformations that result in a rearranged, stable atomic structure with less distorted grains. Therefore, the initial step in most annealing techniques is to heat the steel above its critical temperature to form austenite.

The term annealing is broken down into a number of related operations. Their overall purpose is to reduce hardness, refine grain structure, restore ductility, remove internal stresses left over from some industrial forming process, or to improve machinability or some other property. These operations—full anneal, normalizing, stress relief, process anneal, and spheroidizing—are shown on a simplified equilibrium diagram for steel (see Figure 5-29). A *full anneal* consists of heating steel to the temperature above the line marked A_1, depending on its carbon composition, holding it at that temperature to obtain a homogeneous structure of austenite, and slowly cooling it in a furnace, followed by further cooling in still air to room temperature. The result is a transformation of austenite to coarse pearlite, which is ductile, soft, and stress-free.

Because it is time consuming and expensive, primarily due to the use of furnace cooling, full annealing in many cases is replaced with normalizing. *Normalizing* of steels through using higher temperatures does not require furnace cooling. Instead, all cooling is in still air and a fine pearlite structure is obtained. Steel is normalized to obtain greater hardness than that of a full anneal. Note that normalizing does not approach equilibrium cooling conditions as closely as a full anneal.

A *stress relief anneal* requires temperatures around 600°C, which are below the critical temperature at which austenite begins to form upon heating. The primary objective of this technique is to relieve residual stresses in all metals and steels as a result of a welding, cold-working, or casting operation. These stresses are eliminated or reduced, even though there is no change in the metal's microstructure.

Process anneal is allied with cold working in which the metal, after heating, is cooled slowly in a furnace down to room temperature. The cooling phase distinguishes this process from normalizing. The end product has higher ductility and lower strength than if normalized. Process annealing is used to "soften" metals, particularly steel sheet and wire products, for further cold working. An electrical transmission line made of copper would become brittle after several drawing operations and without a process anneal would fracture (Figure 5-29). This technique, as well as stress relief, does not involve a phase transformation. The structure of the metal involved, however, is affected. A process called recrystallization is involved, which will be explained later in this module.

Spheroidizing anneal is used to improve the machinability of high-carbon steel. Hypereutectoid steels provide good wear resistance. A spheroidizing anneal helps to toughen them by providing more ductility. The hard and brittle cementite network present in hypereutectoid steel also makes machining to close tolerances difficult. Heating the metal for a longer duration near the critical temperature, followed by slow cooling, is one technique used to produce a spheroidal or globular form of cementite in the ferrite matrix. This whole structure is called spheroidite or spheroidized pearlite; it has a lower hardness, higher ductility, and higher toughness than the original metal (Figure 5-29).

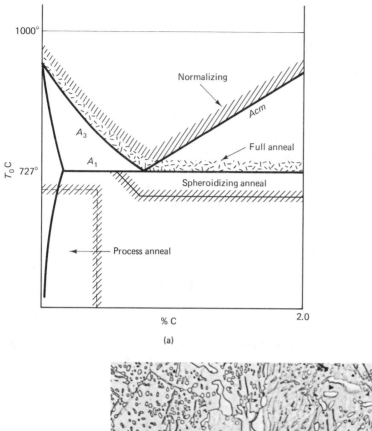

(a)

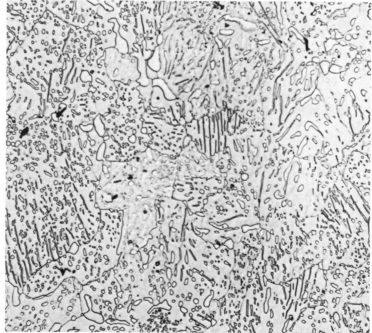

Figure 5-29 Simplified steel equilibrium diagram. (a) Annealing techniques. (b) Spheroidized 1045 steel, 400×. (Buehler Ltd.)

Sec. 5.6 Thermal Processing: Near Equilibrium Conditions **187**

5.6.1 Grain Refinement

Grains that have become plastically deformed as a result of cold working can be given enough energy through thermal processing to permit an orderly rearrangement of the atoms to take place with less deformed grains. The thermal process involved is annealing, as previously discussed, and a certain stage of annealing in which new crystals form, known as recrystallization, is worthy of separate comment.

Recrystallization is the formation of a new, strain-free grain structure from that existing in a cold-worked metal upon heating the metal up through a critical temperature for a suitable period of time. The general term annealing, when referring to cold-worked metals (mostly all nonferrous metals), can be described as consisting of three stages called recovery, recrystallization, and grain growth. The recovery stage (see Figure 5-30), the lowest temperature region, provides the deformed atoms with sufficient energy to rearrange themselves, thus reducing distortion and its attendant stress. A stress relief anneal, as described previously, requires temperatures in this region. As temperatures increase, a point is reached where the nucleation of new crystals appears in the microstructure. These new crystals have the same structure and composition of the original undeformed crystals before the metal was cold worked. The grain boundaries and slip planes with the maximum distortions caused by the cold working contain the majority of these new crystals, and eventually they grow and spread throughout the metal structure. A short period of time after the appearance of the new crystals the properties of the metal change rather drastically. Hardness and strength drop off, while ductility increases. As the temperatures increase, the region of grain growth is reached, which points out that control of grain size can be exercised to some extent by a wise choice of annealing temperatures. Recrystallization temperatures vary with metals and with the degree of cold working. The number and size of crystals that develop in the recrystallization region result from the amount of cold working that is done to the metal. If a metal is not cold worked prior to heating, then no recrystallization will occur and the grains will grow in size. To transfer sufficient energy to the deformed crystals, a minimum plastic deformation of about 7% is required before any change in grain size takes place.

From our knowledge of cold working, we know that a highly cold worked metal contains very little capacity for further plastic deformation. Consequently, the metal is heated to a temperature that permits the atoms to realign and diffuse to a more stable position. The temperatures over which a marked softening occurs (a drop in tensile strength) are in the recrystallization zone. This range or zone of temperatures falls between one-third to one-half of a metal's melting temperature expressed in absolute units (Kelvin and Rankine). The average recrystallization temperature is not the same as the critical or transformation temperature. It depends on the amount of cold working, as well as the duration of the heating. The more a metal is cold worked, the more energy the atoms possess. Therefore, they need less energy transfer from an external source than a lesser cold worked metal. The recrystallization temperature is used to distinguish between cold working and hot working of metals. The cold working of copper at 95°C is at a higher temperature than the hot working of zinc at 20°C. This is so because the recrystallization temperatures of pure copper and zinc are about 120° and 10°C, respectively. A point to remember is that these changes do

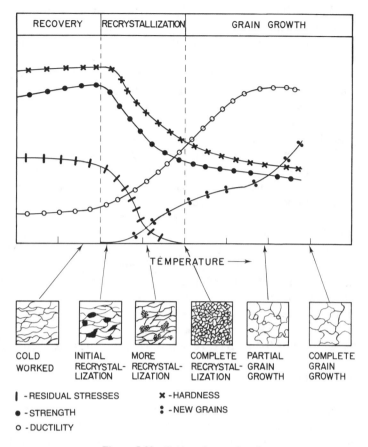

Figure 5-30 Cold work anneal cycle.

not come about as a result of any phase change. When a metal is not allotropic, several processes are needed to change its properties. First, it must be cold worked to bring about a minimum deformation of the crystal structure. Second, it must be heated sufficiently to produce recrystallization. Third, after slow cooling the emergence of new, stress-free, small, uniformly dimensioned grains will result in the desired properties. Finally, it must be realized that if a metal is subjected to a temperature while in service that is higher than its recrystallization temperature, it will recrystallize, lose its strength and hardness and possibly cause a failure of some magnitude.

5.6.2 Grain Size

Heat treating alters the size of grains, which produces corresponding changes in properties. A grain or individual crystal varies in size from one metal to another. Zinc crystals in a galvanized steel sheet and crystals in some brasses can be seen with the naked eye, but these are exceptions to the general rule. Most crystals are only visible with the use of a microscope.

Even within one metal, the grain size may vary from one region to another. Grain size has a direct influence on the properties of the metal. For example, as grain size *decreases,* the yield strength of the metal increases. Another fact is that fine-grained metals are stronger and tougher under low or room temperatures than coarse-grained metals. At high temperatures, the reverse is true; coarse-grained metals are stronger and tougher than fine-grained metals. Grain size ratings, as contained in the American Society for Testing and Materials (ASTM) specifications, are made by comparison with a standard chart of sizes numbered from 1 to 10. Steel is ordered on the basis of either fine grain or coarse grain. Grain size must be specified for metals that are to be cold worked. If the grain size is too coarse, the worked surface will be uneven, and if too fine, the metal will be too hard to work satisfactorily. Several different methods are used in industry to estimate grain size, in addition to the ASTM comparison method.

Temperature alone causes grain growth and not grain size reduction or refinement. If you wish to reduce the size of grains to refine them, you must destroy the original grains and produce new ones with the desired size. One way of doing this is by heating an allotropic metal sufficiently to produce a phase change with its initially small grains. In steel, the fcc grains of austenite are large; but on transforming to bcc crystals, the grains are small. Another way to refine grains is to plastically deform the metal by cold working. This refinement in grain size is not the primary purpose of cold working. However, due to the smaller size of the grains, the cold-worked metal is found to possess different properties than before it was cold worked.

5.6.3 Grain Boundaries

We know that the atoms that make up the grain boundaries are the last to form. As a result, they are not arranged in the same orderly way as the same atoms in the interior of the crystal. We also know that any time we observe a disordered arrangement of atoms we look for distortion, increased stress, and resistance to further plastic deformation or slip—all of which result in stronger material in the grain boundaries (Figure 5-31).

The more small grains present in a metal, the more grain boundary per unit of volume of metal. The conclusion drawn is that fine-grained metals are stronger than coarse-grained metals at room temperatures. At high temperatures, the heat supplied to the fine-grained metal atoms that make up the grain boundaries increases their energy. This reduces their strained condition, which allows them to spread out and become less densely packed, which results in a decrease in strength. Based on this reasoning and coupled with knowledge gained from experience, coarse-grained metals are required for high temperature applications where resistance to creep is critical. Can you recall the definition of creep? A good example where knowledge is needed is in the selection of the correct metal for the manufacture of turbine blades that revolve at very high speeds and temperatures, with extremely small clearances between the moving tips of the blades and the stationary parts of the turbine.

Cold working results in the formation of more grain boundaries due to the large number of dislocations coming together with their disorderly arrangement of atoms. With more grain boundaries come smaller grains, reduced ductility, increased hard-

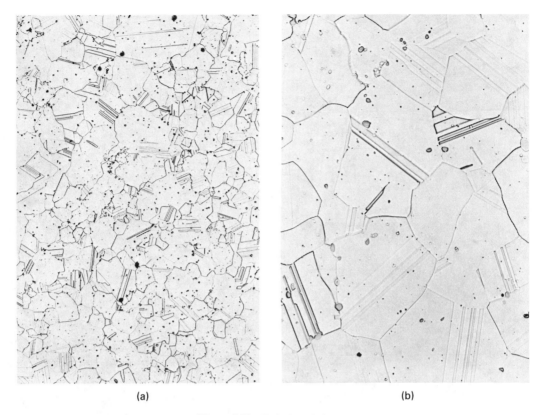

(a)   (b)

Figure 5-31   Grain boundaries. (Buehler Ltd.)

ness, tensile strength, and electrical resistance. These conditions describe the terms *work* or *strain hardening* produced by cold working. A good example of cold working is the drawing of copper wire. *Drawing* has two meanings in materials science. In heat treatment, it means tempering. In cold working, the process of drawing refers to the pulling of metal through dies, as in the manufacture of wire. A copper wire transmission line is cold worked to increase its strength so that it can support itself without excessive sagging over a moderate span.

5.6.4 Grain Shape

Cold and hot working produce a distorted crystal structure. That is, the shape of the grain is deformed or becomes elongated in the direction of the metal flow. Figure 5-32a and b compare hot- and cold-worked grains of copper. Such deformed metals no longer possess uniform properties in all directions (isotropic). Instead, the metal generally shows higher strength properties in certain directions, with a corresponding decrease in those same properties in other directions (anisotropic).

To summarize the preceding discussion, the set of photomicrographs in Figure 5-32c illustrates the concepts of cold working, recrystallization, and grain growth.

Figure 5-32 (a) Hot worked, then cold worked copper specimen ($100\times$). (b) Hot worked copper specimen. (c) 70/30 brass $300\times$. Reduction: (1) 0%, (2) 5%, (3) 25%, (4) 75%.

The samples are from annealed, commercially available 70/30, (70% Cu, 30% Zn) nonleaded brass, initially 0.128 in. thick. Reductions were accomplished by small reduction rolling passes while maintaining the strip at room temperature at all times. Alongside each photomicrograph the average surface hardness (Rockwell 30T) has been recorded. All photomicrographs of the strip samples are shown at $300\times$ for the best possible comparison of microstructures.

5.7 THERMAL PROCESSING: NONEQUILIBRIUM CONDITIONS

By varying the amount of carbon, in the form of *cementite,* dissolved in the single-phase solid solution of iron and carbon (austenite) and cooling the austenite under equilibrium conditions, the austenite was transformed into a multiphase mixture known as carbon steel. Depending on the amount of carbon, steel alloys of ferrite and pearlite or cementite and pearlite are formed, which have different hardness and strength properties. But we now reach a limit. Low-carbon steel (0.25% carbon) has a strength of about 44,000 psi; eutectoid steel (0.8% carbon) has about 112,000 psi. To obtain steel with greater strengths, we must depart from equilibrium conditions primarily in the cooling of the austenite. This departure, in the form of a rapid cooling, is known as *quenching.*

The degree of cooling depends mostly on the quenching medium used. For drastic quenching, water or brine is used. Less severe cooling rates are achieved with such media as oil, molten salt baths, still air, or molten sand. Most liquid media require agitation of the liquid to help reduce the gaseous layer formed adjacent to the hot metal as a result of the vaporization of some of the liquid adjacent to the hot metal whose temperature is above the boiling point of the liquid quenching medium.

Martensite (Figure 5-33) is too hard and brittle to be generally useful. To reduce martensite's brittleness, the heat-treating technique of *tempering,* also called drawing, is called on. In tempering, the martensite is heated to various temperatures to allow

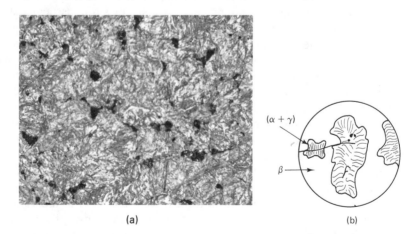

(a)

(b)

Figure 5-33 Martensite. (a) 1095 steel. (b) Microstructure of martensite ($100\times$). (Buehler Ltd.)

the carbon atoms to continue their diffusion through the structure. In allowing the carbon atoms to proceed once more on their paths through the atomic structure, varying microstructures and crystal sizes are formed with just the correct hardness and strength desired for some industrial application. One application may require a good amount of toughness with very little brittleness; another may demand greater hardness. As in many situations, we cannot have the best of two worlds; in this case, to gain greater strength and toughness, we have to trade a little hardness.

To summarize what we have said about equilibrium and nonequilibrium conditions of cooling, our initial step was to heat the basic components of steel (into the *austenite* forming region) to produce austenite, a single-phase solid solution of carbon in gamma (γ) iron. Once we formed austenite, we could cool it under equilibrium conditions to produce a two-phase mixture of ferrite and cementite called *pearlite*. The carbon in the cementite distorts this structure, producing qualities of strength and hardness. The ferrite contributes ductility. If we depart from equilibrium conditions and cool the austenite more quickly, we can form pearlite with greater hardness. If we cool it even more rapidly, we produce a transformation product called *martensite,* which is extremely brittle, strong, and hard. It far surpasses the hardness and strength of pearlite. Steel with 100% martensite has limited use due to its sensitivity to fracture by impact. Therefore, we call on another heat-treating process known as *tempering* to heat the martensite to selected levels to transform some or all of the martensite to other microstructures with differing properties. Tempering is known by another name, *drawing,* due to the fact that, through controlled heating, hardness can be drawn from the material. Actually, tempering, like any process that supplies energy to atoms through heating, depends on the diffusion of these atoms. The rate of diffusion depends not only on the temperature, but on the time allowed for diffusion to take place.

In discussing nonequilibrium conditions in the thermal processing of steel, we make use of an isothermal transformation (IT) or transformation–temperature–time (TTT) diagram for a particular steel composition. This diagram is derived by plotting hundreds of isothermic cooling curves for samples of a particular steel composition. The isothermal cooling curve sketched in Figure 5-34 shows the percentage of austenite and the percentage of the austenite transformed plotted against time. The isothermal temperatures are all below the critical A_e temperature at which the austenite becomes unstable and begins to transform. The data thus collected from these isothermal curves are then used to construct the isothermal transformation diagram for that particular steel which shows the beginning and ending of the austenite transformation as well as the transformed product at the various isothermal temperatures.

This procedure is illustrated in Figure 5-35 using just one set of data. A characteristic S-shaped curve is formed that represents the transformation zone or region at the various temperatures. Figure 5-36 is a copy of such a diagram for eutectoid steel indicating the Brinell and Rockwell hardness readings alongside photomicrographs of the various products formed from austenite at the various temperatures. Figure 5-37, another sketch of a TTT diagram, shows the S-shaped austenite-to-pearlite transformation zone, the austenite-to-martensite transformation zone, and a subdivision of the pearlite into coarse and fine pearlite, and bainite, a transformed product of austenite that has characteristics of both pearlite and martensite. In addition,

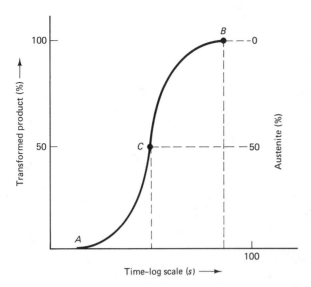

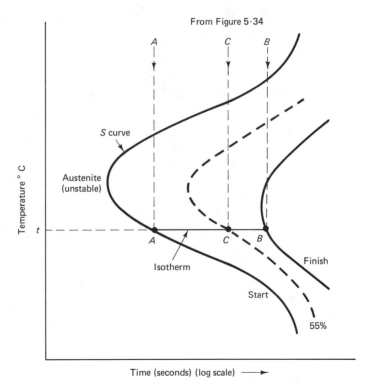

Figure 5-34 Isothermal transformation curve at temperature $t°C$.

Figure 5-35 Isothermal transformation diagram (IT or TTT).

Sec. 5.7   Thermal Processing: Nonequilibrium Conditions   **195**

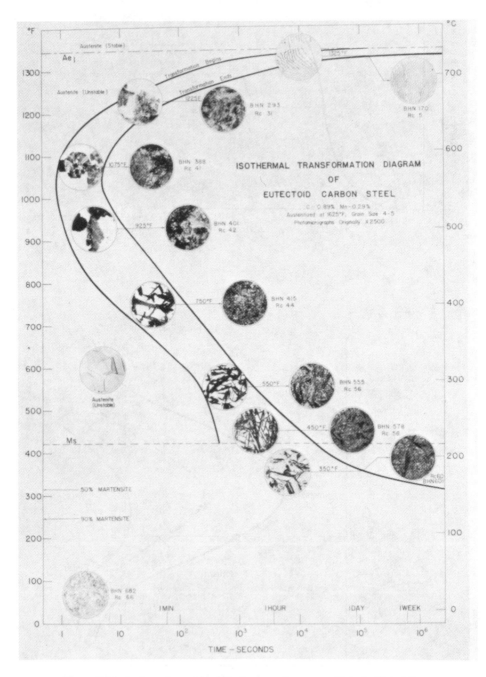

Figure 5-36 Isothermal transformation diagram of an eutectoid steel. (United States Steel Corporation.)

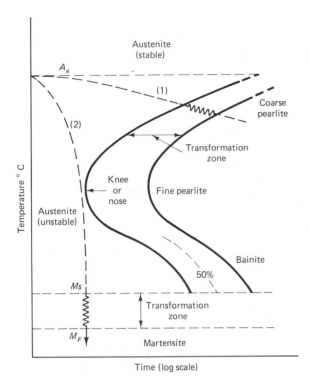

Austenite
(stable)

A_e

(1)

Coarse
pearlite

(2)

Transformation
zone

Knee
or
nose

Fine pearlite

Temperature °C

Austenite
(unstable)

Bainite

50%

M_s

Transformation
zone

M_F

Martensite

Time (log scale)

Figure 5-37 IT diagram showing transformation zones, two cooling curves, and products formed.

two cooling curves are plotted. The first curve, labeled 1, represents the transformation of austenite to coarse pearlite. Such a curve has a relatively small slope, which indicates a slow cooling rate. The second curve has a much steeper slope and consequently a much faster cooling rate. This latter curve is shown passing to the left of the leftmost portion of the pearlite transformation zone, known as the knee or nose. As a consequence, all the austenite transforms into martensite.

It should be evident at this point that the goal of thermal processing under nonequilibrium conditions is to produce steels with greater strengths than those produced by solution hardening under equilibrium conditions (very slow cooling). The mechanism used to accomplish this goal is a phase transformation. Such a mechanism is possible because steel is allotropic. The next few pages describe briefly some major thermal processing techniques used by industry to strengthen steel under nonequilibrium conditions.

5.7.1 Conventional (Customary) Heat, Quench, and Temper Process

The previous statements on the formation of austenite, quenching to form martensite, and the final heating to temper the martensite describe the conventional process of making steel parts with desired properties. Study Figure 5-38 for the customary quench/temper cooling rate. The log scale along the bottom indicates time. One serious drawback to this process is the possibility of distorting and cracking the metal as a result of the severe quenching required to form the martensite without transforming any of the austenite to pearlite.

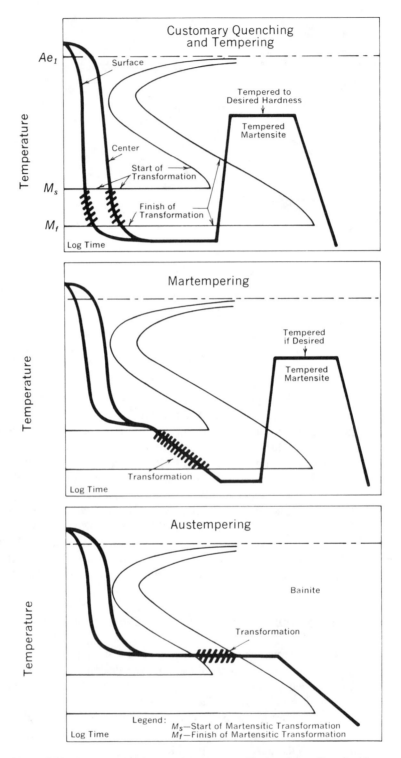

Figure 5-38 Comparison of heat-treating processes. Heating and cooling operations superimposed upon a typical isothermal transformation diagram on which the temperature range has been indicated at which the transformation to the hard product bainite or martensite occurs. (United States Steel Corporation.)

As a metal object is quenched, the outer area is cooled quicker than the center. Thinner parts are cooled faster than parts with greater cross-sectional areas. What this means is that transformations of the austenite are proceeding at different rates. As we cool a metal object, it also contracts and its microstructure occupies less volume. Put these statements together and the conclusion is that extreme care is called for to prevent undue distortion and/or fracture. Extreme variations in the size of metal objects complicate the work of the heat treater and should be avoided in the designing of metal parts. This means there is a limit to the overall size of parts that can be subjected to such thermal processing.

5.7.2 Martempering

To overcome these restrictions, two other thermal processes are used. The first, *marquenching* or *martempering,* permits the transformation of austenite to martensite to take place at the same time throughout the structure of the metal part. This is shown graphically on the diagram in Figure 5-38. By using an interrupted quench, the cooling is stopped at a point above the martensite transformation region to allow sufficient time for the center to cool to the same temperature as the surface. Then cooling is continued through the martensite region, followed by the usual tempering.

5.7.3 Austempering

A second method of interrupted quenching is called *austempering.* Figure 5-38 shows this process graphically. The quench is interrupted at a higher temperature (200° to 375°C) than for marquenching to allow the metal at the center of the part to reach the same temperature as the surface. By maintaining that temperature, both the center and the surface are allowed to transform to bainite and are then cooled to room temperature. The advantages of austempering are (1) even less distortion and cracking than marquenching, primarily due to the higher transformation temperatures, and (2) no need for final tempering. However, austempering has the disadvantage of requiring more time, even though it requires no tempering treatment. Also, parts with a large thickness cannot be handled. Sections with a maximum thickness of $\frac{1}{2}$ in. are cooled sufficiently fast to permit the transformation of austenite to bainite without the formation of pearlite.

5.8 PRECIPITATION HARDENING (AGE HARDENING)

Nonferrous metals do not undergo the significant phase transformations possible with steel. Consequently, most thermal processing of nonferrous metals is used to relieve stresses in a single-phase microstructure and/or to produce recrystallation. These techniques do not result in significant strengthening of a metal's structure. Instead, the most effective thermal processing technique for increasing the strength of such metals is *precipitation hardening* or *age hardening.* Precipitation hardening involves two steps. The first is *solution treatment,* the heating (annealing) of an alloy that exists at room temperatures as a two-phase solid solution to produce a single-phase solid solution. The second step is the subsequent rapid cooling (*quenching*) of this single-phase solid solution to form an unstable supersaturated solid solution. A further heating process may be necessary in the case of some alloys to expedite the strengthening of the metal.

To explain how this strengthening comes about, it is necessary first to review the previous material under phase diagrams in this Module specifically concerning the effects of solid solubility curves (solvus lines), using Figure 5-22. Such a review at this point will assist the reader in understanding our explanation of precipitation hardening. As stated previously, the partial solubility of one solid metal in another is determined by the slope of the solid solubility curve (solvus line). The solvus must slope such as to indicate that there is greater solubility of one metal in another at a higher temperature than at a lower temperature.

Figure 5-22, repeated as Figure 5-39, is a schematic of the equilibrium diagram for silver (Ag)–copper (Cu) alloys. At each end (the silver-rich and the copper-rich), the solvus indicates the type of partial solubility desired for precipitation hardening. Using the silver-rich end of the diagram (see Figure 5-40), an alloy consisting of about 92.5% Ag and about 7.5% Cu (sterling silver) is shown as a vertical dashed line. At room temperature this alloy exists as a metal consisting of two distinct phases (α and β). α represents a silver-rich solid solution with some dissolved Cu. β is a solid solution of almost pure Cu with very little dissolved Ag. As the alloy is heated, it crosses the solvus at about 760°C. The β phase dissolves at this temperature and diffuses uniformly to form a solid solution of α. If, then, this same alloy were slowly cooled, the β phase would precipitate out of the α solid solution, because the solubility

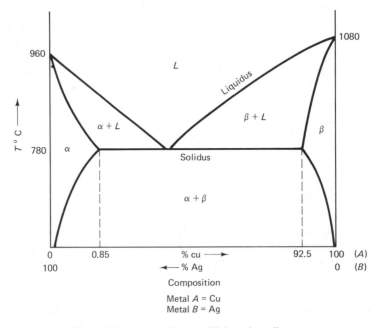

Figure 5-39 Ag-Cu alloys equilibrium phase diagram.

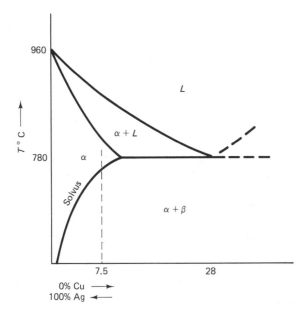

Figure 5-40 Ag-rich end of Ag-Cu equilibrium diagram.

of β (copper) in the α (silver) decreases from about 8.5% at 780°C to less than 1% at room temperature, as shown by the solvus line.

Now if this alloy after being heated (annealed) to around 770°C were cooled quite rapidly (quenched), there would be insufficient time to permit the Cu atoms to diffuse, which would produce equilibrium-type precipitation of β atoms. The result is a retained α phase that is unstable, with its excess Cu atoms trying to precipitate out to form the β-phase crystal structure. The unstable, supersaturated solid solution formed is quite ductile.

With the metal in such a crystal structure, it can be worked, straightened if distorted, or machined much more easily than when the metal was in the stable, original form at room temperature. Once the metal is worked the strengthening process can continue. The Cu atoms are now given more energy along with sufficient time to precipitate by allowing the temperature to return to room temperature. Other alloys may require heating them to some temperature below the solvus line to cause this precipitation. In the case of sterling silver this alloy can be heated to around 300°C for about 30 minutes and the hardness doubles in value. The increase in strength and hardness is explained by the precipitation of the trapped β metal out of the α metal structure at the grain boundaries and at the sites of impurities in the lattice structure of α metal, which produces resistance to the movement of dislocations. If the precipitation occurs at room temperatures within a few days duration, it is known as *natural aging*. If additional heating is required, as in the case with sterling silver, to cause precipitation and full strengthening, the process is called *artificial aging*.

It is possible to stop the precipitation entirely by lowering the temperature sufficiently through refrigeration. Once the precipitation has produced a saturated normal dual phase at or near room temperature, the metal reaches its greatest hardness and strength. As fine precipitates form, they have a tendency to grow with time and with natural-aged alloys this growth of the precipitate will result in a decrease in the strength. Such a phenomenon is known as *overaging*. For this reason, artificial-aged alloys are less subject to overaging since that growing process can be effectively stopped by a simple quench. It must be pointed out that the strength property was used more in this discussion than any other property. We know that other properties as well are affected by distortions of a lattice structure brought about by the precipitation of atoms of another element.

To round out our treatment of precipitation hardening, the popular aluminum–copper alloy known as Duralumin (2017) along with a part of its equilibrium diagram is used to illustrate the steps taken to bring out the strength of the alloy in industry. Figure 5-41 is a schematic of the aluminum-rich end of the Al–Cu equilibrium diagram. Alloy 2017 contains about 96% Al and 4% Cu. Naturally or artificially aged, this alloy finds many applications, one of which is for aircraft rivets. The vertical dashed line in Figure 5-41 represents this particular alloy. Upon solution treatment to a temperature of about 550°C, the single-phase solid solution designated *k* is formed in which all the Cu atoms are diffused into the crystalline structure of the Al atoms. Next it is quenched to form the unstable, supersaturated *k* phase, and the alloy becomes very ductile and can be maintained if it is refrigerated to retard the eventual precipitation of Cu atoms from the *k* phase. Once sufficient energy is received through heating to room temperature or beyond, the aging process, and hence the increase in strength, take place. Rivets made from this alloy taken from refrigerated storage can then be driven easily. Once driven, the aging process begins as room temperature is reached, which brings on the attendant growth in strength and hardness. Thus, we have explained in some detail the statement made in Section 5.1 about the effect of refrigeration on the movement of atoms in an aluminum rivet.

5.9 HARDENABILITY

Not to be confused with hardness, *hardenability* is the measure of the depth to which hardness can be attained in a metal part using one of the standard thermal processing techniques designed primarily to increase hardness. We know that carbon is vital to steel making. Steels are even classified by their carbon content. Hardness, as an indicator of strength, depends on carbon content. The higher the carbon content, the higher the maximum hardness that can be attained at greater depths within a part, assuming other factors remain the same. In addition, the increase in carbon permits a greater degree of hardness. Alloys also play a significant role in producing steels with higher degrees of hardenability. We know that one way to change a metal alloy's microstructure is to purposely add impurities. The impurities are elements that will alloy or join with the main elements or ingredients to form solid solutions—substitutional and/or interstitial. These alloying elements, such as nickel, boron, copper, or silicon, produce effects similar to impurities found in the original metal when mined and/or those added, out of necessity, during the processing of the metal. The presence of atoms of the alloying element (1) distorts the lattice structure, (2) increases the metal's resistance to plastic slip, (3) hinders the diffusion of atoms past these

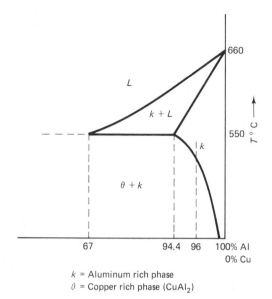

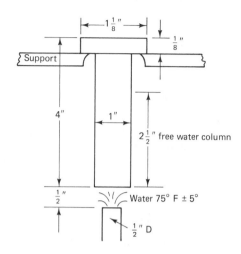

Figure 5-41 Al-rich end of Al-Cu equilibrium diagram.

k = Aluminum rich phase
θ = Copper rich phase (CuAl$_2$)

Figure 5-42 Jominy end quench hardenability test.

areas of distortion, and (4) effectively increases the brittleness, hardness, and strength of the metal alloy. To this list we can add (5) increases significantly the metal's hardenability. Adding alloys or increasing carbon content shifts the location of the S-curve (transformation zone) of austenite to pearlite or bainite to the right on the TTT diagram. The result is that a steel need not be quenched as rapidly to achieve the desired hardness. One disadvantage is that alloys are costly, but benefits usually outweigh such costs. One benefit in using a less drastic quench is the reduction in the danger of cracking and/or warping of steel parts, particularly the parts that are, out of necessity, complex in shape or size or require great dimensional stability.

It is essential to point out here the necessity to follow proper design standards that help reduce such things as residual stresses, warping, and cracking. Such techniques as maintaining, if possible, uniform cross sections or thicknesses and using fillets and rounds make the job of the heat treater less difficult. The Jominy end-quench test performed on steels using standardized specimens and procedures determines the necessary nonequilibrium cooling rates to attain a certain hardenability in a particular steel part (see Figure 5-42).

5.10 SURFACE HARDENING AND SURFACE MODIFICATION

In all our discussions about hardening and strengthening of metal alloys, our objective was to obtain the same hardness and strength throughout the part we were heat treating. It would be impractical, if not impossible in many cases, to harden the whole part, that is, the interior and the exterior, to that hardness demanded by wear-resistant surfaces. Even if we could attain this degree of hardness throughout, the part would have little *toughness* (the ability to withstand impact loads). Little application is found in industry for such a part.

Most steel parts require machining, and this operation is best performed when the part is not hard. Once machining is completed and the part is within the required tolerances, it may be hardened on the surface to provide a strong, hard, wear-resistant surface.

5.10.1 Case Hardening

Hardening of the surface only is known as *case hardening,* and the surface layers that are actually hardened make up the *case.* The inner surfaces that are not hardened are known as the *core* (see Figure 5-43).

The first three case-hardening methods are referred to as chemically modifying methods or diffusion-coating methods. They provide the needed elements that are lacking in the surface layers to permit the hardening reactions to take place. We have learned that carbon is an essential element in steel if we are to harden it (about 0.3% carbon in steel is the minimum for hardening steel). Many hot-working operations result in a removal of the carbon from the surface. This is known as *decarburization* (Figure 5-44).

5.10.2 Carburizing

To harden the steel, the carbon content must be restored. As the carbon content in steel increases, the heat treater's ability to create greater hardness also increases. The purpose of carburizing is to either restore lost carbon and/or increase the amount of the carbon in the outer layers of the metal part. Basically, the metal part is surrounded by a high-carbon-content solid material or gas; and by supplying the necessary energy through high furnace temperatures, the carbon atoms diffuse from this high-carbon-content medium into the crystal structure of the metal part to be case hardened.

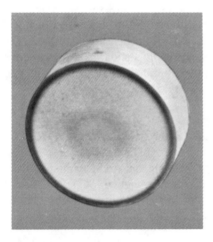

Figure 5-43 Polished section of 8620 steel bar carburized to a depth of 0.060; measured to a 0.40 percent carbon content. (Republic Steel Corporation.)

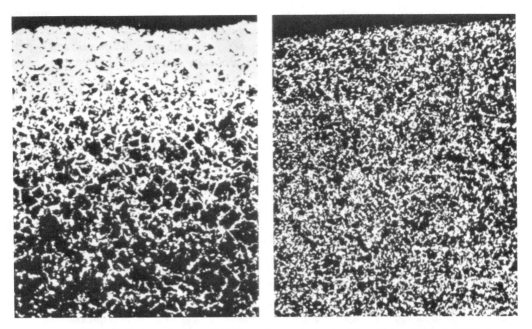

Figure 5-44 Decarburizing. (left) Photomicrograph showing "decarb" to a depth of .020″ in the surface of a hot rolled bar of 1050—magnified 100 times. (right) Same bar after carbon correction annealing. Note the restoration of carbon on outer surface and also the refinement of grain structure. (Republic Steel Corporation)

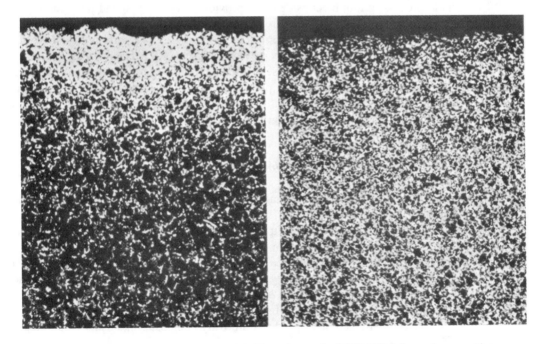

Figure 5-45 Carbon correction. (left) Photomicrograph of AISI 5046 before carbon correction—magnified 100 times. (right) Same bar after carbon correction annealing with surface carbon restored. (Republic Steel Corporation.)

After carburization, the steel part is ready for a nonequilibrium thermal processing, which transforms the high-carbon-content austenite into martensite. This is followed by various quenches and tempers, depending on the degree and depth of hardness as well as the degree of grain refinement desired in both the case and the core (Figure 5-45).

5.10.3 Nitriding

In our study of the various phase changes produced by heat treating steel, we learned that some iron reacts with carbon to produce a very hard, brittle compound, iron carbide (FeC), that precipitates out during a phase change, giving the steel hardness and strength. Aluminum alloyed with copper forms another compound ($CuAl_2$) with similar results. In addition to these elements, there are several other elements, such as chromium (Cr) and molybdenum (Mo), that react with nitrogen in steel to produce compounds called nitrides that are also hard and brittle like carbides. The purpose of nitriding is to provide the necessary nitrogen in the steel so that these hard nitrides may form, producing the necessary hardness. The rich nitrogen medium may be ammonia gas. As with carburizing, heat is called on to provide the energy for the nitrogen atoms to diffuse into the steel case. The heat also helps produce the necessary chemical changes, and therefore no additional heat treating is required.

5.10.4 Cyaniding and Carbonitriding

These methods also provide cases that are rich in carbon and nitrogen. *Cyaniding,* an old process, might be considered as providing both carbon and nitrogen. Cyanide (CN) decomposes in a liquid bath of molten salts containing sodium cyanide, and the nitrogen reacts with iron to form hard iron nitride. The carbon diffuses to assist in further hardening operations, as previously discussed. *Carbonitriding* is very similar to gas carburizing in that it uses a medium consisting of a mixture of gases rich in carbon and also nitrogen. The presence of nitrogen in the austenite also permits slower transformation to martensite, and therefore lower temperatures are required, followed by less drastic quenching. These conditions result in less distortion and less danger of cracking.

5.10.5 Flame Hardening and Induction Hardening

These last two methods do not change the chemical makeup of the steel case. Through the application of heat, the outer layers of the metal part are first transformed into austenite, followed by quenching to transform the austenite into martensite. The steel to be case hardened must contain all the hardening ingredients. Any decarburized surface would have to be removed by machining or grinding prior to any induction or flame hardening. *Flame hardening* uses oxyacetylene torches as sources of heat, as seen in Figure 5-46. *Induction hardening* uses heat generated by the resistance of the metal to the passage of high-frequency induced electrical currents.

Surface-hardened steel products are truly *composite materials* containing two

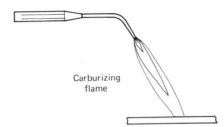

Carburizing
flame

Figure 5-46 Flame hardening.

or more integrated components. Reinforced concrete is a composite material whose integrated components are steel rods strong in tension and a concrete matrix strong in compression. But both combine to work together in an integrated system providing strength properties not obtainable by either one acting alone. With surface-hardened steel parts, the integrated components are the hard case and the tough core, together providing superior performance not obtainable by their use alone.

Even steels that are through-hardened, such as fine pearlite or tempered martensite steels, are strong and relatively hard as a result of the diffusion of fine particles of a second harder phase throughout a soft ferrite matrix. The relatively soft matrix material, the solvent component of an aluminum–copper alloy, though deformable, cannot deform independently of its harder phase. The point is that there is a striking similarity among all man-made materials—steel, aluminum alloys, prestressed or reinforced concrete, fiber-reinforced plastics, or surface-hardened metal products. They all contain integrated components working together to provide the right mix of material properties needed for the almost endless uses in today's highly technological society. Module 8 on composite materials will provide further details.

5.11 FERROUS METALS

Iron and its many alloys, including cast irons and a nearly limitless variety of steels, comprise the ferrous metals group. Even with the wide acceptance of aluminum and polymeric materials, the iron-based alloys dominate all other materials in the weight consumed annually for manufactured products. Ten times more iron (mainly in the form of steel) is used than all other metals combined. Figures 5-47 and 5-48 show the processes that iron ore, coal, and limestone undergo in the production of iron. Coke is made from coal; many other products also come from this source.

5.11.1 Cast Iron

As shown in the iron–iron carbon equilibrium diagram (Figure 5-25), cast iron has between 2% to 4% carbon, compared with less than 2% for steel. Other elements in cast iron are silicon and manganese, plus special alloying elements for special cast irons. Many cast-iron products are used as they are cast, but others require changes in properties, which are achieved through heat treatment of the cast parts.

5.11.1.1 Gray cast iron. This type of cast iron is a supersaturated solution of carbon existing in a *pearlite* (two-phase iron structure) matrix. The carbon is mostly in the form of *graphite* flakes (soft form of carbon known as *elemented carbon*). It is the familiar metal used as the engine block of most automobiles and for other internal combustion engines. Figure 5-49 shows photomicrographs of two "as-cast" gray cast-iron specimens, one of low strength and one of medium strength. The amount of carbon, 3.2%, exceeds the solubility limits of iron, and the carbon precipitates out of solution with ferrite (carbon precipitates out in graphite form). Silicon (2%) is important in the *graphitizing* of gray cast iron. The graphite promotes machinability and lubricity of this metal. The dampening ability of this alloy provides excellent absorption of vibrations and noise, which leads to its selection as piano sound boards and machine parts. These combined properties have also made gray iron a popular gearing material.

The ASTM system of designation for gray iron places it into classes 20 to 60 based on the minimum tensile strength for each class. For example, class 30 would have a minimum tensile strength of 30,000 psi (207 MPa), while a class 60 gray iron would be 60,000 psi (414 MPa). This classification is often preceded by 48 (ASTM A48 Class 40), which designates the specification used to determine the mechanical properties of representative samples. Brinell hardness numbers (BHN) range from

a flowline of steelmaking

From iron ore, limestone and coal in the earth's crust to space-age steels — this fundamental flowline shows only major steps in an intricate progression of processes with their many options.

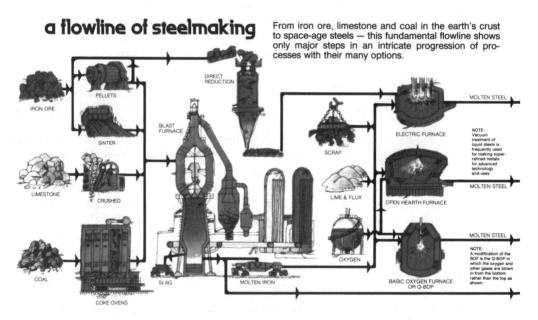

Figure 5-47 A flowline of steelmaking. (American Iron and Steel Institute.)

160 to 200 for ASTM 48 Class 20 to 212 to 248 for ASTM 48 Class 60. *Machinability* ratings go to 110, as compared to a 112 carbon steel that is rated as 100% machinable. Gray cast iron is also available with alloys of nickel, chromium, and molybdenum to improve resistance to wear, corrosion, and heat while improving strength. Flame and induction hardening allow for increased surface hardness with a slightly tougher core.

5.11.1.2 White cast iron. Through slow cooling in sand molds, chilling of specific portions of a casting, and alloying, graphitic carbon is stopped from precipitating out of solution with the ferrite to produce a white cast iron. The name white iron comes from the white color produced with the fracture of the alloys. Figure 5-50 shows a photomicrograph of white cast iron to contrast with the photomicrograph of gray cast iron in Figure 5-49, which reveals graphite flakes. The carbon composition of 3.5% for unalloyed white iron has 0.5% silicon. The structure is an intermetallic compound of carbon and iron known as *cementite*, plus a layered two-phase solution of ferrite and cementite known as *pearlite*. The castings are very brittle, with Brinell hardness values from over 444 to 712. White cast iron has very good compressive strength above 200,000 psi (1380 MPa), with tensile strength around 20,000 psi, and good wear resistance; it finds applications as rolls for steel making, stone and ore

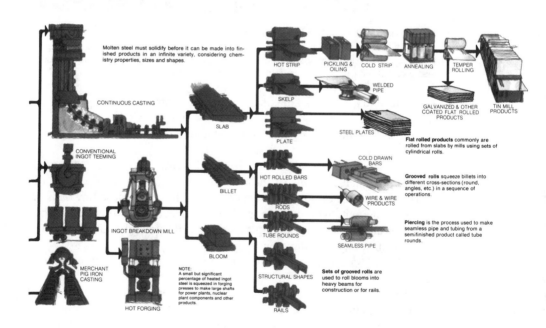

Figure 5-48 Molten iron rushes from a blast furnace through a series of clay-lined runners into a "submarine." Most of this iron will be charged in a molten state into basic oxygen furnaces for refinement into steel. Slag is tapped from a blast furnace several times during a cast and passes in an opposite direction into huge pots loaded on railroad cars for delivery to the slag dump. (Bethlehem Steel Corporation.)

crushing mills, and brickmaking equipment. This is generally the lowest-cost cast iron. Heat treatment can reduce brittleness and, as with gray iron, other properties are possible with alloying; but both of these add cost to the castings.

5.11.1.3 Nodular or ductile cast iron.

With 3.5% C and 2.5% Si the addition of small amounts of magnesium (Mg), sodium (Na), cerium (Ce), calcium (Ca), lithium (Li), or other elements to molten iron will cause tiny balls or *spherulites* to precipitate out. As the name implies, the spherulitic structure improves the elongation or ductility while yielding superior tensile strength (150,000 psi or 1034 MPa) and machinability (similar to gray iron).

ASTM specifications for nodular iron indicate minimum tensile strength, minimum yield strength, and minimum percentage of elongation in 2 in. For example, as ASTM 120-90-02 would have a minimum tensile strength of 120,000 psi (828 MPa), minimum yield strength of 90,000 psi (621 MPa), and 2% minimum elongation in 2 in. (50.8 mm). In internal combustion engines, nodular iron finds application as crankshafts, rocker arms, and pistons. It is also used for cast gears, pumps, ship propellers, and with caustic handling equipment. While it is more expensive in terms of weight than gray iron, its specific strength makes ductile iron more economical, yet it lacks the dampening ability and thermal conductivity of gray iron.

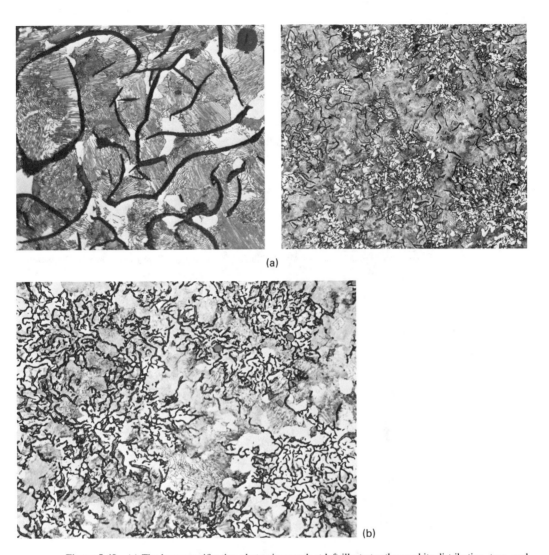

(a)

(b)

Figure 5-49 (a) The low magnification photomicrograph at left illustrates the graphite distribution, type, and size. At 1000× (at right), the pearlite colonies and small ferrite grains adjacent to the graphite flakes are clearly distinguishable. The large round gray particles are manganese sulfide inclusions. The surface of the specimen contains products of transformation of a faster cooling rate. (b) The distribution of fine graphite flakes in this sample results in an increase in strength of the iron. (Buehler Ltd.)

5.11.1.4 Malleable iron. The annealing of white-iron castings causes nodules (large flakes) of soft graphitic carbon to form through the breakdown of hard and brittle cementite (Fe_3C). Two basic types of malleable iron are possible by varying the heat-treatment cycle. *Pearlitic malleable* iron is strong and hard, whereas *ferritic malleable* iron is softer, more ductile, and easier to machine. Malleable iron has

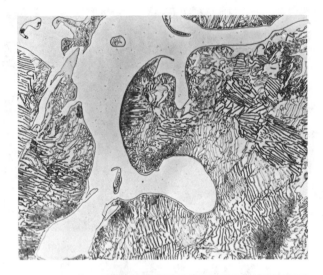

Figure 5-50 This specimen shows a hypereutectoid structure of pearlite and massive cementite. The dark areas are pearlite colonies surrounded by a network of cementite. At higher magnification, the alternate lamellae of alpha ferrite and Fe_3C are clearly resolved. (Buehler Ltd.)

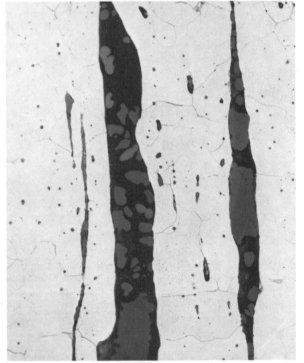

Figure 5-51 The matrix of this specimen consists of ferrite grains similar to that of ingot iron. The elongated stringers are inclusions of slag composed largely of FeO and SiO_2. At higher magnification, small, dark particles within the ferrite grains are visible. These are finely dispersed impurities, apparent only after etching. (Buehler Ltd.)

2.2% carbon and 1% silicon. In pearlitic malleable iron, 0.3% to 0.9% of the carbon is combined as cementite and allows for selective hardening of portions of a casting.

According to ASTM specifications A-47-52 and A197-47, three grades are available: 35018, 32510, and cupola malleable iron. The 35018 and 32510 grades are ferritic, with the latter lower in silicon and consequently more ductile. Cupola malleable iron has higher carbon and lower silicon content than the other grades, which yields lower strength and ductility. Basic properties for the three grades are as follows:

TABLE 5-1 PROPERTIES OF 3 CAST IRON GRADES

Grade	Minimum tensile strength, psi (MPa)	Minimum yield strength, psi (MPa)	Minimum elongation % in 2 in. (50.8 mm)
35018	53,000 (365)	35,000 (241)	18
32510	50,000 (345)	32,500 (224)	10
Cupola	40,000 (276)	30,000 (206)	5

Applications of the ferritic grades include machined parts (120% machinability rating), automotive power trains, and hand tools such as pipe wrenches that take hard beatings. Applications for the stronger and harder pearlitic malleable iron include parts that require high surface hardness (up to R_c 60 or BHN 163–269), such as bearing surfaces on automobiles, trucks, and heavy machinery.

Table 10.11 provides a properties comparison of cast irons and wrought iron. *Wrought iron*, as shown in the photomicrograph (Figure 5-51), is an iron of high purity (less than 0.001 parts carbon) with the slag (iron silicate) rolled or wrought into it. The ferrite matrix encloses iron silicate fibers shaped in the direction of rolling, which makes it an easy material to form. It is not a common metal today, but before the development of cast iron and steel making, a cruder form of wrought iron served society as weapons, tools, and architectural shapes. The Eiffel Tower in Paris was constructed of wrought iron in 1872.

5.11.2 Steel

As the most widely used engineering material, steel is available in an almost limitless variety. Several groups can be used, such as cast steel and wrought steel. Wrought steel covers the largest group and is the steel most common to consumers. Steel is cast into ingots when it comes from such steel-making processes as the open hearth furnace or basic oxygen furnace. These ingots are further processed while in the hot "plastic" state to produce a variety of *wrought* or *hot-rolled steel* (HRS) products such as bars, angles, sheet, or plate. Further working of HRS sheet or bar stock at below the recrystallization temperature of the steel is known as *cold working* or *cold finishing*. *Cold-rolled steel* (CRS) is a harder steel because its grains have work hardened. Further classifications of steel are the carbon steels and alloyed steels.

The classification of steels takes a variety of forms. A very common system developed by both the Society of Automotive Engineers (SAE) and the American Iron and Steel Institute (AISI) uses four or five digits and certain prefix and suffix letters to cover many steels and steel alloys. As seen in Figure 5-52, the numbers reveal the major alloying element, its approximate percentage, and the approximate amount of carbon in *hundreds of one percent*, commonly called points of carbon. Table 5-2, shows the major groupings of carbon and alloy steels under SAE–AISI classification. The alloy and carbon contents given in this 4- or 5-digit system are

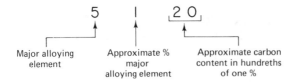

Major alloying element

Approximate % major alloying element

Approximate carbon content in hundreths of one %

Examples:

Shown above – chromium steel alloy with about 1% chromium and 0.20% (0.002 parts or 20 points) carbon

1015 – plain carbon steel with 0.15% (0.0015 parts or 15 points) carbon

E52100 – chromium steel alloy with about 20% chromium and 1 point carbon produced in an electric arc furnace

Figure 5-52 SAE-AISI steel designation.

approximations. Complete specifications are available from SAE, AISI, or from handbooks such as the *Machinery's Handbook*. For example, the chromium steel alloy 5120 has the following composition: C = 0.17% to 0.22%; Mn = 0.70% to 0.90%; Cr = 0.80% to 1.10%; P = 0.040%; and Si = 0.20% to 0.35%. Prefixes can indicate the process in making the steel, such as E for electric arc furnace; and suffixes further clarifiy, for instance, H for hardenability guaranteed. Other societies such as ASTM and ASME (American Society of Mechanical Engineers) have specifications for speciality steels, such as tool steels for dies, cutting tools, and punches or structural bolts and plates.

5.11.2.1 Carbon steel. This group of steels, also known as *plain carbon* and *mild* steel, dominates all other steels produced and is essentially iron and carbon with other elements that naturally occur in iron ore or result from processing. These elements are held to certain maximum levels: manganese (Mn), 1.65%; silicon (Si), 0.60%; and copper (Cu), 0.60%. Carbon steel may be cast or wrought. Typically, cast steels have more uniform properties since wrought steel develops *directional properties* as a result of rolling it into shape.

5.11.2.2 High-strength low-alloy steel. HSLA steel is a product of recent technology that aimed to obtain strong, lightweight steel at a price competitive with carbon steels. While the price per pound of HSLA steel is greater than carbon steel, thickness is reduced due to a higher strength of 414 kPa (60,000 psi) versus 276 kPa (40,000 psi) for carbon steel; consequently, overall cost may be better for the HSLA, while significant weight savings are realized. The transportation industry, especially the automotive section, has employed HSLA in numerous structural applications. While not as malleable as carbon steel, sheet HSLA steel could not be used in auto bodies, but a modification resulted in a dual-phase steel acceptable for the small bending radii required on auto bodies.

5.11.2.3 Alloy steel. The classification of alloy steel is applied when one or more of the following maximum limits are exceeded: Mg, 1.65%; Si, 0.60%; Cu, 0.60%; or through the addition of specified amounts of aluminum (Al), boron (B), chromium (Cr up to 3.99%), cobalt (Co), niobium (Nb), molybdenum (Mo), nickel

TABLE 5-2 SAE–AISI SYSTEMS OF STEEL CLASSIFICATION

Digit designation	Types of steel
$10 \times \times$	Plain carbon
$11 \times \times$	Sulfurized (free-cutting)
$12 \times \times$	Phosphorized
$13 \times \times$	High manganese
$2 \times \times \times$	Nickel alloys
$30 \times \times$	Nickel (0.70%), chromium (0.70%)
$31 \times \times$	Nickel (1.25%), chromium (0.60%)
$32 \times \times$	Nickel (1.75%), chromium (1.00%)
$33 \times \times$	Nickel (3.5%), chromium (1.50%)
$34 \times \times$	Nickel (3.00%), chromium (0.80%)
$30 \times \times \times$	Corrosion and heat resistant
$4 \times \times \times$	Molybdenum
$41 \times \times$	Chromium–molybdenum
$43 \times \times$	Nickel–chromium–molybdenum
$46 \times \times$	Nickel (1.65%), molybdenum (1.65%)
$48 \times \times$	Nickel (3.25%), molybdenum (0.25%)
$5 \times \times \times$	Chromium alloys
$6 \times \times \times$	Chromium–vanadium alloys
$81 \times \times$	Nickel (0.30%), chromium (0.30%) molybdenum (0.12%)
$86 \times \times$	Nickel (0.30%), chromium (0.50%) molybdenum (0.20%)
$87 \times \times$	Nickel (0.55%), chromium (0.50%) molybdenum (0.25%)
$88 \times \times$	Nickel (0.55%), chromium (0.50%) molybdenum (0.35%)
$93 \times \times$	Nickel (3.25%), chromium (1.20%) molybdenum (0.11%)
$98 \times \times$	Nickel (1.10%), chromium (0.80%) molybdenum (0.25%)
$9 \times \times \times$	Silicon–manganese alloys

$\times$'s indicate that numerals vary with the percentage of carbon in the alloy.

(Ni), titanium (Ti), tungsten (W), vanadium (V), zirconium (Zr), or others. Alloy steels are grouped as low, medium, or high alloyed steel with high alloy steels encompassing the stainless steel group. Table 5-2 shows the SAE–AISI classification systems used for certain alloy steels. Elements added to steel can dissolve in iron to strengthen ferrites or alpha iron (bcc) and form with carbon in the austenite or gamma iron phase (fcc) to produce carbides to improve hardness.

Chromium is effective in increasing strength, hardness, and corrosion resistance. Copper forms in austenite to reduce rusting. Manganese is an austenite former that, much like carbon, increases hardness and strength. Vanadium forms with ferrite to improve hardness, toughness, and strength. Molybdenum combines in carbide to improve high-temperature tensile strength and high hardness. Silicon dissolves in ferrite to improve electromagnetic properties, plus toughness and ductility. Nickel is an austenite former that improves high-temperature toughness and ductility, plus rust resistance. Aluminum is effective as a ferrite former in reducing grain size, thus giving improved mechanical properties. Cobalt dissolves in austenite to improve magnetic properties and high-temperature hardness. Tungsten dissolves in ferrite to increase both hardness and toughness at elevated temperatures.

Table 10.11 provides a comparison of selected alloys.

5.11.2.4 Stainless steel.

This group of high-alloy steels contains at least 10.5% chromium; it is more correctly called corrosion-resistant steels (CRES). The 10.5% of chromium does not ensure that the steel will not rust because a sufficiently high carbon content or other alloys may negate the passivity of the chromium (see Figure 5-53). As with other steels, stainless may be wrought or cast. Wrought stainless is grouped by its structure as ferritic, martensitic, austenitic, or precipitation hardening (PH). Cast stainless may be classified as heat resistant or corrosion resistant.

The ASTM and SAE have developed the Unified Numbering System (UNS), a five-digit designation with an S prefix to replace the AISI designations for stainless steels. Stainless steels in the S30000 (AISI 300) series are nickel–chromium steels, while the S40000 (AISI 400) series have chromium as the major alloy. Series S20000 (AISI 200) are austenitic alloys with manganese and nitrogen replacing some of the nickel, far less expensive alternatives, when high formability and good machinability

Figure 5-53 Stainless steel. Cookware of nickel-chrome stainless. (International Nickel Company.)

are not required. *Austenitic stainless steel* is a single-phase solid solution that has good corrosion resistance.

Martensitic stainless steels in the S40000 (AISI 400) series have high carbon content up to 1.2%, with 12% to 18% chromium. The higher carbon content allows formation of more gamma iron, which quenches to a hard martensitic (up to 100%) steel, but the high carbon content reduces some of the corrosion resistance. If an austenitic stainless steel is heated sufficiently so that carbon precipitates out of solid solution as chromium carbide which leaves less than 12% chromium in some segments of the alloy, it promotes intergranular corrosion.

The *ferritic stainless steels* have low carbon (0.12% or less) content and high chromium content (14% to 27%) in a solid solution and do not harden by heat treatment; they are in the S40000 (AISI 400) series and unlike martensitic and austenitic are magnetic. The ferritic grades have good formability, machinability, and corrosion resistance above martensitics. Specific properities of each type of stainless steel are found in standard references on steels. Table 10.11 shows selected stainless steels and their properties.

5.11.2.5 Other steel alloys.

Beyond and including high-alloy steels and stainless steels discussed previously, there is a wide variety of specialty alloys. The ASM publishes a *Metals Handbook* of several volumes; *Volume I, Properties and Selection of Metals,* gives in-depth coverage on most available metals. Included in the steel alloys is a range of tool steels, high yield strength steels (HY), magnetic and electrical steels, ultrahigh-strength steel, maraging steels, low expansion alloys, and ferrous powdered metals (PM).

5.12 NONFERROUS METALS

While the ferrous alloys are used much more than other metals, nonferrous metals comprise three-fourths of the known elements. However, the problem of extracting certain metals from the earth or ocean and the practical commercial value limits the number of metals used in significant quantity. The most commonly used nonferrous metals include aluminum, copper, zinc, nickel, chromium, tin, magnesium, beryllium, tungsten, lead, molybdenum, titanium, tantalum, and the noble or precious metals such as gold, platinum, silver, and rhodium. With demands for superior properties through advances in technology, each metal receives careful consideration for its potential value in unique circumstances. Catalytic converters used in automobile exhaust systems contain platinum to aid in the removal of pollutants in our environment. Ruthenium is an example of a metal that has little practical application alone, but as an alloy with platinum it is valued in thin-film circuitry (ceramic substrates with printed circuits) for solid-state electronics. In addition to the properties of a material, selection for design bears heavily on the processability of the material. Table 5-3 provides a simplified comparison of metals to common processes.

TABLE 5-3 PROCESSABILITY OF COMMON METALS

	Steel	Iron	Aluminum	Copper	Nickel	Magnesium	Zinc	Titanium	Tin
Castability									
Centrifugal	■		■		■				
Continuous									
Investment	■				■				
Permanent mold	■							■	
Die casting								■	
Formability									
Cold	■		■		■				
Hot	■		■		■				
Machinability	■		■		■		■		
Powder metal Compacting	■		■		■				
Weldability									
Gas	■		■	◢	■				
Inert arc	■		■		◢	■		■	
Electrical Resistance	■		■		◢			■	

☐ — Common on most alloys

◢ — Used on some alloys

Adapted from *Machine Design*, March 15, 1976.

5.12.1 Aluminum

The most abundant of metals, aluminum, unfortunately is locked up with other elements in the form of bauxite ore. In 1852, this metal was only rarely available in usable form and was prized by Emperor Napoleon as dinnerware; it sold for $545 a pound. The development by Charles Hall in Ohio and by Paul Heroult in Paris, France, of an electrolytic process for economical extraction of bauxite occurred simultaneously in 1886. Even though the price had dropped to 24 cents per pound by 1964, the cost of electrical energy keeps up the price of virgin aluminum. About 4% of our metal-processing energy is consumed in the production of aluminum. However, recycling of aluminum drastically reduces the cost of the metal.

Figure 5-54 shows the steps in the production of aluminum from bauxite. Aluminum has many favorable properties, including excellent thermal conductivity, good strength, corrosion resistance, and light weight (one-third steel's weight). Figures 5-55 & 56 show typical products which stem from aluminum properties. Table 5-3 shows that it is processable by the major methods. Wrought aluminum and cast

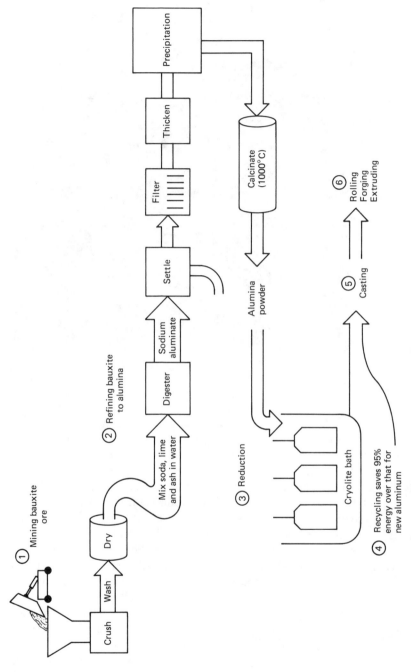

Figure 5-54 Making aluminum. (Adapted from The Aluminum Association flow charts).

① Mining bauxite ore

Crush

Wash

Dry

② Refining bauxite to alumina

Mix soda, lime and ash in water

Digester

Sodium aluminate

Settle

Filter

Thicken

Precipitation

Calcinate (1000°C)

③ Reduction

Alumina powder

Cryolite bath

④ Recycling saves 95% energy over that for new aluminum

⑤ Casting

⑥ Rolling Forging Extruding

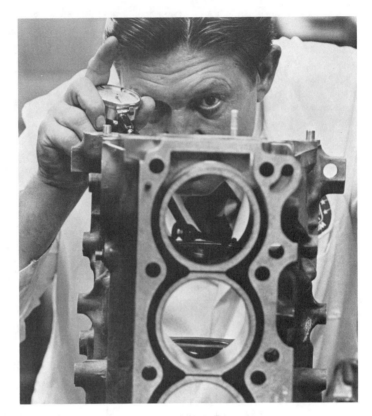

Figure 5-55 Energy-saving transportation uses, like this all-aluminum engine block, make aluminum important today in modern engineering and design. (Reynolds Metal Company.)

aluminum are designated by two different numbering systems. The Aluminum Association uses four digits to specify wrought alloy groups. The meaning of each digit is shown in Figure 5-57. Wrought and cast designations are similar; the last digit in cast aluminum indicates cast (0) or ingot (1 or 2). Temper designation follows the four digits; they indicate thermal treatment (T through T10), solution heat treated (W), as fabricated (F), annealed (0), or strain-hardened (H).

Aluminum of high purity (1000 series) or at least 99% finds applications as thin foils for electronics applications; as plate and bar stock for chemical apparatus, fuel tanks, cooking utensils; and as paint pigments and in gasoline production. Aluminum alloyed with copper (2000 series) extends the tensile strength (greater than mild steel), improves machinability, and proves good specific strength, but has reduced corrosion resistance and weldability. Aluminum–manganese alloys (3000 series) provide good strength, workability, formability, and weldability, plus the high corrosion resistance makes this a popular alloy for such applications as highway signs, furniture, cooking utensils, architectural shapes, and truck panels.

The aluminum and silicon alloys (4000 series) have high wear resistance and relatively low coefficients of thermal expansion; they permit good castings and forg-

Figure 5-56 Energy-saving and low-maintenance requirements are two important reasons aluminum building products, like the siding being applied here, are used today. (Reynolds Metal Company.)

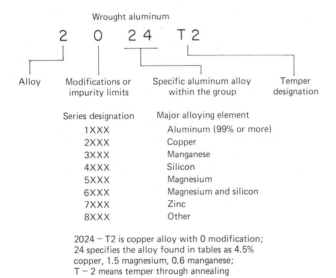

Wrought aluminum

2 O 2 4 T 2

Alloy | Modifications or impurity limits | Specific aluminum alloy within the group | Temper designation

Series designation	Major alloying element
1XXX	Aluminum (99% or more)
2XXX	Copper
3XXX	Manganese
4XXX	Silicon
5XXX	Magnesium
6XXX	Magnesium and silicon
7XXX	Zinc
8XXX	Other

2024 – T2 is copper alloy with 0 modification;
24 specifies the alloy found in tables as 4.5% copper, 1.5 magnesium, 0.6 manganese;
T – 2 means temper through annealing

Cast aluminum

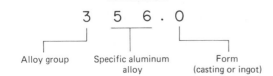

3 5 6 . 0

Alloy group | Specific aluminum alloy | Form (casting or ingot)

Figure 5-57 Aluminum Association number system.

ings and find applications as welding and brazing rods, anodized architectural shapes, marine equipment, and automotive pistons. Aluminum alloyed with magnesium (5000 series) yields moderate to high strength, good corrosion resistance in marine environments, and good weldability and formability for uses as extrusions, ship and boat parts, and automotive structural parts. The 6000 series alloys of aluminum, magnesium, and silicon have excellent corrosion resistance, plus good weldability, machinability, and formability; they find uses as bridge rails, automotive sheets, piping, and extrusions. The aluminum–zinc series 7000 alloy is heat treatable with tensile strength up to 606 MPa (88,000 psi), has good corrosion resistance in rural atmosphere, but is poor in marine environments. Uses for the 7000 alloys are chiefly as aircraft structures and other equipment requiring high specific strength. The 8000 series includes alloys of combinations not included above, for example titanium or zirconium. Table 10.5.3 provides comparisons on various aluminum alloys.

Corrosion resistance in aluminum is a result of the natural occurrence of an aluminum oxide film that prevents the metal from further corrosion. Through an *anodizing* process, it is possible to build up the oxide layer and even add color to the aluminum surface. Alloying, cold working, and heat treatment can reduce the corrosion resistance of aluminum. For example, intergranular corrosion can occur in a precipitation-hardened aluminum because of the lack of homogeneity (i.e., some areas have high copper content, which become anodic to copper-depleted areas).

5.12.2 Copper

Both as an alloying element in such alloys as steels and aluminum and as a base for copper alloys, copper is a very valuable metal. Included in the copper-base alloys are brass (copper and zinc), bronze (copper and tin), cupronickel (copper–nickel), leaded copper, leaded brass, aluminum bronzes, nickel silver (copper, nickel, and zinc), and high copper. Table 5-3 shows that copper and its alloys are easily processed by most common methods except certain types of welding. Several systems of designation exist for copper and its alloys but are beyond the scope of this text. Besides its value as an alloy, copper has its greatest value as an electrical conductor. High purity of over 99.9% is sought to attain the best conductivity and formability. Due to its scarcity and weight, the trend is to move away from copper as electrical conductors when possible. Copper and copper alloys develop a protective film that will not corrode in water or nonoxidizing acids. It will corrode if liquids, solids, or gases break down the protective film or if other conditions cause electrolysis to develop. In brass the presence of oxidizing chemicals can result in dezincification in which the zinc dissolves leaving a porous metal sponge. Table 10.11 lists several copper alloys with corresponding designations, compositions, and properties.

5.12.3 Nickel

This metal gains widest use as an alloying element. It provides resistance to both atmospheric and chemical corrosion, while also imparting other properties to alloying systems such as formability and strength. Table 5-3 shows that several of the common methods of processing are not recommended for nickel. While alloyed in steel, copper,

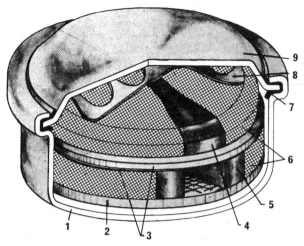

1 Cell cup	4 Negative electrode	7 Sealing washer	**Figure 5-58** Schematic cutaway of pressed-
2 Bottom insert	5 Positive electrode	8 Contact spring	plate button nickel-cadmium battery cell. (In-
3 Separator	6 Nickel wire gauze	9 Cell cover	ternational Nickel Company.)

and other systems, nickel is also electroplated onto objects for decorative and corrosion-resistance purposes. After development at the end of the nineteenth century, nickel–cadmium batteries have undergone tremendous refinement to the point that these rechargeable batteries are found throughout society in hand electronic calculators, aircraft engine starters, emergency power equipment, and portable medical equipment. Figure 5-58 shows the typical construction of nickel–cadmium batteries, which uses nickel hydroxide as the active positive plates and cadmium as the negative plates with an electrolyte of water and potassium hydroxide. Common alloys of nickel include *Monel*[R] (nickel–copper), *Inconel*[R] (nickel–chromium), *Hastelloy*[R] (nickel–molybdenum–iron), *Duranickel*[R] (nickel–aluminum), and illium (nickel–chromium–molybdenum–copper). Nickel is magnetic up to 360°C and is alloyed with iron, aluminum, cobalt, and copper in the powerful permanent alinco magnet. Appendix 10.11 shows the composition and properties of certain nickel alloys.

5.12.4 Magnesium

About 6 million tons of magnesium is available for processing in a cubic mile of sea water, which makes this a highly plentiful metal. But the need for electrolysis for extracting the metal from mineral deposits or sea water increases its cost. With a specific gravity of 1.75, magnesium is one and a quarter times lighter than aluminum and four times lighter than steel. Its ability to dampen noise and vibration compares with cast iron, while its light weight and good strength make it attractive to electronics chassis designers. Table 5-3 shows the methods for processing magnesium. Caution is required with the chips in machining because it burns and corrodes easily. Magnesium is often used as an alloying element for aircraft engine parts and in wheels for race cars. Appendix 10.11 provides compositions and properties of various magnesium alloys.

5.12.5 Titanium

As with aluminum and magnesium, titanium (Ti) is a very abundant metal in its ore form. As a raw material element, it ranks fourth in availability of the structural metal elements. Unfortunately, the extraction processes are costly. Table 5-3 shows common processes used on titanium. Its alloys offer superior specific strength (up to 26.5×10^3 in.) in high temperatures (over 590°C) and low temperatures (-253°C), which makes it a popular structural metal in ultrahigh-speed aircraft and accounts for its use on the Space Shuttle. The *superplastic* nature of titanium allows it to deform over 2000% without nicking or cracking when it is heated to around 925°C in a process known as superplastic forming. Advances in such new technologies as *superplasticity forming* and *diffusion bonding* (see Chapter 8) have furthered the potential of titanium. Titanium is nonmagnetic and has a lower linear coefficient of expansion and lower thermal conductivity than steel alloys or aluminum. Like zirconium and beryllium, titanium is allotropic and exists in a hcp structure (alpha) at below 885°C and a bcc structure (beta) above that temperature (see Table 3-1). Various alloying elements alter the effect of the structure and stabilize the alpha or beta phases. For example, aluminum stabilizes the alpha structure, raising the temperature of the alpha–beta transformation. Appendix 10.11 shows some titanium alloys, compositions, and properties.

5.12.6 White Metals

This general term covers zinc, tin, lead, antimony, cadmium, and bismuth, which all have relatively low melting temperatures, as seen in Table 5-4. *Zinc* is a readily available and inexpensive metal that can be easily applied to steel to serve as the sacrificial protection coating, as discussed previously. Zinc is also die cast for housing and decorative trim; used as additives in rubber, plastics, and paint; and used in wrought forms that can achieve high surface hardness of over 70 R_C. *Tin* has long been used as a coating for "tin cans" made of tin plate and also finds application as a valuable alloying element with lead as soft solder, in tin babbit, and antimonial tin solder. Solder balls as small as 2 mils (.002) hold semiconductors onto metal-pinned ceramics which are used for computers. *Pewter* (alloyed) with lead was a popular alloy for cooking and eating utensils, plus ornamental casting, in colonial America. The modern tin-based pewter is lead free with 91% tin, 7% antimony, and 2% copper and is often used to replicate colonial dishes. *Lead* is not only an important alloy in solder, but when evenly dispersed in leaded steels at 0.15% to 0.35%, it provides

TABLE 5-4 MELTING TEMPERATURE OF WHITE METALS

Metal	°C	Metal	°C
Antimony (Sb)	631	Lead (Pb)	327
Bismuth (Bi)	271	Tin (Sn)	232
Cadmium (Cd)	321	Zinc (Zn)	420

built-in lubrication to ease machining and produce tightly curled chips. Lead is also important in electrical batteries as terminals and, alloyed with antimony, for grid plates (Figure 5-59). Its resistance to many corrosive chemicals, x-rays, and gamma rays and its sound dampening capacities coupled with its high density find unique applications in the medical, chemical, and nuclear industries. It is compounded with ceramic glazes. Small portions of lead built up in the human body cause lead poisoning (plumbism). Consequently, it is desirable to eliminate lead from paint on children's furniture, gasoline, and soldered tin cans, and to reduce exposure by industrial workers. *Cadmium* is harder than tin and serves as a corrosion-resistant coating electroplated on steels, especially fasteners. A coating of 0.0008 mm has the equal protection of a 0.025-mm zinc coat. It also serves as an alloy in copper to improve hardness.

Figure 5-59 Lead-acid batteries. Experimental small vehicles to run on electric power rely on twelve lead-acid maintenance-free traction batteries. (General Motors.)

5.12.7 Other Metals

Beryllium is a high-cost metal that serves as an alloying element in copper for age-hardening to enhance oxidation resistance and refined grains. *Refractory metals* are those with melting points above 1980°C and include tungsten (W) (3370°C), tantalum (Ta) (2850°C), columbium (Cb) or niobium (Nb) (2415°C), and molybdenum (Mo) (2620°C), which are commonly alloyed with many of the metals discussed above to impart strength and hardness at high temperatures, such as type 6-6-2 molybdenum–tungsten high-speed steel. Because of the high temperatures and fabrication difficulties, refractories are often formed as powdered metal parts. Numerous non-ferrous alloys known as *superalloys* have recently been developed to meet the requirements of high-technology innovations, such as jet and rocket engines, for high strength and corrosion and creep resistance at extemely high temperatures (exceeding 1100°C). These superalloys use iron and nickel, nickel, or cobalt as the base metal. Examples of such alloys are Inconel[R] (nickel base), Incoloy[R] (iron–nickel base), and S-816 (cobalt base). Possessing such high strength, most of the alloys are cast or use powder metallurgy techniques in their manufacture. Where machining is required, chipless-type machining methods are mandated.

5.13 METAL POWDERS

Through a variety of techniques, powders of selected materials are produced for powder metallurgy (P/M) applications. While there is evidence P/M technology existed at least 5000 years ago, only recently has the technology made significant strides. Developments from the eighteenth century through World War I primarily resulted from the inability of metallurgists to fuse such metals as platinum and tungsten, so they turned to powder compacted and heated to allow diffusion bonding. This last step is known as *sintering*, and P/M parts are often referred to as sintered metals. Current powder metallurgy deals with most metals and a host of processes. Figure 5-60 shows the flow of the raw powder through compacting (*briquetting*) to finished P/M parts. Some parts undergo secondary operations to improve tolerances or finishes, to apply lubricants or coatings, or for heat treatment. A major advantage of the P/M part is the reduction or elimination of scrap and machining, so if too many secondary operations are required this advantage can be reduced. Once the part is compacted and sintered, it may undergo secondary operations such as *coining* to squeeze the part to closer tolerances, *infiltration* in which a lower melting point metal is added to close pores and increase density, *impregnation* (Figure 5-61), which adds a lubricant for self-lubricating bearings, or *machining* to achieve a profile not practical through the pressing operation. Not all metal powder is compacted and sintered. Powder iron is added to human and livestock food as a dietary supplement. Iron powders are also used in nondestructive testing (NDT) and magnetic particle testing, for welding electrodes, and in oxyacetylene cutting and scarfing (exothermic reaction increases the heat of flame to melt oxides formed on high alloys or refractories). Platinum

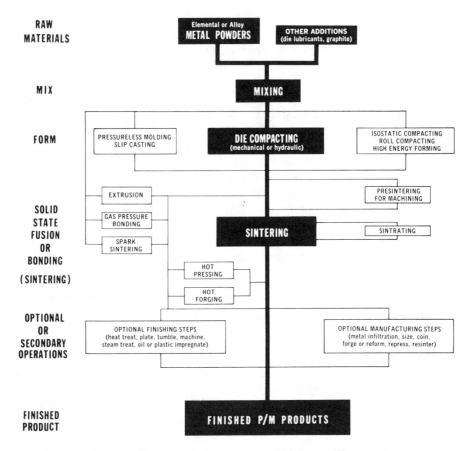

Figure 5-60 Schematic flow diagram of P/M process. (Hoeganaes.)

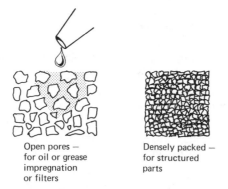

Open pores —
for oil or grease
impregnation
or filters

Densely packed —
for structured
parts

Figure 5-61 Impregnation.

Sec. 5.13   Metal Powders

227

powder is a catalyst used in making gasoline, copper powder goes into marine paints, carbon powder is used in automotive ignition wires and in making xerographic paper copies, and aluminum powder aids in paper manufacture.

In addition to the advantages given, P/M allows the manipulation of a part's density through control of particle size and degree of compaction. Particles range from 0.2 to 2000 μm in size. Smaller particles can pack tightly for increased density; or if open pores are required, as in filters or oil impregnated, self-lubricating bearings, larger particles would be compacted together (Figure 5-62). P/M structural parts of steel powder have tensile strengths from 310 MPa (45,000 psi) to over 1200 MPa (175,000 psi), which makes them competitive or superior in terms of strength with cast iron, steel, and nonferrous alloys in certain applications. While conventional P/M processing can produce parts up to 35 lb, most are smaller and weigh less than 5 lb. The most common P/M materials are iron, steel, copper and copper alloys, stainless and other alloy steels, nickel alloys, aluminum alloys, and refractories such as tungsten carbides, but most other metals find certain unique P/M applications.

APPLICATIONS AND ALTERNATIVES

Metals and humanity. Metals are vital to many aspects of society. Since early times they served in weaponry, transportation, art, construction, and most facets of our lives. Tracing major technological developments often means tracing metals technology. From the early Bronze Age and Iron Age to the development of high-alloy steels for jet engines and rocket engines, humanity has relied on metals to make significant advances. So much has the United States relied on metals that we face threats to our well-being when sources of certain metals are endangered. For over a three-year period 89% to 92% of the chromium used by this country was imported. The major sources of this vital alloying element are Southern Africa and Russia. Because of human rights violation, the United States might prefer not to trade with South Africa, and in the event of hostilities, Russia could not be relied on as a chromium supplier. Therefore, a heavy reliance on chromium makes the United States vulnerable, just as reliance on the Oil Producing and Exporting Countries (OPEC) for our major source of energy (oil) created economic problems. The solution is to reduce reliance on chromium supplies. Stainless steels with aluminum as a major alloy hold promise for some relief from chromium, as do newly developed alloy systems from manganese–molybdenum or manganese–nickel–molybdenum, which produce cost effective heat-treatable steels. The United States relies heavily on imports for many of its vital metals, such as nickel, manganese, cobalt, tin, and titanium.

In addition to looking for alternative metals, other answers to sound metal economics include recycling metals and discovering new sources. Research into better ways of recycling the enormous amount and variety of materials used in automobile manufacturing produced the water elu-

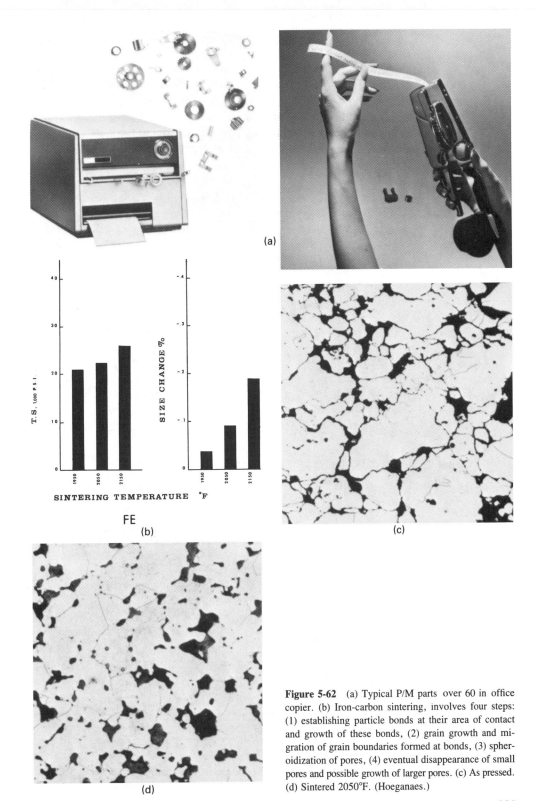

Figure 5-62 (a) Typical P/M parts over 60 in office copier. (b) Iron-carbon sintering, involves four steps: (1) establishing particle bonds at their area of contact and growth of these bonds, (2) grain growth and migration of grain boundaries formed at bonds, (3) spheroidization of pores, (4) eventual disappearance of small pores and possible growth of larger pores. (c) As pressed. (d) Sintered 2050°F. (Hoeganaes.)

triation technique for metal recovery from nonmagnetic automobile shredder rejects. Figure 5-63 depicts a typical shredding operation using water elutriation. The top portion of the diagram shows the car fed in at the left for grinding and magnetic separation on the right. In the middle right, nonmagnetic materials feed through the surge bin into the water elutriator to yield three separations: (1) a float of combustibles, (2) a middling of

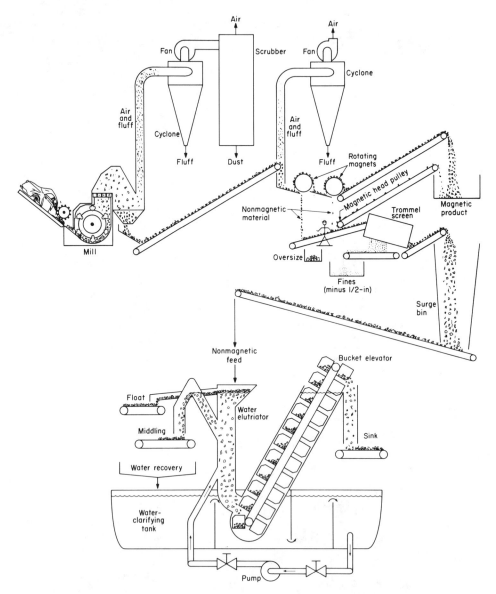

Figure 5-63 Diagram of a typical shredding operation with water elutriation system. (Bureau of Mines, U.S. Department of the Interior.)

heavy plastic, rubber, glass, coated wire, and some light metal; and (3) a sink product comprised of 95% to 98% metal and some dense rubber and heavy rocks. Other techniques can follow elutriation to separate the nonferrous metals into aluminum alloys, zinc alloys, copper alloys, stainless steels, and nonmetals.

In addition to metal recycling, other sources of minerals have received study. The earth's moon and our ocean's floors contain important minerals. The United States imports 98% of its manganese ore, much of which goes into steel making. The ocean floor holds metal nodules (clumps of metal created from metal ions joining) that contain high-grade manganese along with nickel, copper, cobalt, and other metals. The commercial feasibility to ocean-mine the nodules has been demonstrated, but problems of ownership and international law could continue to plague this source of ore.

5.14 SELF-ASSESSMENT

5-1 Name four basic methods for changing a metal's properties.

5-2 What other name does the term work hardening have?

5-3 Name one distinguishing characteristic of a phase.

5-4 The cooling curve for a pure metal is readily identified by a certain portion of this curve. Describe this segment of the curve and tell what it represents.

5-5 Using the variable time, describe what the term "equilibrium conditions" means for phase changes in an allotropic metal.

5-6 Describe the differences between the words eutectic and eutectoid.

5-7 Carbon steel is an alloy of carbon and iron. Is stainless steel an alloy of carbon steel? Is it an alloy steel?

5-8 How does the composition of ferrite differ from pearlite?

5-9 Write the chemical symbol for cementite and attach another name to this ingredient of steel.

5-10 Wrought iron is mainly pure iron mixed with a slag composed mostly of iron silicate. Specify an excellent application for wrought iron. Could wrought iron be alloyed with another metal? If so, which metal would you choose and for what reasons?

5-11 Sketch a cooling curve for an alloy.

5-12 Plain carbon steels (i.e., low, medium, and high carbon steels) contain up to a maximum of about 1% carbon. These steels provide much of the steel that goes into welded parts, bolted and riveted structures, springs, and free machining parts. Why is it necessary to develop more sophisticated specialty steels containing alloying elements other than carbon?

5-13 Using Figure 5-22 and an alloy of about 60% A at a temperature above T_E, what solid is formed upon cooling when this alloy reaches the liquidus line?

5-14 What cystalline phase of steel is referred to as a very fine platelike (lamellar) mixture of ferrite and cementite?

5-15 What is the maximum solubility of carbon (in percent) in the interstitial solid solution of carbon dissolved in gamma iron?

5-16 A carbon steel alloy containing 0.20% carbon when cooled from the austenite region crosses the A_3 line in Figure 5-25. What phase begins to form at the austenite grain boundaries at this point?

5-17 Referring to Figure 5-29, determine approximately what temperature you would heat a hypereutectoid steel to in order to fully anneal it?

5-18 Name two annealing processes that do not involve phase change.

5-19 The recrystallization temperature for Monel metal of about 600°C is the approximate temperature at which a highly cold worked Monel part completely recrystallizes in an hour. Other recrystallization temperatures are low carbon steel, 540°C; zinc, 10°C tin and lead, −4°C. Working of tin, lead, or zinc at room temperature (about 20°C) is considered hot working. Explain.

5-20 Describe how to reduce grain size in a piece of metal.

5-21 In grinding metal for sharpening purposes, operators are cautioned not to grind the metal too long, otherwise the temper in the metal would be drawn. What is meant by this warning and how do you prevent losing the metal's temper in this case?

5-22 Figure 5-36 is a diagram of the transformation of austenite in eutectoid steel. How would you redraw this diagram if the steel were a hypereutectoid steel?

5-23 Hardness and hardenability are related properties. Discuss their differences.

5-24 If a metal part needs to be hard and at the same time possess a high degree of toughness, how are these two properties acquired through thermal treatment?

5-25 Modern steels are sometimes referred to as examples of man-made composite materials. Explain the significance of this statement.

5-26 The production of metals and the casting of metal parts take place under the influences of the earth's atmospheric pressure and gravitational force. Speculate as to how the crystalline structures of metals would change, and hence their properties, if these processes could be accomplished in a vacuum.

5-27 Colbalt, chromium, titanium, tungsten, manganese, and platinum are some of the strategic metals the United States must import in order to produce the vital specialty alloy steels needed in our advanced technological society. Most of these metals are found in the Third World and communist bloc nations. Stockpiling aside, what does this situation signify to the many materials scientists, engineers, and technicians working in the metals industries?

5-28 What is the base metal in all ferrous metals and what raw materials are used in obtaining this metal from ore?

5-29 (a) What is the carbon content of cast iron? (b) Explain the designations ASTM 48 Class 35 and 32510.

5-30 List one application, the key structural difference, and the best quality for each of the following cast irons: (a) ductile cast iron, (b) gray cast iron, (c) white iron, (d) malleable iron, (e) wrought iron.

5-31 (a) How is steel different from cast iron? (b) How is CRS superior to HRS?

5-32 Explain the following designations: (a) SAE 1020, (b) ASTM S 40000, (c) AISI 400, (d) SAE E 52100, (e) 6016 T 2, (f) CRES.

5-33 (a) What drawback does HSLA present to auto body makers? (b) What steel is normally used for auto bodies?

5-34 What is the major function of the following in alloy steel? (a) copper, (b) vanadium, (c) chromium, (d) aluminum, (e) lead, (f) nickel, (g) tungsten.

5-35 Describe the major advantage for each of the following stainless steels: (a) martensitic, (b) austenitic, (c) ferritic.

5-36 Name three noble metals and give an application of each.

5-37 Explain why aluminum, magnesium, and titanium are not inexpensive when they are so readily available as raw materials.

5-38 Explain the corrosion-resistance mechanism of copper and aluminum. Which metal offers the widest range of processability?

5-39 Give three applications of nickel and explain why nickel is used in these cases.

5-40 What is (a) brass, (b) bronze, (c) Monel, (d) Inconel, (e) refractory metals?

5-41 What metal is used for superplastic forming, and why?

5-42 (a) Provide three reasons for the use of powder metal. (b) List three applications of P/M.

5-43 (a) What common characteristic do white metals possess? (b) Provide a typical application for each of these metals. (c) Why is lead being eliminated from paint, plumbing, gasoline, ceramics cookware, and similar uses?

5.15 REFERENCES AND RELATED READING

ALUMINUM ASSOCIATION. *The Story of Aluminum*. New York.

————. *Uses of Aluminum*. New York.

AMERICAN SOCIETY FOR METALS. *Metals Handbook, Vol. 1, Properties and Selection of Metals*. Metals Park, Ohio, 1961.

AMERICAN SOCIETY OF MECHANICAL ENGINEERS. *ASME Handbook: Metal Properties*. New York: McGraw-Hill, 1954.

ANNUAL BOOK OF ASTM STANDARDS PART 11—METALLOGRAPHY; NON DESTRUCTIVE TESTING. Philadelphia: American Society for Testing and Materials, 1983 or current edition.

ANNUAL BOOK OF ASTM STANDARDS PART 10—METALS—MECHANICAL, FRACTURE, AND CORROSION TESTING; FATIGUE EROSION; EFFECT OF TEMPERATURE. American Society for Testing and Materials, 1983 or currrent edition.

AVNER, SIDNEY H. *Introduction to Physical Metallurgy*, 2d ed. New York: McGraw-Hill, 1974.

BETHLEHEM STEEL CORPORATION. *Modern Steels and Their Properties*, 6th ed. Bethlehem, Pa., 1967.

BOSSCKER, JAMES P., *Elements of Metallurgy* (Video Modules), Atlanta: Association for Media-Based Continuing Education for Engineers, 1980.

BRADY, GEORGE S., and HENRY R. CLAUSER. *Materials Handbook*, 11th ed. New York: McGraw-Hill, 1977.

BUREAU OF MINES. *Mineral Commodity Summaries 1983*. Washington, D.C.: U.S. Department of Interior, 1984.

CLAUSER, HENRY R. *Industrial and Engineering Materials*. New York: McGraw-Hill, 1975.

Design Guidebook. Princeton, New Jersey, 1974.

FLINN, RICHARD R., and PAUL K. TROJAN. *Engineering Materials and Their Applications*. Boston: Houghton-Mifflin, 1975.

HOEGANOES CORPORATION. *Creating with Metal Powders*. Riverton, N.J., 1968.

"H.S.L.A. STEELS," *Machine Design*, Apr. 16, 1984.

HURD, PAUL S. *Metallic Materials*. New York: Holt, Rinehart and Winston, 1968.

INTERNATIONAL NICKEL COMPANY. "A Quick Refresher on Stainless Steel." New York.

JOHNSON, CARL G., and WILLIAM R. WEEKS. *Metallurgy,* 5th ed. Chicago: American Technical Society, 1977.

LEAD INDUSTRIES ASSOCIATION. *Lead.* New York.

Machinability of Sintered Iron, Riverton, N.J.: Hoeganaes, 1968

The Many Uses of Metal Powders, Riverton, N.J.: Hoeganaes, 1976

Materials Engineering: Material Selector Edition. Cleveland, Ohio: Penton/IPC, Reinhold Publishing.

Metallography of Sintered Ferrous Metals, Riverton, N.J.: Hoeganaes, 1970

METAL POWDER INDUSTRIES. *Powder Metallurgy.* New York.

MEYERS, MARC A. and K. K. CHAWLA. *Mechanical Metallurgy–Principles and Applications,* Englewood Cliffs, N.J.: Prentice-Hall, 1984.

NEELY, JOHN E. *Practical Metallurgy and Materials of Industry.* New York: Wiley, 1979.

OBERG, ERIK, and FRANKLIN D. JONES. *Machinery's Handbook,* 22nd ed. New York: Industrial Press, 1984.

PETERS, A. T. *Ferrous Production Metallurgy.* New York: Wiley, 1982.

Properties of Some Metals and Alloys, Riverton, N.J.: Hoeganaes, 1968.

ROSS, ROBERT B. *Metallic Materials.* London: Chapman and Hall, 1968.

TITANIUM METALS CORPORATION OF AMERICA. *How to Use Titanium.*

Periodicals

The Journal of Materials Science and Engineering

Machine Design: Materials Reference Issue

Metals Progress

NATIONAL AERONAUTICS AND SPACE ADMINISTRATION, *NASA Tech Briefs*

6

POLYMERIC MATERIALS

6.1 PAUSE AND PONDER

Considerable effort by industry goes into the containers used to hold products. Sometimes the containing method is an integral part of the product, such as the body of an automobile. With packaging the container holds the product for later dispensing; it may serve as an eyecatcher to consumers and protect the product from exposure to contaminants as with a beer bottle. Cabinets used for TV sets and microwave ovens protect the internal parts and add decorative value, while providing a mounting surface for the controls used to operate these devices. Companies constantly strive to improve the design of their containers in order to meet governmental regulations and improve product safety and quality, to keep up with design trends and consumer desires, or to accommodate a new product.

One typical example of a change in containers occurred with toothpaste tubes when the traditional metal tube was replaced by a plastic. Because of the health problems involved in tubes that used lead as part of the alloy for tubing, a suitable replacement was required. The elastic nature of plastics prevents this type of tubing material from rolling up, a desirable characteristic in collapsible tubing containers. Figure 6-1a shows the plastic composite developed in answer to this container problem. Note the complexity of this laminated plastic sheet. The tubing is primarily polyethylene laminated to other materials. The 0.7 mil (0.0007 in.) aluminum foil with its ductility counteracts the elasticity of the polyethylene, so the tube is collapsible and thus capable of rolling up as it is emptied.

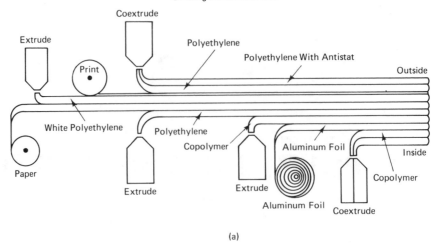

Building a Glaminate Web

(a)

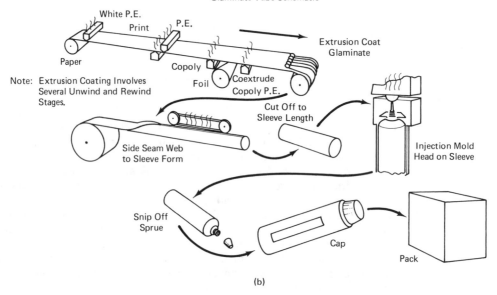

Glaminate Tube Schematic

(b)

Figure 6-1 Polymer packaging (a) Building plastic composite (Glaminate®) web. (b) Web built, then processed into toothpaste tube. (American Can Co.)

A copolymer adhesive film bonds the polyethylene to the aluminum. A sheet of paper is laminated behind the plastic label to ensure opacity. Figure 6-1b shows the processes involved in making the complete tube.

As you study this module on polymers, consider how the many plastics, rubbers, and woods serve us as containers. Keep these questions in mind: What polymers are used for specific packages, cabinets, and body coverings that you encounter daily? What properties do the polymers possess that cause their usage? How would you use polymers as an improvement for containers on everyday products that now use some other material? What may prevent the use of polymers as a substitute material? Consider their cost, desirable properties, raw material availability, and recycling characteristics.

6.2 POLYMERICS

Section 2.3.3 provides an introduction to polymeric materials (often classified simply as polymers). The emphasis in this module is on the main polymeric materials used in engineering, manufacturing, and construction and excludes such natural polymers as skin, hide (leather), bone, or natural fibers such as wool, cotton, silk, and jute. Plastics, woods, elastomers, and adhesives are all polymers; while plastics, adhesives and synthetic rubber are man-made, wood and natural rubber are naturally occurring materials that are used in industry with various alterations to their natural state. Polymers can include many types of industrial materials, as seen in Figure 6-2.

6.3 NATURE OF POLYMERS

The basic makeup of polymers consists of smaller units (*mers*) joined together either naturally or synthetically to produce polymers or macromolecules. This section on structure deals with the structures common to most polymers, and most of the ex-

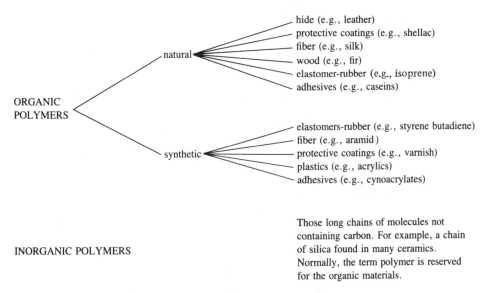

Figure 6-2 Industrial polymers.

amples cited are plastics, due to their abundant use in industry. The discussion will concentrate on the synthetic methods of producing polymers known as *polymerization* and will not deal with the biological process by which the natural or biological polymers are grown.

The initial step in polymer production is achieved by petrochemists or petro-chemical engineers, engineering technicians and engineering technologists, or others with chemistry in their education and training. This initial step involves breaking down such raw materials as crude oil, coal, limestone, salt, natural gas, and air. With cellulose plastics, the raw material is cotton or wood; however, they are not broken down, as are most other polymer ingredients. Whether it be the cracking of natural gas, distillation of coal or crude oil, or the esterification of cotton or wood, processing the basic organic (carbon base) raw materials yields the element carbon together with hydrogen. Other basic elements are also obtained in the initial stages, including oxygen, nitrogen, silicon, fluorine, chlorine, and sulfur. Such basic ingre-dients of polymers as ethylene and naphtha are but two of the many products obtained in the cracking and distilling processes. From 100 gallons of crude oil, at a market value of around $45, it is possible through distillation and further processing to produce such products as gasoline, motor oil, propylene, and a large variety of polymers such as styrene butadiene rubber, polyethylene plastics, polypropylene fibers for fabrics, and others, which combine to make products with a market value of over $4000. While oil is the major raw material for plastics, plastics take only a very small percentage (about 1.3%) of the world's petroleum production. Yet, shortages in oil cause problems in the plastics and polymers industry, so alternate raw materials continue to receive study.

6.3.1 Polymerization

Scientists observed nature's methods of joining elements into chains and duplicated that natural process to produce macromolecules or polymers. Module 2 covers the principles of bonding, which provide a basic understanding of polymerization. *Po-lymerization* is the linking together of smaller units (monomers) into long chains. The repeating units (mers) of polymer chains may be identical, as in the case of polyethylene (Figure 6-3a), polystyrene, and polyvinyl chloride, in which case they are labeled *homo*polymers (Figure 6-3b). *Co*polymers contain two different types of monomers, such as polyvinyl chloride mixed with vinyl acetate to produce polyvinyl acetate (Figure 6-3c), and *ter*polymers like ABS (acrylonitrile butadiene styrene) contain three types of monomers. Isomers (Figure 6-3d) are variations in the molecular structure of the same compositions. It is important to note that copolymers and terpolymers consist of units from each contributing mer and are not an alloy of mers. If polymerization only permitted the production of homopolymers, then the properties of polymers would be severely limited.

Most polymers are produced by unsaturated hydrocarbons, which means they have one or more multiple covalent bonds, such as ethylene,

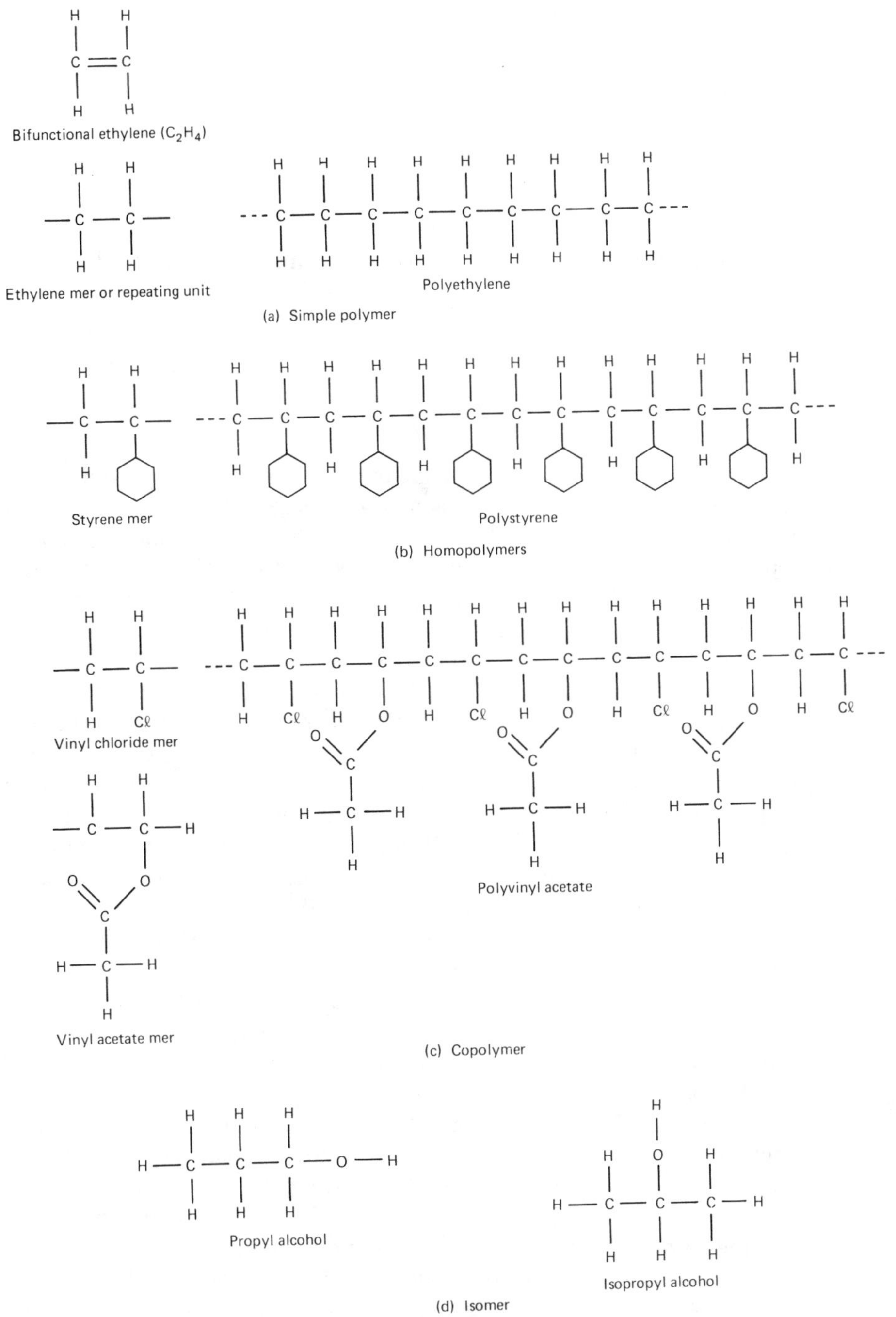

Figure 6-3 Types of polymers (a) Simple polymer. (b) Homopolymers. (c) Copolymers. (d) Isomers.

or adipic acid that is used in nylon synthesis; it has two hydroxyl monomers.

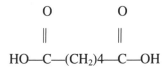

Saturated hydrocarbons have all single bonds. To achieve the polymerization process, monomers must be capable of reacting with at least two neighboring monomers or be *bifunctional*. Because of copolymerization, terpolymerization, or other multicomponent polymerizations, a large variety of polymers is available. One case is styrene, a brittle polymer that has limited toughness and poor chemical resistance. But through copolymerization of acrylonitrile with styrene it is possible to obtain a chemically resistant, more rigid, and stronger plastic. Copolymerization of butadiene with styrene yields an elastomer; and terpolymerization of acrylonitrile, butadiene, and styrene produces ABS plastics that are tough, elastomeric, and have good chemical resistance.

The main polymerization processes are addition (chain reaction) polymerization and condensation (step reaction) polymerization. The addition process is the simpler of the two. By use of heat and pressure in an autoclave or reactor, double bonds of unsaturated monomers (Figure 6-4) break loose and then link up into a chain. The products of addition polymerization, also referred to as chain reaction polymerization, include polypropylene (PP),* polyethylene (PE),* polyvinyl acetate (PVA),* polyvinyl chloride (PVC),* acrylonitrile butadiene styrene (ABS),* and polytetrafluoroethylene (PTFE).* Chain polymers mostly fit into the thermoplastic (soften when heated) group. Most plastics that cannot be resoftened when heated (thermosetting polymers) come from condensation polymerization, also known as step reaction polymerization. This group gets its name from the by-product (condensate) of the polymerization, which is often water, but it may be a gas. Phenolic (PF),* polyester (PET),* silicon (SI),* and urethanes (PUR)* are typical thermosets from the step reaction synthesis, while nylon (PA)* and polycarbonates (PC)* are thermoplastic resins synthesized through step polymerization. Figure 6-5 diagrams the addition process for the production of some typical polymers, and Figure 6-6 shows the flow for production of nylon through condensation polymerization.

The term *resin* covers both solid and semisolid organic polymers. Resin is often considered as the uncompounded ingredients or monomers that are mixed but not yet polymerized. For example, thermosetting resins or pellets are molded into thermo-

*ASTM abbreviation.

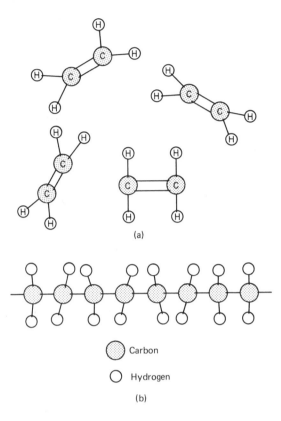

Figure 6-4 (a) Monomer of ethylene (C₂H₄). Carbon atoms with four valence electrons have four arms (shared electrons) for covalent bonding. Two shared electrons hold another carbon atom in a double bond. Two other shared electrons, each covalently-bonded, hold one shared electron of hydrogen which has only one valence electron for single bonds. Double bonds are not stable and form *reactive sites*. (b) Polyethylene, chain of ethylene monomers. During polymerization heat and pressure break the hold of one shared electron on each of the double-bonded (reactive sites) carbon atoms thereby leaving each carbon atom of the monomers free to grasp another carbon atom (covalent bond) from other ethylene monomers which form into a chain of thousands of ethylene monomers (polyethylene). The single bonds satisfy the carbon bond arrangement and produce the most stable saturated polymers since they have no reactive sites.

setting plastic or elastomeric parts. Sometimes the term resin is used synonymously with plastics (e.g., acetal resins instead of acetal plastics or thermoplastic resin instead of thermoplastic plastics).

6.3.1.1 Polymer chain lengths.

Considerable variation in polymer chains results during polymerization. The chain lengths play an important role in processing polymers and their final properties. While we cannot see the length of the polymers, the molecular weight provides a good indication. Refer to Module 2 for an introduction to molecular structure and bonding.

Molecular weight provides a weight for a single unit or molecule in a compound. To determine the *molecular weight of the compound,* it is necessary to multiply the weight of a molecule by the number of molecules or monomers. We said polyethylene typically has about 2000 monomers, so we multiply 2000 times the molecular weight of one monomer to obtain 56,000:

Molecule or monomer	Molecular weight	Number of monomers	Molecular weight of compound
C₂H₄	28	× 2000	= 56,000

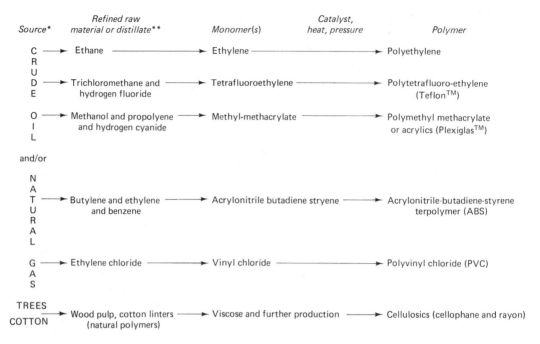

Source*	Refined raw material or distillate**	Monomer(s)	Catalyst, heat, pressure	Polymer
C R U D E	Ethane	Ethylene		Polyethylene
	Trichloromethane and hydrogen fluoride	Tetrafluoroethylene		Polytetrafluoro-ethylene (Teflon™)
O I L	Methanol and propolyene and hydrogen cyanide	Methyl-methacrylate		Polymethyl methacrylate or acrylics (Plexiglas™)

and/or

N A T U R A L	Butylene and ethylene and benzene	Acrylonitrile butadiene stryene		Acrylonitrile-butadiene-styrene terpolymer (ABS)
G A S	Ethylene chloride	Vinyl chloride		Polyvinyl chloride (PVC)

TREES
COTTON — Wood pulp, cotton linters (natural polymers) → Viscose and further production → Cellulosics (cellophane and rayon)

*LRG (liquefied refinery gas) is obtained from refining crude oil and can be further refined to yield the butadiene, ethylene, methane, and propylene. Coal and coke also serve as raw materials for production of gases used to produce polymers.

**In addition to petroleum raw materials which provide the carbon and hydrogen base, other elements such as oxygen, nitrogen, sulfur, chlorine, hydrogen, and fluorine are mixed to obtain the monomer.

Figure 6-5 Typical polymers produced through addition (chain) polymerization.

Involved chemical analysis is required to determine polymer chain lengths, and such analysis reveals that chains are not uniform in length. Consequently, the value given for a particular resin would be either its *average molecular weight* or *degree of polymerization* (DP), which indicates the number of repeating mers in the material. Long-chained polymers have a greater molecular weight, which causes more entanglement of the chains and thereby increases *viscosity* (resistance to flow). The degree

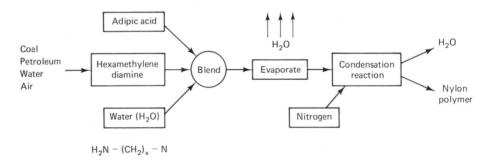

$$H_2N - (CH_2)_x - N$$

Figure 6-6 Graphical illustration of condensation (step) polymerization of a polyamide (nylon).

of viscosity affects the processing of the polymer, so manufacturers must take it into account.

6.3.2 Crystalline Structures

In addition polymerization neighboring chains are held together by weak secondary (intermolecular) bonds known as van der Waals forces, as described in Module 2. As the chains grow during polymerization, they intermingle in a random pattern that will lead to an *amorphous structure* (unorderly). Bear in mind that growth of chains is three dimensional and not simply flat as normally diagrammed. Visualize a bowl of cooked spaghetti with its uneven strands entangled; that gives a good idea of the arrangement of an amorphous polymer structure.

Actually, polymers are semicrystalline in varying degrees, with the amorphous structured polymers having only slight regularity, while other polymers may have a high degree of crystallinity (Figure 6-7). Metals achieve crystallinity due to the uniform nature of the unit cells of their space lattices. Crystallinity in polymers can alter strength and toughness, whereas the molecular motion of short chain segments and side groups cause amorphous polymers to be stiff, hard, and brittle at room temperatures. Increases in temperature cause greater molecular motion through thermal mixing of the atoms and molecules, thereby causing an increase in volume of the material. The thermal mixing from higher temperatures increases the spacing between molecular segments and permits increased flow of the materials. A discussion of glass transition later in this module deals further with this concept of molecular motion and flow in polymers. The degree of crystallinity is determined by structural regularity, compactness, and amount of flexibility, which allows packing of chains. Reduction of random chain lengths provides regularity. Stronger secondary forces allow greater compacting.

Other chemical factors affect crystallinity, including configuration and tacticity but will not be discussed here. See references at end of module for more information.

Polyethylene serves as a good example of a polymer that is capable of a high degree of crystallinity because (1) the linear structure of the chains is conducive to packing, and (2) its molecular pattern is flexible, which provides easy packing even

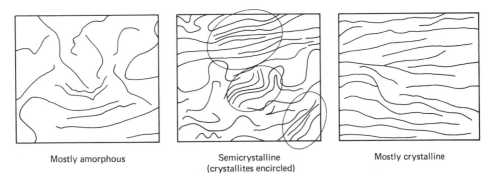

Mostly amorphous   Semicrystalline   Mostly crystalline
                          (crystallites encircled)

Figure 6-7 Polymers semicrystalline structure.

though it has weaker secondary bonding. Secondary bonds (van der Waals forces) normally promote crystallinity.

Polyethylene (PE) has the potential for a wide range of properties because of technology's ability to control its molecular weight and crystallinity. Low-density polyethylene (LDPE) has a molecular weight below 10,000, while ultrahigh-density polyethylene (UHMWPE) has molecular weights much above 1.0×10^6 (see Figure 6-8a).

Crystallinity of polymers is also achieved through processes such as extrusion and drawing. *Oriented polymer fibers* are obtained by drawing an amorphous polymer

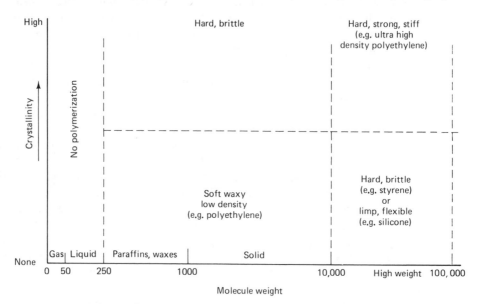

Properties: Relation to molecular weight and crystallinity

(a)

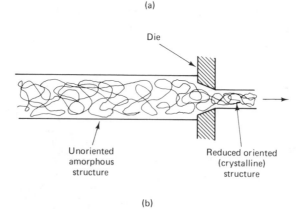

(b)

Figure 6-8 Polymers and crystallinity. (a) Properties: relation to molecular weight and crystallinity. (b) Effect of pulling amorphous polymer through die.

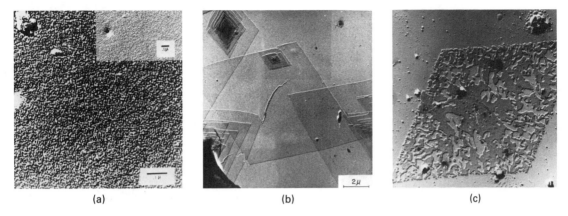

(a)   (b)   (c)

Figure 6-9 Crystallinity in polymers. (a) Electron micrograph of amorphous polyethylene. (b) Single crystal of polyethylene. (c) Single crystal of linear polyethylene annealed at 120°C for 30 minutes. (Dr. Phillip H. Geil, University of Illinois.)

through dies, which improves their strength or *orientation* in the direction of drawing (see Figure 6-8b). This concept can be illustrated by stretching a polyethylene sandwich bag. The method of manufacturing orients the polymers in one direction. By pulling on the bag in perpendicular directions it is possible to notice the greater resistance from the oriented direction; also, as the bag is stretched it becomes stronger due to further orientation.

The cooling rate and processing during cooling affect the crystal patterns. Spherulites form as a result of supercooling in which nuclei generate a spherical pattern that grows until several spherulites melt at boundaries. When large spherulites are allowed to grow, weakness results. Heat treatment of polymers can change their crystal structure; for example, the annealing processes reheat the polymer for a specified time to permit crystal thickening. In Figure 6-9, electron micrographs show polyethylene in the amorphous state, then as a single linear crystal, and then as an annealed crystal.

6.3.3 Branched and Network Structures

As seen in Figure 6-10c, some polymers do not simply grow linear chains (a and b); rather the chains develop branches much like those of a tree. As a result, the branched polymers are not as symmetrical and consequently will not achieve the degree of crystallinity obtained in the more regular linear polymers. Numerous techniques are employed by polymer chemists and materials technologists to vary both the length of linear chains and the degree and uniformity of branching.

Networks of polymers develop and primary bonds form between chains. As contrasted to the weaker secondary forces in linear polymers, *crosslinking* is the development of covalent bonds between chains, as seen in Figure 6-10d and e. The crosslinked network polymers do not soften when heated, as do the *thermoplastic*

(a) Single linear polymer

(b) Multiple linear polymers

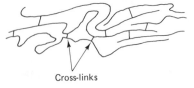

(c) Single branched polymer

(d) Network — crosslinked branched polymers
(e.g., phenolic, epoxy, and silicones)

Cross-links

(e) Crosslinked linear polymers
(e.g., polyesters and diene elastomers)

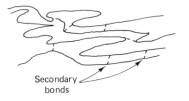

Secondary bonds

(f) Thermoplastic polymer

Figure 6-10 Various polymer structures. (a) Single linear polymer. (b) Multiple linear polymers. (c) Single branched polymer. (d) Network: crosslinked branched polymers, e.g. phenolic, epoxy, and silicones. (e) Crosslinked linear polymers, e.g. polyesters and diene elastomers. (f) Thermoplastic polymer.

polymers that are held together by van der Waals forces (Figure 6-10f). Heating of thermoplastics causes these weaker bonds to lose their hold, movement of the mers, and a softening of the polymer. Thermosetting polymers are like hard-boiled eggs. Once cooked, the egg cannot be softened to flow into another shape. But a thermoplastic material such as a candle can be continually reheated and molded indefinitely into new shapes. Normally, polyethylene is a thermoplastic material, but through electron irradiation crosslinks are developed. Epoxy can exist as a thermoplastic polymer, or through the connections of end groups it will become a thermosetting polymer. Sulfur is used in rubber to form numerous crosslinks in the *vulcanization* process. The more crosslinks, the harder the rubber is. A reference to crosslinking is the *netting index,* which designates the number of crosslinks per 100 linear bonds.

The number of double bonds in a polymer provides more sites for reaction than those with less double bonds. Natural rubber has hundreds more double bonds than butyl rubber, which has more tightly placed single-bonded molecules. A hard rubber has a netting index of 10 to 20, while a hard thermosetting plastic such as phenolic is around 50.

6.4 PLASTICS

The Society of the Plastics Industry defines plastics as

> Any one of a large and varied group of materials consisting wholly or in part of combinations of carbon with oxygen, hydrogen, nitrogen, and other organic and inorganic elements which while solid in the finished state, at some stage in manufacture is made liquid, and thus capable of being formed into various shapes, most usually through the application, either singly or together, of heat and pressure.

This definition is very broad and rather awkward, but points out that plastics come from such wide ranges of raw materials and processes, have so many varied properties, and take such diverse forms that they almost defy definition. Even natural and synthetic rubber fit this description.

6.4.1 Classifications of Plastics

Several methods are used in the classification of plastics. The older and perhaps obsolete system of thermosetting plastics and thermoplastic plastics groups is still commonly used; however, a number of plastics and some day possibly all plastics can fit into either grouping. Using a grouping of thermosetting or thermoplastic is useful as a *processing classification* since it indicates what types of processes a certain plastic can undergo. To illustrate, a thermoplastic sheet can be used as finished product or heated to be reshaped. Once a thermoset part has been molded, it can no longer be reheated for further shaping. Classifying plastics as either *linear* or *crosslinked network* provides a more descriptive system that is useful to the designer in determining general properties. For example, linear plastics can continually be remolded and usually have low heat resistance, whereas the crosslinked network plastics have greater heat and chemical resistance. Another grouping system involves the *nature of the material:* rigid, flexible, or elastic. Closely connected with the nature of the material classification are the *uses:* general purpose or engineering. *General purpose* would include the bulk of plastics that we encounter daily, such as cellulosic, acrylics, and vinyls. *Engineering plastics* are those that substitute for traditional engineering materials such as steel and wood. Some engineering plastics include polycarbonates, nylons, and polyurethanes. Most engineering plastics are considered *structural plastics* capable of bearing supporting loads. Much overlap exists for plastics in these classification systems.

Cellular (foam) plastics are listed by the Society of Plastics Industry (SPI) as a plastics group. Cellular plastics are those that have been foamed or had gas and/or air entrapped in the resins to reduce the density of the finished product (Figure 6-11a). Foaming agents include gases (nitrogen, carbon dioxide, or air) and liquids (chlorinated aliphatic hydrocarbons, alcohols, or ethers). Hollow glass or resin spheres are also used to create voids. Whipping action introduces air, carbon dioxide is generated from chemical reactions in the resin, and volatile liquids vaporize into gas

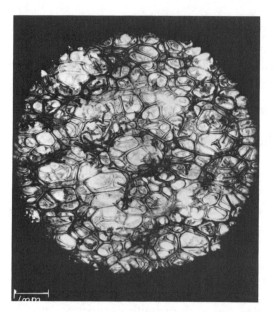

(a)

Figure 6-11 Cellular foam polymer. (a) Photomicrograph enlargement of foamed polymer reveals cells created by gas. (b) Foamed urethane: special soft front end of Calspan/Chrysler RSV weighs only 31 pounds. (Monsanto.)

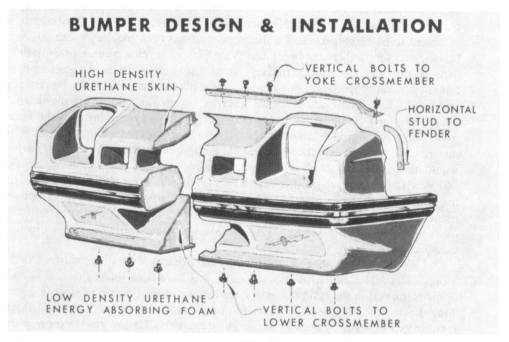

BUMPER DESIGN & INSTALLATION

HIGH DENSITY
URETHANE SKIN

VERTICAL BOLTS TO
YOKE CROSSMEMBER

HORIZONTAL
STUD TO
FENDER

LOW DENSITY URETHANE
ENERGY ABSORBING FOAM

VERTICAL BOLTS TO
LOWER CROSSMEMBER

(b)

through *endothermic* reactions (absorbed heat when resin bonds break). Water is produced through *exothermic* chemical reactions (heat given off) and it volatilizes to produce cells. Practically all thermosetting and thermoplastic resins can be foamed.

Advantages offered by cellular plastics include weight reduction, increased bulk, improved thermal insulation values, greater shock absorption, increased strength-to-weight ratio, and ability to duplicate wood in terms of texture, density, and feel. Cost savings come with foaming. The addition of glass microspheres or bubbles to *sheet molding compounds* (SMC)* can reduce weight while improving impact strength because the hollow spheres dissipate energy through shock-absorbing action (dampening). The tiny bubbles, which are from 10 to 200 micrometers (μm) in diameter also improve mechanical and chemical properties while facilitating processing. Cellular plastics are available in many densities (1.6 to 96 kg/m^3) as rigid, semirigid, or flexible compounds, and with or without color. The auto industry has accepted many cellular plastics for their numerous favorable properties, especially their weight-saving value. The polyurethane (PUR)* foams produced through reaction injection molding (RIM)* composited with reinforcing fibers (FR)* serve as hoods, seat frames, or doors.

Cellular plastics are widespread and include polystyrene foams for building construction, packaging and appliance insulation; cellular polyethylene for electrical wire insulation; polyimide foams with high heat resistance for aircraft; epoxy foams for flotation devices; cellular silicone for electrical and electronic encapsulation, chemical resistant fillers, and structural insulation; flexible and rigid polyvinyl chloride (PVC)* for cushioning, upholstery, carpet backing and toys; and foamed acrylics for decorative materials. Syntactic glass-filled foams of polyester are used in sports equipment for composite tennis rackets, bowling balls, floating golf balls, helmets, skateboards, and skis.

Fiber-reinforced plastics (FRP), along with many laminated and filled plastics, comprise a large group of plastics that fit diverse engineering and general-purpose applications. For both woven fabrics and as reinforcing fibers in FRP, the *aramid fiber* is a recently developed plastic of superior qualities. Under tensile loads this polymer equals steel when compared to similar sizes but has a superior strength-to-weight ratio. The superior strength develops as a result of the stringing out of the carbon atoms, which line up rather than coiling like a spring in most polymers. The built-in stiffness of the C—C bond greatly resists stretching under tensile loads. See Module 8 for further discussion of this remarkable fiber with the trade name Kevlar.

Table 10.9 provides comparisons of various plastic classification systems. In using the table it is important to realize that these are general data, and any given plastic may have considerable grade variations due to fillers, additives, and structure. The table also provides the basic molecular structure, chemical formula, American Society for Testing and Materials (ASTM) abbreviations, some selected trade names, and typical uses of the plastics listed. Note that carbon (C) is the backbone for all those plastics shown except silicone, which has an inorganic silicon (Si) and oxygen

---

*American Society for Testing and Materials (ASTM) abbreviations. These abbreviations are frequently used in journals, tables, and handbooks.

(O) backbone. In the polyolefins group, ionomers come from ethylene gas combined with inorganic compounds consisting of sodium (Na), zinc (Zn), magnesium (Mg), or potassium (K) salts that bond both covalently and ionically. Ionomers are resilient, oil resistant, and tough, while having very high transparency.

6.4.2 Properties of Plastics

In earlier discussion the internal structure of polymers was shown to dictate properties. The nearly infinite variations of structure in plastics provide a wide range of properties that continue to grow annually as the demands of technology prompt new discoveries in the manipulation of polymeric structures. These properties can be considered in terms of mechanical, chemical, thermal, electrical, and general. Additives and fillers so significantly affect the properties of plastics that one must turn to a handbook or manufacturers' specifications for a given plastic to ascertain its properties.

6.4.2.1 Additives (modifiers).   Most plastics are truly composites because of the many additives compounded with the resin to enhance its properties. Beyond the catalysts (also known as accelerators or initiators) that are added in minute quantities to start polymerization, additives serve such functions as to increase processability, reduce oxidation (corrosion), add color, reduce molecular weight, increase flexibility, reinforce, retard flammability, or increase electrical conductivity. Most polymers will absorb foreign elements that can cause degradation. *Stabilizers* are added to prevent this. Ultraviolet radiation and oxygen (ozone) can alter the chemical bonds of plastics so that *free radicals* develop within chains. The free radicals are segmentations within the polymer that can easily combine with other elements to cause a breakdown of the plastics structure and severely affect properties. Crosslinks may form to produce a more brittle plastic or cause the disruption of chain structure, which will limit tensile strength. *Carbon black,* produced through the burning of *carbonaceous* materials such as gas or oil, will block out ultraviolet light while also strengthening the plastic. Such metals as barium, cadmium, and lead in compounds are used as stabilizers. Amine and phenol chemicals serve as sun screens as they interact with photons.

Colorants. Both organic *dyes* and inorganic (metal-based) pigments add color to plastics, and some serve dual roles such as stabilizing. Generally, the inorganic pigments will withstand high temperatures without charring or fading. *Pigments* disperse rather than dissolve in plastics and reduce the transparency of the material. The pigments also hide flaws such as air bubbles, making it difficult to judge the quality. While nearly an infinite range of color, transparency, translucency, and opacity is possible, the Food and Drug Administration allows only certain types of colorants for plastics that make contact with food and drugs.

Plasticizers. Plasticizers are additives that increase flexibility, while crosslinking agents such as organic peroxides cause hardening to produce free radicals. Plasticizers reduce the attraction (secondary valence bonds) between polymer chains and are normally solvents such as alcohol. Polyvinyl chloride (PVC) can be very hard and rigid for water and sewage piping. The addition of a plasticizer such as carbon alphates alcohol produces a resin of low volatility that is used in the slush or dip molding of such very flexible producet as tool handle coatings, rain boots, or doll parts. *Flame retardants* have become more important in the United States as our search for greater product safety broadens. Elements such as boron, nitrogen, chlorine, antimony, and phosphorus reduce the flammability of plastics by preventing oxygen reactions and improving charring. *Charring* is seen when wood burns and leaves a protective residue or ash that slows burning; this is known as *ablation*. Flame retardants can cause problems by reducing flexibility, tear resistance, tensile strength, and heat deflection. Fluorocarbons, polyvinyls, and polyimide plastics can offer low or nonburning properties without the use of flame retardants.

Fillers. Fillers improve plastics by increasing bulk, tensile strength, hardness, abrasion resistance, and rigidity; they improve electrical and thermal properties, appearance, and chemical resistance while either increasing or decreasing specific gravity. Common fillers are woodflour, quartz, glass spheres, talc, wollastonite ($CaSiO_3$), calcium carbonate ($CaCO_3$), carbon black, clay, and alumina trihydrate (ATH). For example, glass spheres, clay, or calcium carbonate are added to sheet molded compounds (SMC) or bulk molded compounds (BMC) to decrease cost while increasing the rigidity of polyester resins.

Other additives include *antistatic agents, coupling agents* to aid bonding between plastics and inorganic materials in composites, *foaming agents* to produce cellular plastics, *heat stabilizers* for processing and end product durability, *lubricants* to decrease friction and resin melt during processing, *mold release agents, preservatives* to retain physical and chemical properties and in some cases prevent growth of bacteria and algae, and *viscosity depressants* to reduce viscosity during processing.

6.4.2.2 Mechanical properties.

As discussed previously, (see Figure 6-8a), variations in properties are achieved through variations of molecular weight and crystallinity, With mechanical properties, an increase in crystallinity or density, plus increase in molecular weight, produces corresponding increases in tensile strength, hardness, creep resistance, and flexural strength, but decreases the impact resistance and percentage of elongation. Additives, fillers, and reinforcers also change mechanical properties to a great extent.

Table 6-1 lists a few selected plastics that reveal the range of *tensile strength* for plastics and also includes other engineering materials for comparison. Because many different varieties and grades are available within a certain plastics group, there is a wide range of tensile strength for plastics of the same name. For example, some flexible cast epoxies have a tensile strength lower than 10 MPa while a cycloaliphatic epoxy casting resin exceeds 100 MPa. The values in Table 6-1 are for specimens tested at standard temperature and pressure (STP). Many plastics rapidly lose strength at relatively low temperatures (20°C) while others, such as polyimides, can maintain full strength near 500°C over short periods while also withstanding cryogenic temperatures around −300°C. Figure 6-12 shows a service application of ABS encountering cold temperatures and contact with food. Table 6-1 shows that other engineering materials have far greater tensile strengths than the engineering plastics, but the low

TABLE 6-1 RANGES* OF TENSILE STRENGTH AT ROOM TEMPERATURE FOR SELECTED PLASTICS COMPARED TO OTHER MATERIALS

Tensile Strength in MPa (ksi)

Column scale: 0 10(1.5) 20(2.9) 30(4.4) 40(5.8) 50(7.3) 60(8.7) 70(10.2) 80(11.6) 90(13.1) 100(14.5) 150(21.8) 175(25.4) 200(29.0) 225(32.6) 250(36.3) 500(72.5) 750(108.8) 1000(145.0)

Plastics	Range (X marks across scale)
ABS[e]	XXXXXX (≈30–60)
Acetals[e]	XXXX (≈50–70)
Acrylics	XXXXX (≈40–70)
Alkyds	XXXXXXXXX (≈20–70)
Amino	XXXXX (≈50–80)
Cellulose acetate	XXXXX (≈20–50)
Cellulose butyrate	XXXXXXX (≈20–50)
Diethylene glycol bisallyl	XXX (≈30–50)
Epoxy	XXXXXXXXXXXXXXXXXX (≈20–100)
Fluorocarbons—TFE[e]	XXXXX (≈10–40)
Nylon[e]	XXXXXXXX (≈60–100)
Phenolic	XXXXXX (≈40–70)
Phenylene oxide[e]	XXX (≈50–70)
Poly(amide/imide)[e]	XXXXX (≈150–200)
Polycarbonate[e]	XXX (≈60–80)
Polyester—TS**	XXXXXXXXXXX (≈20–80)
Polyester—TP**	X (≈60)
Polyethylene—LDPE	XX (≈0–20)
Polyethylene—HDPE[e]	XX (≈10–30)
Polyethylene—UHMWPE[e]	XXXXX (≈20–50)
Polyimide[e]	XXXXXXXX (≈60–100)
Polypropylene[e]	XXXX (≈20–50)
Polystyrene	XXXXXXXXXX (≈20–70)
Polysulfone[e]	XXX (≈60–80)
Polyurethane[e]	XXXXXXXXXXXXX (≈10–80)
Silicones	XXXXXXXXX (≈20–60)
Vinyl—PVC	XXXXX (≈10–40)
Vinyl—PVDC	XXXXXX† (≈30–60) XXXXXXXXXXXXX‡ (≈150–350)

Comparison Materials

Material	Range (X marks across scale)
Aluminum alloys	XXXXXXXXXXXXXXXXXXXXXXX (≈100–500)
Alumina	XXXXX (≈150–250)
Brass	XXXX (≈200–250)
Cast iron	XXXXXXXXXXXXXXXX → (≈150–500+)
Fiber reinforced plastic	XXXXXXXXXXXXXXXXXXXXXXXXXXX → (≈100–1000+)
Glass	XXXXXXXXXXXXXXXXXXXX E & S fibers → (≈70–1000+)
Steel—plain carbon	XXXXX → (≈250–500+)
Stainless steel	XXXX → (≈200–500+)
Wood	X (≈0)

*Ranges reflect the varieties of the named plastic. Does include certain filled varieties but not fiber-reinforced plastics (FRP). Ranges for other materials are general.

**TS, thermoset; TP, thermoplastic; e, engineering plastic.

†Unoriented.

‡Oriented.

Figure 6-12 Plastics in appliances. ABS (acryonitrile butadiene styrene) serves as refrigerator door food liner because of the toughness at low temperature, stain and odor resistance. (Shell Chemical Company.)

density of plastics often yields a competitive or higher strength-to-weight ratio. Using specific strength as the comparison factor,

$$\text{Specific strength (meters)} = \frac{\text{tensile strength}}{\text{density (weight)}}$$

polyester pultrusion, fiber-reinforced plastics (FRP), have specific strengths around 35,000 m while stainless steel has a specific strength around 25,000 m. Other FRPs, dealt with in Module 8, have even higher specific strength ratios (e.g., up to 90,000 m for epoxy reinforced with carbon fibers). Table 6-6 can be used with the tensile strengths given in Table 6-1 to calculate other specific strengths for comparisons.

Plastics compared to steel and most other engineering materials are much softer. Table 4-2 shows various hardness scales in an approximate comparison for matching hardness of selected materials. Hardness comparisons of most common plastics are made with the Shore Durometer A and D method and Rockwell M (R_M) tests. The problem with hardness tests of plastics is that they do not closely correlate to wear or abrasion resistance as do most other materials: polystyrene has a Rockwell M value of 72 but scratches easily; diethylene glycol bisallyl carbonate marketed as CR39 (allyl diglycol carbonate) by PPG Industries has a Rockwell M of 95 to 100 but has abrasion resistance approximating glass. This makes it a competitor with glass because it has optical properties approaching glass. Methyl methacrylate under trade names such as Plexiglas and Lucite has average hardness of R_M 93 to 98 but abrasion resistance 30 to 40 times less than CR39[R] using a modified Taber test. *Abrasion resistance* and *wear resistance* are measured by Taber and other tests that determine the *percentage of haze* (loss of clarity) due to marking, or *percentage of material lost* through rubbing with abrasives such as aluminum oxide. Table 6-2 shows the results of Taber tests on several plastics and their hardnesses.

Toughness or *impact resistance* is generally better in thermoplastics than thermosets. As seen in Table 6-3, polycarbonate is the toughest of the transparent rigid plastics at 12 to 14 foot pounds per inch and far exceeds polystyrene (10.4 ft lb/in.) and acrylics (0.8 to 1.6 ft lb/in.). Polyurethane with an impact strength of 5 to 8 ft

TABLE 6-2 TABER TEST RESULTS

Plastic	Hardness	Abrasion Resistance (mg loss per 1000 cycles)
Nylon 6/6	R_R 114	5
Acetals	R_M 95	16
ABS	R_R 100	84
Polysulfone	R_R 120	20
Polyimides	R_M 98	0.08

lb/in. has gained acceptance as automobile bumpers and hoods. Both polyethylene and polyvinyl chloride (PVC) have very broad ranges of impact strength, as you may know from the thin-walled tough polyethylene milk jugs and PVC piping that have begun to replace much of the copper and cast-iron pipe used for water and sewage. The values given in Table 6-3 are notched specimens. Notching of most plastics severely limits their toughness, so design should avoid sharp corners or conditions in which parts subjected to impact might develop scratches or cuts. Instead of the notched Izod test, which may not give the most reliable results, some designers prefer

TABLE 6-3 COMPARATIVE IMPACT STRENGTH OF SELECTED PLASTICS

	Notched Impact Strength (Foot Pounds Per Inch)										
Plastic	0	2	4	6	8	10	12	14	16	18	20
ABS[e]		XXXXXX									
Acetals[e]		X									
Acrylics	XX										
Alkyds		XXX									
Aminos—melamines		X									
Cellulose acetate		XXXXXXXX									
Epoxy	X										
Fluorocarbon—TFE[e]			XXX								
Nylon 6/6[e]		XX									
Phenolic	X										
Phenylene oxide[e]				X							
Poly(amide/imide)[e]		XXX									
Polycarbonate							XXXXXX				
Polyester—PBT		X									
Polyester—PET	X										
Polyethylene—LDPE										XXX→	
Polyethylene—HDPE	XXX XXX XXX XXX XXX XXX XXX XXX XXX XXX										
Polyimide—GRP[e]										XXX	
Polypropylene[e]		XXX									
Polystyrene	X										
Polysulfone[e]		XX									
Polyurethane[e]				XXXXXXXX							
Silicones	X										
Vinyl—PVC	XXX XXX XXX XXX XXX XXX XXX XXX XXX										

[e]Engineering Plastic

 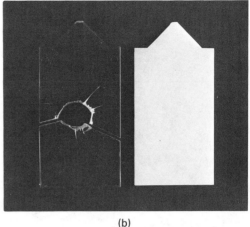

(a)   (b)

Figure 6-13 Tough plastics. (a) Nylon resin, among the toughest of *engineering plastics,* serves as motorcycle sprockets with wear characteristics that surpass conventional sprockets made of steel and aluminum. (b) While polycarbonate (left) is considered a tough plastic, the super-tough nylon (right) resisted falling dart that penetrated polycarbonate of the same thickness. (Dupont.)

data obtained from impacting with a falling weight on sheet plastic (ASTM D 3029), film (ASTM D 1709), and pipes and fittings (ASTM D 2444). Figure 6-13 reveals certain plastics compete well with metals where toughness is required.

6.4.2.3 Viscoelasticity.

Viscoelasticity is a property unique to polymers (plastics, elastomers, adhesives, and wood) and not found in metals or ceramics. Because of the viscoelastic property of plastics, engineering technologists must become involved in rheology (study of flow and deformation of matter) to determine why plastics will react to mechanical stresses in a different manner than metals. The property *viscoelasticity* incorporates two properties: viscosity and elasticity. *Viscosity* refers to the nature of a liquid's resistance to flow. Motor oil used in an automobile engine is rated by its viscosity; heavy oil such as 40 SAE (Society of Automotive Engineers) is thick syrup and is used in hot weather, while lighter oil (10 SAE) that flows much like water is used in temperatures below zero. As with oil, plastics become less viscous or flow easier with increased temperatures. This makes them easier to process but also means they lose strength with heat.

Elasticity was covered in Module 4 and refers to the ability of a material to return to its original size and shape once a load is removed. With plastics the two factors come into play so that a sudden impact on the plastic would not result in immediate strain or immediate and full recovery when the stress is removed. Rather, there is a viscous flow internally as the force is absorbed through shearing of molecular bonds, much the way a shock absorber works on an automobile. The shock absorber is a combination of *dashpot* (force gradually released through a slow bleeding of fluid) and spring. Even after all load is removed, the deformation is not fully recovered as with a metal that experiences elastic loading. The effect of heat on viscoelastic

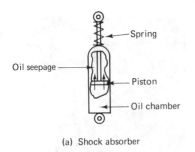

(a) Shock absorber

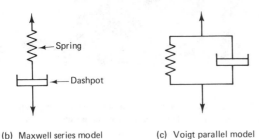

(b) Maxwell series model   (c) Voigt parallel model

Figure 6-14 Viscoelasticity. (a) Shock absorber. (b) Maxwell series model. (c) Voigt parallel model.

properties is discussed under thermal properties. Figure 6-14 illustrates the mechanical models (Maxwell and Voigt) that simulate viscoelasticity.

6.4.2.4 Thermal properties.

Many plastics lose their strength at relatively low temperatures. Continuous service temperature comparisons of plastics reveals most common plastics can endure temperatures no more than 150°C when under low or no stress. Slight increases in stress would reduce the continuous service temperature even more. Newer plastics, such as the polyimide thermosetting resins, resist intermittent heat up to 500°C with low stress and up to 250°C for thousands of hours. Graphite reinforced polyimides can withstand flexural stress to nearly 69,000 Pa at 250°C. Plastic composites have replaced aluminum pistons in some race cars because the aluminum with high thermal conductivity acts as a heat sink and begins to soften at around 150°C, while plastic composites of polyimide and epoxy with glass and graphite reinforcement withstand temperatures approaching 290°C and tremendous mechanical stresses, yet they offer weight savings over metals.

Glass transition temperature or glass point (T_g) is the point at which polymers act as glass or become frozen liquids. In cooling, the thermal mixing of the atoms and molecular chains slows, and the volume of the plastic decreases since they pack closer together. In packing, many polymers do not form crystals as do metals; instead they freeze into an amorphous glassy structure. Of course, highly crystalline polymers such as high-density polyethylene (HDPE) would not follow this pattern, and the degree of crystallinity of plastics would control T_g. The result of cooling to the glass point is that noncrystalline polymers become brittle. This occurs at or above room temperature for certain plastics, for example, polystyrene and acrylic at about 100°C. In crystalline plastics with low T_g, such as polypropylene at -10°C, flexibility and impact strength are maintained. In reheating an amorphous thermoplastic, it passes from the brittle glassy structure through its T_g point, and then turns into a tough rubbery form progressing to a softer and more pliable stage into a viscous liquid upon reaching the melting point (T_M).

Exact melting points are determined optically in crystalline and semicrystalline plastics as the *birefringence* (double refraction) is lost due to the change of structure from crystalline to glassy state. The earlier discussion on mechanical properties covered viscoelasticity. One can see how the T_g will affect this property. A linear amorphous plastic will be quite brittle and not exhibit viscous flow below its T_g but be quite rigid and elastic. As the degree of crystallinity increases in plastics, they become more viscous and less capable of recovering from deformation under load as they possess less elasticity.

Time becomes an important factor in the loading of plastics. A stressed linear amorphous polymer at temperatures above T_g over sufficient time will begin to deform without additional loading or heat as the molecular chains become untangled through viscous flow; this is *stress relaxation*. A network thermoset at relatively low temperatures that is under constant stress over a long period of time will begin to deform and continue deforming with progressively less stress; this is *creep*. The softening temperature of polymers can also be increased through an increase in average molecular weight or polymer chain lengths. Thermoplastics are far more prone to creep than thermosets.

Thermal conductivity is low in most plastics, which makes them valuable as thermal insulators. In the cellular or foamed state, air cavities, which do not carry heat or cold, improve even further the insulating properties. However, the rate of *thermal expansion* is quite high for most plastics, generally ten times as much as metals. Table 6-4 provides a comparison of the rate of thermal conductivity and coefficient of thermal expansion for selected plastics and compares them with other materials.

The range of *flammability* for plastics is quite wide. For instance, phenolics, polyimides, and fluorocarbons without fire-retardant additives are considered as nonburning, whereas the cellulosics are highly flammable. Certain plastics such as polyurethane and vinyls give off highly toxic fumes when burned. The vinyl siding used on housing ignites above 370°C, whereas wood siding ignites at about a 100° lower temperature. Generally, thermoplastics are more flammable than thermosets. Thermoplastics can be ground into chips and melted for recycling. While thermosets have the potential for recycling through the use of chemical solvents, the economics may not be justified. However, thermosets can be ground up and used as a composite filler. The high thermal energy possible from thermosets, nearly 3 MJ/kg (megajoules per kilogram), has value as an energy source in which scraps are burned to generate heat for steam turbine generators in the production of electrical energy. As with wood, plastics develop a char layer that serves as an *ablative* or protective shield, which insulates the unexposed area. Nylon and phenolic resins were used on the NASA spacecraft as ablative heat shields for reentry into the earth's atmosphere. While the polymers did burn when exposed to temperatures over 6500°C for several seconds, the charred ablative shield protected the spacecraft because of its thermal insulating properties.

TABLE 6-4 COMPARISONS OF THERMAL PROPERTIES FOR SELECTED PLASTICS AND OTHER MATERIALS

Coefficient of Thermal Expansion	10^{-6} m/m/K				
	60	120	180	240	300
Plastics					
ABS[e]	X				
Acetals[e]		X			
Acrylics	XXXXXXXXX				
Cellulose—CA			XXXXXXXXXXXX		
Epoxies	XXXXXX				
Fluorocarbons—PTFE[e]		X			
Nylon		X			
Phenolic	XXX				
Polycarbonate	XXXXXXXXX				
Polyester—PBT		XX			
Polyimide	X				
Polyethylene—LPDE			XXXXXX		
Polyethylene—HDPE[e]			XXXXXX XXXXXX XXX XXX XXX XXX XXX		
Polypropylene		XXXX			
Polystyrene—SAN		X			
Polysulfone[e]		X			
Polyurethane	X				
Silicone[e]		XXXXXXX			
Vinyls (PVC)	XX				
Alumina	X	...			
Aluminum	XXX		. . . .		
Cast iron	XX	. .			
Copper	XX				. . . .
Glass	XX				
Rubber	X				
Silver	X				
Steel	X				
Stainless steel	X				
Wood	|	. . . .			
Zinc	X	. . . .			

Thermal Conductivity	W/(m · K)
	0 . . . . 10 . . . . 50 . . . . . . 100 . . . . . . 150 . . . . . . 200 . . . . . . 300 400

[e]Engineering Plastic

6.4.2.5 Chemical Properties.

Plastics do not corrode in the same manner as metals, but they are subject to deterioration and chemical attack. Whereas the corrosion of metals is determined by weight loss, a thermoplastic's deterioration is measured by weight gain as the attacking chemical combines with the plastic. The result is usually discoloration, swelling, or crazing (fine cracks) with corresponding loss of tensile, impact, and flexural strengths. Many of the composites used in space are fibers reinforced with plastics resins. The ability of the resins to resist intense ultraviolet radiation is still a topic of research. Long term space exposure of polymer composites on the space station will provide a real test (Figure 6-15). Scientists are also concerned about higher orbits that involve the van Allen radiation belt. Large space structures such as solar collectors or space stations will be subjected to the bombardment of atomic and subatomic particles.

As with mechanical properties, *chemical resistance* will decrease in plastics as temperatures increase. Table 6-5 provides data for plastics tested at room temperature for their degree of resistance to weak acid, organic solvents, and water absorption. Some common strong acids include hydrochloric and nitric, and weaker acids are lactic, boric, and citric. Common organic solvents include gasoline, alcohol, and acetone. The *weatherability* comparison involves several chemical stresses, including heat, moisture, ultraviolet light, and chemicals in the air such as ozone (allotropic form of oxygen) and hydrochloric acid (see Figure 6-16). Notice the fluorocarbons with their inertness have great resistance to weak acids, organic solvents, water absorption, and weathering. Acetal homopolymer resins have great resistance to acetone at room temperature, but at elevated temperatures (65°C) they are unacceptable for service nor do they perform well when in environments with strong acids. The copolymerization of acetal resins greatly improves their resistance to inorganic liquids at elevated temperatures, so they are utilized as plumbing products such as valves and pumps. High-density polyethylene (HDPE) has great resistance to acids, water absorption, and weathering, which, coupled with its low cost, makes it a good candidate for numerous packing applications such as blow-molded bottles used for the home and industry; it also serves as electrical wire insulation, oil storage tanks, and recreational body vehicles. Acrylics (Lucite and Plexiglas), as seen in Table 6-5, have great resistance to *weathering*, which provides good service as exterior

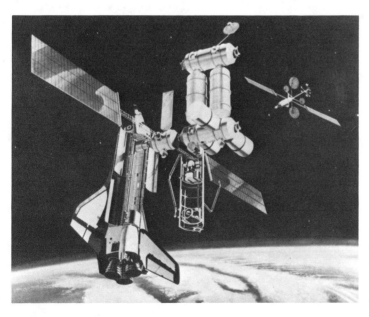

Figure 6-15 An artist's concept drawing for the U.S. manned Space Station. A number of polymer composites are used in constructing the station. A Space Shuttle is shown docked to the Space Station.

(Exposures at room temperatures except weatherability)

	Weak Acid Resistance			Organic Solvent Resistance			Water Absorption Resistance			Weatherability		
	Little	Fair	Great	Little	Fair	Great	Little	Fair	Great	Little	Fair	Great
ABS[e]	XXXXXXX			XXXXXX			XXXX			XXXXX		
Acetals[e]	XXXX			XXXXXXXXX			XXXX			XXXXXXXX		
Acrylics	XXXXXXXX			X			XXXXXX			XXXXXXXXX		
Alkyds	X			X			XXXXXXXX			XXXXXXXX		
Aminos—Melamides	XXXX			XXXXXXXXX			XXXXXXXXX					
Cellulose acetate	XXXXXX			XXXXXX			X			XXXXXX		
Cellulose butyrate	XXXX			XXXXX			XX			XX		
Epoxy	XXXXXXXXX			XXXXX			XXXXXXXX			XXXXXXXXX		
Fluorocarbons (PTFE)[e]	XXXXXXXXX			XXXXXXXXX			XXXXXXXXX			XXXXXXXXX		
Nylon 6/6[e]	XXXX			XXXX			X			XXXX		
Phenolic	XXX			XXX			XXXX			XXXX		
Phenylene oxide[e]	XXXXXXXXX			XXXXX			XXXXXXXX			XXXXXXXXX		
Poly(amide/imide)[e]	XXXXXXX			XXXXXXX			XXXX			XXXXX		
Polycarbonate	X			X			XXXXXXXXX			XXXXX		
Polyester—(TS)	XXXX			XXXX			XXXXXX			XXXXXXXX		
Polyester—(TP)	XXXXXXXX			XXXX			XXXXXXXXX			XXXXXXXXX		
Polyethylene—(LDPE)	XXXXXXXXX			XXXXX			XXXXXXXX			XXXXXXXX		
Polyethylene—(HDPE)[e]	XXXXXXXXX			XXXXX			XXXXXXXXX			XXXXXXXXX		
Polyethylene—(UHMWPE)[e]	XXXXXXXXX			XXXXX			XXXXXXXXX			XXXXXXXXX		
Polyimide[e]	XXXXX			XXXXX			XXXXXXXXX			XXXXXXXXX		
Polypropylene[e]	XXXXXXXXX			XXXXXXX			XXXXXXXXX			XXXXX		
Polystyrene	XXXXXXX						XXXXXXXX					
Polysulfone[e]	XXXXXXXX			XXXXXXXXX			XXXXXXX			XXXXX		
Polyurethane[e]	XXXX			XXXXXXXXX			XXXXXXX			XXXXX		
Silicones	XXXX			X			XXXXXX			XXXXXXXX		
Vinyl—(PVC)	XXXXXXXXX			XXXXX			XXXXXXXXX			XXXXXXXXX		
Vinyl—(PVDC)	XXXXXXXXX			XXXXX			XXXXXXXX			XXXXXXXX		
Comparison Materials												
Aluminum	XXXXXXXX			XXXXXX			XXXXXXXX			XXXXXXXX		
Alumina	XXXXXXXX			XXXXXXXXX			XXXXXXXX			XXXXXXXXX		
Brass	XXX			XXXXXXXXX			XXXXXXXX			XXXXXXXXX		
Cast iron	X			X			XXXXXXXX			XXXXXXX		
Fiber reinforced plastics vary with plastic matrix from little to great												
Glass	XXXXXXXXX			XXXXXXX			XXXXXXXXX			XXXXXXX		
Steel-plain carbon	X			X			X					
Stainless steel	XXXXXXXXX			XXXXX			XXXXXXX			XXXXXXXXX		
Wood	XXXXXXXXX			XXXXXXXXX			X			XXXX		

[e]Engineering Plastic

Figure 6-16 Chem suit used when entering chemical tanks uses face shield of optical grade plastic made of CR-39℠ monomer which resists solvents, acids, scratching, and impact. (PPG Industries.)

windows since their impact resistance is better than soda-lime glass and they transmit light (92%) equally as well as fine optical glass. Certain plastics with little *water absorption resistance* such as nylon 6/6 are dried before processing so the absorbed moisture will not cause corrosion of machinery. Moisture absorption causes swelling in plastics, which creates problems in holding dimensional accuracy. When high accuracy must be maintained, plastic of low moisture absorption is required. Polyester used as the resin in fiber glass has great water absorption resistance and serves well in marine applications such as boat hulls and surf boards.

6.4.2.6 Density.   The density (kilograms per cubic meter) of plastic is generally lower than other engineering and general-purpose materials. Table 6-6 shows that only magnesium and some wood have as low mass per unit volume as plastics. Even glass-reinforced plastics (GRP) are lighter than aluminum and steel, while their strength approaches or in some cases equals these popular engineering metals. CR-39℠ (allyl diglycol carbonate) (Figure 6-16) developed by Pittsburgh Plate Glass for competition with optical glass has a density of about one-half that of glass. This plastic is a practical substitute for glass in larger stylish eyeglasses and for industrial safety glasses because it combines light weight with impact resistance and is superior to acrylics in wear resistance.

6.4.2.7 Specific gravity.   Specific gravity is the ratio of the mass of a measured volume of any material to the mass of an equal volume of water at a standard temperature. This ratio is often used for weight comparisons. Specific gravity ratios for plastics range from below 0.06 for foams to over 2.0 for fluorocarbons, as compared to 0.5 for softwoods, 0.7 for hardwoods, 2.2 to nearly 4 for glasses and ceramics, nearly 3 for aluminum, and around 8 for steels.

6.4.2.8 Optical properties.   A number of plastics have optical properties comparable to glass while offering impact strength superior to most glasses; in addition, they process more easily and when broken do not produce sharp splinters.

TABLE 6-6 MASS DENSITY COMPARISONS OF SELECTED PLASTICS AND OTHER MATERIALS (Kilograms per cubic meter, pounds per cubic inch)

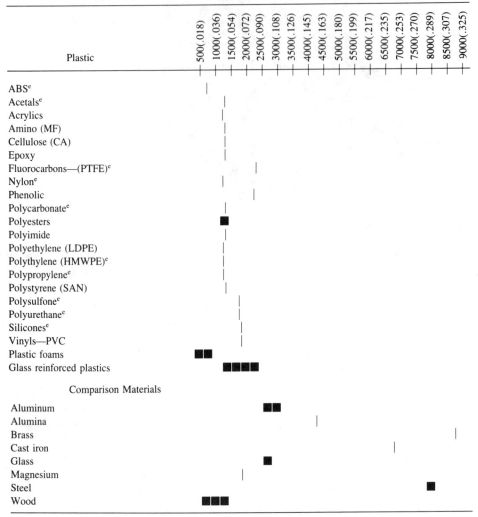

Plastic — scale values: 500(.018), 1000(.036), 1500(.054), 2000(.072), 2500(.090), 3000(.108), 3500(.126), 4000(.145), 4500(.163), 5000(.180), 5500(.199), 6000(.217), 6500(.235), 7000(.253), 7500(.270), 8000(.289), 8500(.307), 9000(.325)

ABS[e]
Acetals[e]
Acrylics
Amino (MF)
Cellulose (CA)
Epoxy
Fluorocarbons—(PTFE)[e]
Nylon[e]
Phenolic
Polycarbonate[e]
Polyesters
Polyimide
Polyethylene (LDPE)
Polythylene (HMWPE)[e]
Polypropylene[e]
Polystyrene (SAN)
Polysulfone[e]
Polyurethane[e]
Silicones[e]
Vinyls—PVC
Plastic foams
Glass reinforced plastics

Comparison Materials

Aluminum
Alumina
Brass
Cast iron
Glass
Magnesium
Steel
Wood

[e]Engineering Plastic

Opacity develops in plastics with increase in crystallinity, whereas an amorphous structure produces transparency. Low-density polyethylene film such as sandwich bags is clear, while HDPE for detergent bottles is opaque. Acrylics are available in an unlimited range of colors and are as transparent as the finest optical glass (light transmission equals 92%) or can be made opaque or through a full range of translucencies. Acrylics have a *refractive index* of 1.49 compared to 1.52 for soda-lime glass, 1.47 for borosilicate glass, and 1.46 for 96% silica glass. (Refractive index indicates the ratio of the speed of light through the materials compared to light traveling through a vacuum. Higher values indicate greater bending of the light.) The limits of transparency of some other plastics follow:

Acrylics	92%	Polyethylene	80%
Cellulose	88%	Polypropylene	90%
Ionomer	92%	Polystyrene	92%
Polycarbonates	90%		
Amino	29%*	ABS*	33%

*Translucent.

Translucent plastics transmit light but objects cannot be clearly seen.

Certain plastics, such as acrylics, polyesters, cellulosics, and polystyrene, show colorful stress concentrations when viewed with a polarized light filter. This photoelastic effect is used as a design tool. Photoelastic plastic is shaped to a specific design and stressed under polarized light to determine what areas will receive the greatest stress, thus allowing for added material or redesign to handle expected loads (Figures 6-17 and 18). As with glass, many plastics can effectively bend light and serve as *light pipes* for optical fibers used in medical applications, signs, telecommunications, and plastic art.

6.4.2.9 Electrical properties. Advances in materials have fostered many technological developments. The electrical and electronic developments over the past 75 years owe much to the continuing breakthroughs in plastic materials. Superior insulating properties coupled with good heat resistance of silicones and fluorocarbons lead to large reductions in the weight of electrical motors. Epoxies serve to encapsulate

Figure 6-17 Photoelastic plastic. A demonstration model to show stress concentrations. (Measurement Group, Inc., Raleigh, N.C.)

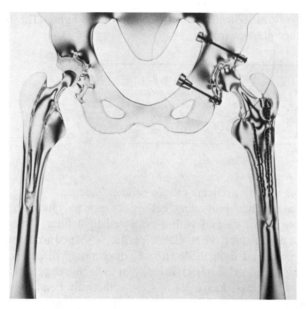

Figure 6-18 Actual model to analyze stress concentration resulting from pins inserted in bone for replacement joint. (Measurement Group Inc., Raleigh, NC.)

electronic components subjected to temperatures as high as 150°C, corrosive chemical environments, and high vibrations and shocks. Phenolics have the oldest history as electrical insulators and find wide use as housings for automotive ignition parts (coils and distributors), switches, receptacles, terminal blocks, and bulb bases.

The *dielectric strength* (maximum voltage that a dielectric can withstand without rupture) of most plastics makes them the logical choice as insulators. They range from 79×10^6 volts per meter (V/m) for fluorocarbons to 12×10^6 V/m for certain phenolics. These values compare to 19×10^6 V/m for porcelain and 12×10^6 V/m for alumina. The *dielectric constant* (ability to store electrical energy) of certain plastics puts them into the condenser category. Condensers in electrical circuits maintain voltage with less fluctuation. At 60 hertz, comparative dielectric constants (K) compare as follows: polyvinyl fluoride, 8.5; cellulose acetate, 4.0 to 5.0; polyethylene (LDPE), 2.25 to 2.35; and polystyrene 2.45 to 4.75; as compared to alumina, 10 to 8; mica, 8.7 to 5.4; and glass, 4.6 to 3.8.

Arc resistance is the ability of a material to withstand the arcing effect of an electric current. Insulators, terminals, and switches are subjected to arcing effects, which can burn through or damage the material's surface. The arc test (ASTM D495) measures the total elapsed time in seconds an electrical current must arc in order to cause failure. The plastic may carbonize and become a conductor, burst into flames, produce thin wiry lines (tracks) between electrodes, or become incandescent (glowing hot) conductors. Arc resistance for unfilled polyimide is 125 seconds; melamines, 110 to 150; FEP fluorocarbon, over 300; macerated fabric and cord-filled phenolic show tracking, thermosetting cast polyester, 100 to 125; and polyethylene (LDPE), 135 to 160.

6.4.3 Cost of Plastics

In the selection of a material as a designer or a consumer, cost is of paramount concern. Assuming properties are equivalent and other features such as appearance are similar when comparing materials, then cost is usually the overriding factor in the selection of one material over the other. Often a trade-off is made when costs are higher for one material than another. For example, epoxy reinforced with graphite fibers offers tensile strength nearly three times that of stainless steel; however, the cost for stainless steel may be 10 or 20 times less than the epoxy composites. If weight is important, as in aircraft, the fact that the stainless steel is over five times as dense as the graphite epoxy may cause the selection of the reinforced plastic. These comparisons are for exotic materials. More common materials such as plain carbon steel, which is used more frequently, find strong competition from low-cost plastics such as polyethylene, polypropylene, and vinyls.

6.5 ELASTOMERS: NATURAL AND SYNTHETIC RUBBERS

Elastomers refer to that group of polymers that exhibit rubbery or elastic behavior and include *natural rubber* and *synthetic rubber*. The American Society for Testing and Materials (ASTM) defines elastomers in ASTM D 1566-66T as . . . "Macromolecular material that returns rapidly to approximately the initial dimension and shape after substantial deformation by a weak stress and release of the stress." Elastomers will stretch from 200% to over 900% their original length at room temperature and return to their original state when the tensile stress is released. Natural rubber known as *hevea rubber* comes from the rubber tree *Hevea brasiliensis*. Columbus brought this bouncy material back to Europe with other curiosities from the New World, but nearly 300 years passed before it was given serious attention. The name rubber resulted from the material's ability for rubbing black pencil marks from paper. Synthetic rubber, as with plastic, is a hydrocarbon with petroleum as its major source. Development of the processing of synthetic rubber into a method suitable for high-volume production did not fully materialize until the beginning of World War II when the United States was cut off from the natural rubber supply in India and was forced to use large quantities of synthetic rubber.

6.5.1 Natural Rubber (NR) and Vulcanization

Natural rubber (NR), in the form of liquid resin secreted from the inner bark of the *Hevea brasiliensis* tree, is known as latex. Latex is not tree sap. It consists of isoprene molecules as seen in Figure 6-19a. Polyisoprene or rubber is formed through a natural polymerization in the tree (Figure 6-19b). The liquid latex dries into a thermoplastic material. The structure is an amorphous mass of coiled and kinked chains with constant

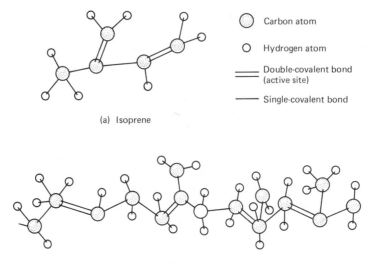

Carbon atom

Hydrogen atom

Double-covalent bond
(active site)

Single-covalent bond

(a) Isoprene

(b) Rubber (natural polyisoprene)

Figure 6-19 Natural rubber.

thermally induced motion of the atoms and chains. Rather than stretch when heated, rubber shrinks because the thermal agitation causes the chains to entangle and draw up. The structure provides the *resilience* for rubber to spring back into shape when compressed or stretched. When a tensile load is applied, the structure changes as the chains straighten (bond straightening) and stretch (bond lengthening) (Figure 6-20), and crystallinity is achieved in varying degrees depending on the amount of stress. With increased crystallinity comes a greater strength, increased rigidity, and increased hardness. The structural changes in rubber make it good for tires because the portion of the tire under high stress is crystalline and provides support for the vehicle it is carrying, while at the same time the portions of the tire not under high stress are still resilient and absorb shock from bumps in the road.

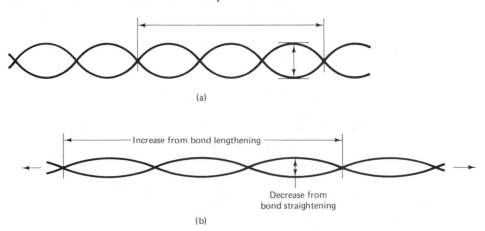

(a)

Increase from bond lengthening

Decrease from
bond straightening

(b)

Figure 6-20 Effect of tension on bonding. (a) No load. (b) Tensile load.

The problem with natural rubber is that it is too soft and has too many reactive sites (double bonds) which cause rapid oxidation and *dry rot*. It is also somewhat plastic and will not recover from high stress. In 1839, Charles Goodyear discovered through the addition of sulfur to natural rubber compounds that it was possible to increase hardness and reduce susceptibility to oxidation and reaction with other chemicals. The sulfur *vulcanizes* the rubber or changes it into a thermosetting polymer by linking together the molecular chains at their double bonds (Figure 6-21). Two sulfur atoms per pair of isoprene mers are needed, and with 5% of the mer pairs crosslinked

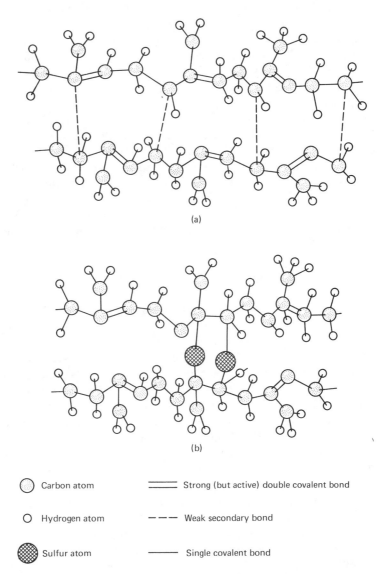

(a)

(b)

⬤ Carbon atom   ═══ Strong (but active) double covalent bond

◯ Hydrogen atom   --- Weak secondary bond

▨ Sulfur atom   ── Single covalent bond

Figure 6-21 Vulcanized rubber. (a) Natural rubber—thermoplastic. (b) Vulcanized (crosslinked) rubber—thermoset.

a flexible resilient rubber exists. Further crosslinking increases hardness. Vulcanization of only sulfur to rubber requires several hours and temperatures around 145°C. However, accelerators and activators added to the compound will result in vulcanization within minutes.

In addition to natural rubber, sulfur crosslinks polybutadiene (BR), polyisoprene (IR), and acrylate (ACM) elastomers. Along with sulfur, other crosslinking agents vulcanize certain synthetic elastomers. Butyl rubber (IIR) and ethylene-propylene copolymers (EPM) rely on sulfur and phenolic agents; zinc oxides form crosslinks in polysulfide (PTR) and polychlorapene (CSM). Vulcanized natural rubber has excellent flexural strength or deformability, tensile strength, and abrasion resistance, but it is attacked by petroleum oil, greases, and gasoline. Its value in auto tires comes from low heat buildup, discussed later. NR has superior overall engineering capabilities compared to synthetics. The choice of a synthetic rubber over natural rubber or of one synthetic over another boils down to specific properties required, price, and processability.

Since natural rubber comes from a renewable source and because it has excellent properties, natural rubber science and technology continue to make strides. Deproteinized natural rubber (DPNR) is an example of developments in improving this natural elastomer. By removal of ingredients with an affinity for water such as protein, polyols, and inorganic salts, fatigue resistance and other mechanical properties are enhanced. Research has been renewed in the United States into the use of guayule shrubs to produce rubber; they were grown in California during World War II. Only 5% to 10% yield was possible then; a 20% yield is required to justify the use of this raw material.

6.5.2 Synthetic Rubber

Even with the improved properties obtained through vulcanization, natural rubber has poor resistance to aging. It is attacked by ultraviolet light, oxygen, and heat because it still has many reactive sites. Due to these shortcomings and since the rubber tree only grows in special environments that were vulnerable to political sanctions, the search for a substitute produced several synthetic rubbers in the early to mid-part of the twentieth century. The number of synthetic rubbers has grown to include many special-purpose elastomers. Improvements in synthetic elastomers have also brought improvement in the additives and fillers and for natural rubber (NR). Almost identical to NR is synthetic polyisoprene (IR), except it has greater stretching ability. Synthetic rubbers have the same raw materials used in plastics. Starting with crude oil and natural gas, many formulations are possible. Styrene, the monomer for styrene-butadiene rubber (SBR), is derived from coal as vinyl benzene, $CH\text{=}CH_2$, a product of benzene and ethylene. Butadiene is derived from petroleum. The hydrocarbon is obtained in fractionation of cracking petroleum used for olefins, polymers, or gasoline. The 1-butene is separated and catalytically dehydrogenated in the vapor phase to produce butadiene. Figure 6-22 reveals the complexity of synthetic rubber production, which produces the raw synthetic rubber or *crumb* that is shipped to the processors who will make semifinished or finished products. During the final processing, the vulcanizing is preceded by *masticating* or kneading; the crumb or

PRODUCTION OF NATURAL RUBBER

PRODUCTION OF SYNTHETIC RUBBER

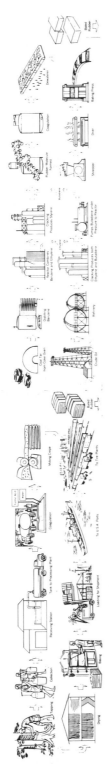

PROGRESSIVE STEPS IN THE MANUFACTURE OF A TIRE

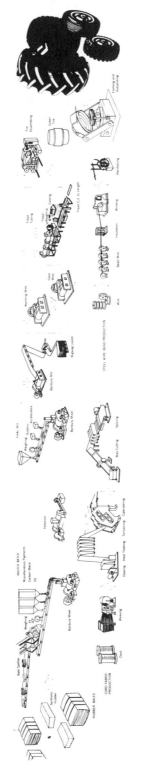

Figure 6-22 Natural and synthetic rubber in tire manufacturing. (The Firestone Tire and Rubber Company.)

rubber bales are ground up to soften them. In some cases, vulcanizing is achieved in the final shaping of the parts as with tire manufacture, and in other cases the products are placed under steam heat and pressure in an *autoclave*.

6.5.2.1 Additives (modifiers).

Additives (modifiers) to synthetic rubber are introduced at the initial processing stage and also during compounding. Soap and water are added to styrene and butadiene in the making of raw styrene butadiene rubber (SBR) to produce an emulsion that keeps droplets of the monomers from separating out. *Accelerators* and *activators* speed up sulfur vulcanizing, while *retarders* prevent vulcanizing before it is required. For those saturated synthetic elastomers that have no double bonds, *peroxides* promote sulfur vulcanizing. *Peptizers* soften raw rubber, *pigmenters* add color, *abrasives* provide abrasing action for such products as erasers, *hardeners* increase rigidity, and *antioxidants* and *antiozonants* prevent aging or dry rot from the sunlight and ozone.

Fillers help to increase bulk to the compound while holding down cost. Inert fillers include talc, chalk, and clay. Reinforced rubbers use many of the same *fillers* as plastics, including metals, glass, and polymer fibers. *Carbon black* also serves as a strengthening filler and imparts hardness.

6.5.2.2 Thermoplastic rubbers (TR) or elastoplastics.

Thermoplastic rubbers (TR) or elastoplastics do not crosslink as do most elastomers, even though their properties are similar to many rubbers. A copolymer of styrene and butadiene was introduced in 1965 as a thermoplastic rubber. The elastoplastics have the ability to soften when heated for processing and, upon cooling, become solid yet maintain their elastic behavior. Vulcanization is eliminated. The thermoplastic behavior develops as a result of the structure in which the chains of one monomer form links in the center of block molecules of other monomers. Figure 6-23 shows the domains in the middle have a hard linking effect, while the other end of the segments has elastomeric behavior. Upon heating through T_g, the hard centers soften to plasticize the rubber for processing. When cooled, the hard blocks reform to again link together the copolymers. This linking is especially effective at low temperatures.

Two-block, copolymer, thermoplastic elastomers manufactured under the trade name of Kraton include styrene-butadiene-styrene (SBS) and styrene-isoprene-styrene

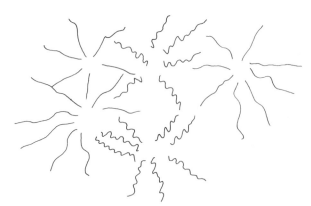

Figure 6-23 Polystyrene domains of thermoplastic styrene butadiene rubber.

Polymeric Materials   Module 6

(SIS). The styrene (S) occurs in two thermoplastic blocks as shown in Figure 6-24a on either end of a rubber block of butadiene (B) or isoprene (I). Rather than having randomly distributed monomers, as do most copolymers, block copolymers form uniformly distributed sections as blocks. Figure 6-24b shows that a network of block chains has formed, which consists of a highly uniform distribution of *polystyrene domains* and the separate butadiene rubber chains. The *rubber network,* which is physically joined, links the domains and cannot move because the domains are immobile, thus providing a physical rather than a chemical crosslinking. The physical crosslinking can be broken when heated, which makes the compound a thermoplastic. Kraton is therefore capable of continuous molding and remolding. The polystyrene domain consists of actual particles that become hard and glasslike at room temperature; they take on cylindrical, spheroidal, and platal configurations.

The major elastoplastics are polyester copolymers, olefinics, polyurethanes, and styrene copolymers. The wide service temperature ranges, low temperature flexibility, plus ease of painting, place thermoplastic rubbers in contention as a substitute for

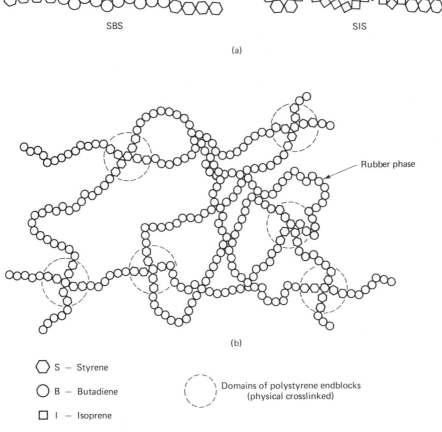

(a)

(b)

◯ S — Styrene

◯ B — Butadiene

☐ I — Isoprene

Domains of polystyrene endblocks (physical crosslinked)

Figure 6-24 Thermoplastic rubber "Network" structure. (a) Block copolymers. (b) SBS network.

thermosetting rubbers. Their ease of processability has given them wide acceptance in the auto industry as flexible bumpers and other exterior panels, steering wheels, hose covers, housings for seat belts and horns, wire covering, and oil seals. Other applications include housewares, toys, sporting equipment, adhesives, rainwear, footwear, and skate wheels. As with plastics, scrap thermoplastic elastomers can be ground and remolded.

6.5.3 Classification and Properties of Elastomers

Designations of elastomers by the American Society of Testing and Materials (ASTM D 2000 standard) and Society of Automotive Engineers (SAE) group them by type and class (Figure 6-25). *Type* reveals the maximum service temperature, which ranges from A (70°C) to J (275°C). *Class* indicates the maximum percentage of swell when immersed in oil and ranges from A with no requirement to K at 10% swell. Appendix 10.5.2 shows the class and type for selected elastomers. ASTM D 1418 establishes abbreviations for elastomers, also shown in the appendix. To illustrate, PTR, AK designates a polysulfide elastomer with a maximum service temperature of 70°C that can swell as much as 10%. Thiokol is the trade name for polysulfide. Further systems of designation indicate hardness and minimum tensile strength (Figure 6-25). ASTM AA625 designates (AA) natural rubber with Shore durometer hardness (6) meaning 60 plus or minus 5 points on the scale and (25) 2500 as the minimum tensile strength.

Appendix 10.5.2 provides a means for comparison of properties for common elastomers. NR is the standard for comparison of the synthetics, as can be seen by its overall superiority with excellent resistance to abrasion, tear, impact, electrical current, and water absorption, plus a tensile strength of 3100 Pa and a wide hardness range of 30 to 100 Shore A. The synthetic rubbers through designed formulation can surpass NR for special properties. NBR, ACM, FPM, and PTR offer excellent re-

Type	Test temp (°C)	Class	Maximum swell (%)
A	70	A	No requirement
B	100	B	140
C	125	C	120
D	150	D	100
E	175	E	80
F	200	F	60
G	225	G	40
H	250	H	30
J	275	J	20
		K	10

(a) ASTM D 2000/SAE J200 type and class

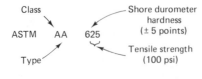

(b) ASTM designation example

Figure 6-25 Elastomer designation. (a) ASTM D 2000/SAE J200. (b) ASTM designation example.

(a)   (b)

Figure 6-26 (a) Gasoline pump hose made of epichlorohydrin rubber resists constant jerking and tugging during use, as well as the oily, greasy environment of service stations. The specialty elastomer helps the hose withstand constant exposure to gasoline and remains flexible in cold weather. (BF Goodrich.) (b) Elastomer bumper. Foamed elastomers such as polybutadiene not only rebound from impact but allow colorants to match bumper to autobody. (General Motors.)

sistance to oil, kerosene, and gasoline, while NR is not recommended for service when exposure to these chemicals is expected. Figure 6-26 shows a familiar application. For high-temperature service, SL and VE are excellent. Neoprene resists deterioration from weathering, oxygen, ozone, oil, gasoline, and greases and has good tear and abrasion resistance; it finds applications as soles and heels for work shoes and liners for chemical tanks and pipelines, and serves both as an adhesive and is applied like paint as a protective coating. Fluoroelastomers including FPM are the space-age elastomers that meet the demands of the aircraft and aerospace industry because of excellent resistance to heat and fluids, but they carry a high price tag.

Sec. 6.5   Elastomers: Natural and Synthetic Rubbers   **273**

NR has only fair *gas permeability,* which means it will allow some gas to pass through it. Butyl rubber (IIR) resists gas permeating and serves well as inner tubes, gas hoses, tubing, and diaphragms.

Hysteresis is an energy loss through heating in elastomers and creates a problem with many synthetic elastomers. If you stretch a rubber band and hold it against your cheek, you can feel the release of heat. The constant flexing of auto tires generates enough heat in such elastomers as SBR that if a tire were made wholly of this compound it would quickly deteriorate. NR, on the other hand, converts the stressing into elasticity and quickly rebounds with less heating. The high hysteresis of SBR has an advantage in preventing slippage on wet surfaces plus better abrasion resistance than NR, so it is used for the tread of tires while the NR goes into the sidewalls.

Elastomers possess extremely diverse properties, as do plastics. Thousands of additives and fillers are available and as many as 30 may be mixed into a single elastomeric compound. The range of hardness includes soft foam rubbers used for pillows to hard rubber (ebonite) battery cases. Figure 6-27 provides a comparison of hardness of selected elastomers and plastics. Foam rubbers have been formulated to respond to the body temperature and pressure. These foams developed for astronauts now serve as orthopedic support for bed mattresses and chairs because of their ability to provide even support, which reduces fatigue and soreness caused by prolonged body contact. Upon impact the foam firms up, and when heated, it softens. Joggers and other athletes use foam inserts in their shoes to reduce shock due to impact of running and jumping.

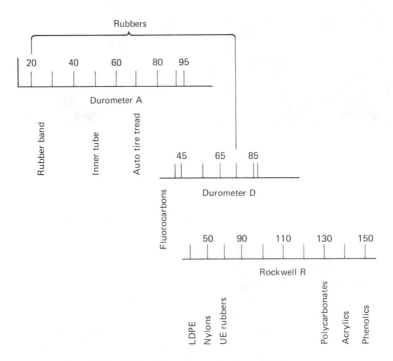

Figure 6-27 Comparisons of polymer hardness.

EPDM is a class of 13 types of materials that are a miracle of rubber technology in the form of very low cost, paintable elastomerics. Through compounding of carbon black, clay, processing oils, and crosslinking agents, the elastomer compound can be molded or extruded as thermoplastics or thermosets. Temperature insensitivity, impact resistance, high deformation recovery, dimensional stability, and distortion resistance make them a good choice for flexible exterior parts on automobiles, including bumpers and fascia.

Heat resistance is an important property in certain elastomer applications. Fluoroelastomers (FPM) can withstand temperatures over 340°C and continuous service temperatures up to nearly 290°C in exposure to oils, steam, and certain fluids. They have been improved to handle low temperatures down to around -35°C, at which point they become brittle. Among the silicon elastomers, room temperature vulcanizing (RTV) compounds have stable properties through a range of temperatures from nearly -53° to 250°C. Ethylene acrylics used for belts, seals, rollers, hoses, and gaskets can withstand hydraulic fluids, engine coolants, and hot oils for over a year at temperatures over 200°C. Chlorinated polyethylene elastomer, which is a random compounding of chlorine with high-density polyethylene, has good weather, hydrocarbon fuel, and oil resistance down to -15°C and up to nearly 150°C of continuous service. Grades of thermoplastic elastomers (TP) such as polyurethane provide good weathering as low as -15°C and as high as 120°C. Appendix 10.5.2 provides general ranges of maximum service temperatures for the groups of elastomers listed.

6.5.4 Cost of Elastomers

The cost of elastomers varies widely and has a tremendous influence on selection of one over another. The fluoroelastomer mentioned previously that is capable of temperatures over 340°C sells for over $4500 per kilogram (kg) on finished parts, whereas the price for SBR would range from $0.60 to $0.80/kg. Natural rubber sells for about 90 cents/kg, SI for $3.75 to $11/kg, and heat-resistant ethylene acrylics sell for $3 to $4/kg. While all fluoroelastomers do not cost as much as $4500/kg, they are generally the most expensive elastomers selling at around $28/kg, and as copolymers with silicones, most fluorosilicones reach prices of $65 to over $90/kg. Of the thermoplastics the following indicate general prices per kilogram: copolyesters, $3.20 to $3.65; olefins, $1.37 to $1.75; and styrenes, down to $0.90 and $1.37.

6.6 WOOD

Along with rock, wood was humanity's first material. The stick as a ready weapon or fuel for the fire required no major processing. Since those early beginnings, wood has been the target of much technology and still is being researched for methods of not only improving the yield of forests, but also ways of better using the tree and the many by-products of wood. Many plants possess woody or cellulosic substances, but the tree is our major source of wood. *Extractives* from the tree that contribute to such properties of wood as decay resistance, color, odor, and density are also useful in a number of industries. These extractives include tannin (tannic acid) used in processing leather and polyphenolics used to make phenolic plastics, plus coloring agents, resins, waxes, gums, starch, and oils. In addition to the fruit and nuts that are harvested, trees serve as the raw material for paper, cellulosic plastics, explosives, rayon fibers, films, lacquers and drilling muds, ethyl alcohol, food flavoring, concrete additives, and rubber additives. Wood was our first fuel, but with the technological development of coal, oil, and gases, wood became only a romantic fuel for most of the technologically advanced people of our world. However, tree products once again are being given serious consideration as a fuel. The ability to renew wood supplies through forestry increases humanity's interest in this material as nonrenewable resources such as oil, coal, and minerals rapidly become depleted.

6.6.1 Structure

Wood develops through photosynthesis as chlorophyll in trees uses carbon, hydrogen, and oxygen to produce sugar, starches, and cellulose. The dry wood used as an engineering material has an approximate composition of the following percentages by weight: cellulose (50%), lignin (16% to 33%), hemicelluloses (15% to 30%), extractives (5% to 30%), and ash-forming mineral (0.1% to 3%). *Cellulose* ($C_6H_{10}O_5$) (Figure 6-28a), a high molecular weight linear polymer, forms fibers that make up cell walls of vessels and ducts. *Lignin* (Figure 6-28b), an amorphous polymer, forms a matrix around the cellulose, much like the plastic matrix in fiber glass. Through

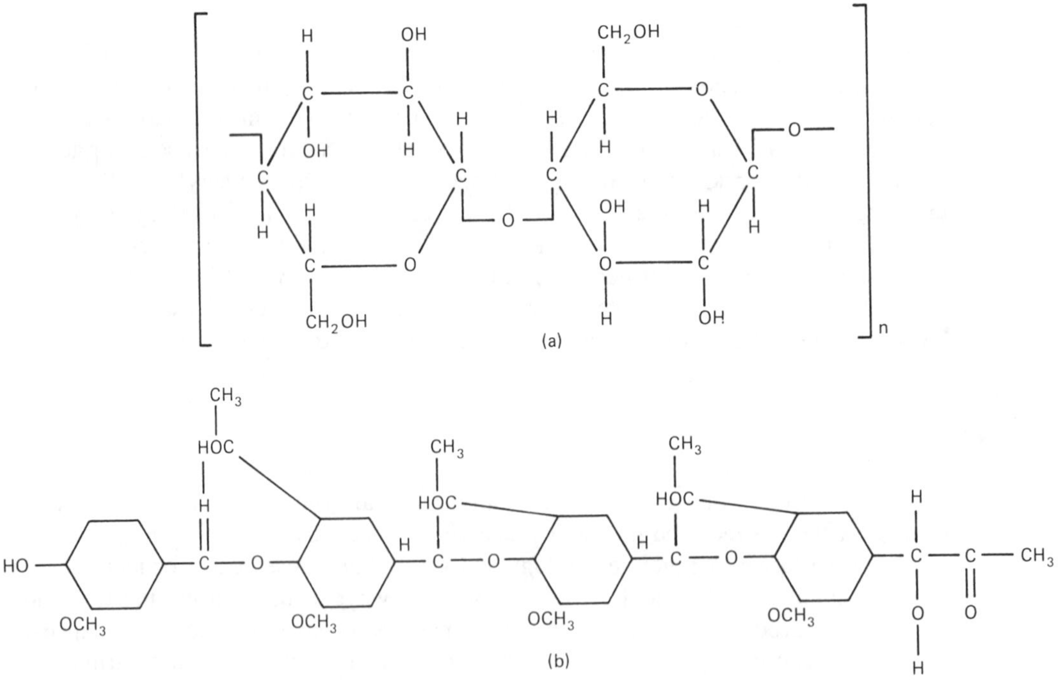

Figure 6-28   (a) Cellulose formula. (b) Lignin formula.

removal of lignin, wood can be broken down into fibers that are used in making paper and other synthetics. The cellulose forms cellular networks of ducts, vessels, fiber rays, and pits, which transport and store the extractives and minerals throughout the living tree. Depending on whether it is a hardwood or a softwood tree, the network differs. Softwoods are not always softer than hardwoods; Douglas fir, a softwood, is about twice as hard as basswood, a hardwood. *Softwood* species (Figure 6-29) include firs, pines, and spruces; they bear cones and have needles or scalelike leaves. *Hardwoods* are broadleaved and not normally evergreen; they include maples, oaks, birches, and mahogany. *Hemicellulose* is another polymer closely akin to cellulose; like cellulose it breaks down into sugars when chemically treated. The ash-forming minerals include calcium, potassium, phosphate, and silica. Extractives are removed by heating the wood in water, alcohol, or other chemicals; they represent a significant commercial value.

Most common engineering materials such as steel and concrete are isotropic, which means that generally the strength is the same in all directions because the materials are homogeneous. Wood is *anisotropic,* meaning it has greater strength in some directions than others. The anisotropy develops as a result of the way a tree grows, with such factors influencing the wood structure as cellular structure, branches, and bending by prevailing wind. A full discussion of all these factors is beyond this text, but the resulting tensile strength and stiffness of wood are greater in the radial direction than in the tangential direction (Figures 6-30 and 6-31). Rays that radiate from the pith outward on the tree tend to tie together the layers of cells (tracheids) growing longitudinally. This directional strength results partly from the complex structure of cell walls, with long polymer chains of cellulose that form layers running

Figure 6-29 Softwood block showing three complete and part of two other growth rings in the cross-sectional view (X). Individual cells can be easily detected in the earlywood (EW), whereas the smaller latewood cells are difficult to distinguish in the latewood (LW). Note the abrupt change from earlywood to latewood. The two longitudinal surfaces (R-radial; T-tangential) are illustrated. Rays which consist of food-storing cells are evident on all three surfaces. 70× (Dr. Wilfred A. Côté, SUNY College of Environmental Science and Forestry.)

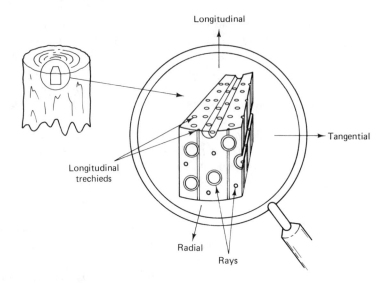

Figure 6-30 Anisotropy of wood. (U.S. Forest Products Laboratory, Madison, Wisconsin.)

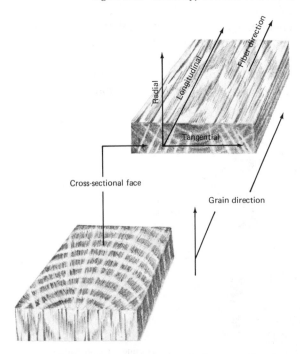

Figure 6-31 Grain direction. (U.S. Forest Products Laboratory, Madison, Wisconsin.)

in varying patterns to reinforce each other. Primary covalent bonding holds together these *microfibrils*, which are almost parallel to the cell axis. Weaker secondary bonds operate in the perpendicular axis.

Lumber consists entirely of dead cells. Figure 6-32 illustrates a cross section of a white oak tree trunk. The *cambium layer* produces all the growth of cells (wood and bark). *Sapwood* is a transitional wood of both living and dead cells. The sapwood's

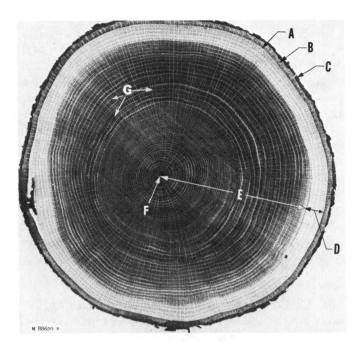

Figure 6-32 Cross section of white oak tree trunk: (A) Cambium layer is inside inner bark and forms wood and bark cells. (B) Inner bark is moist, soft, and contains living tissue; carries prepared food from leaves to all growing parts of tree. (C) Outer bark containing corky layers is composed of dry dead tissue; gives protection against external injury. (D) Sapwood, which contains both living and dead tissue, is light-colored wood beneath the bark; carries sap from roots to leaves. (E) Heartwood (inactive) is formed by a gradual change in sapwood. (F) Pith is the soft tissue about which the new growth takes place in newly formed twigs. (G) Wood rays connect the various layers from pith to bark for storage and transfer of food. (U.S. Forest Products Laboratory, Madison, Wisconsin.)

main role is to conduct sap and store food. *Heartwood* evolves from sapwood as the cells die and become inactive. The heartwood cells no longer transport sap but rather become storage cells. Sapwood of all species is highly susceptible to rotting, while heartwood of some species is protected from decay because it has taken on infiltrating materials that serve as preservatives. Further information on wood structure, plus other information on wood, is available in the excellent source *Wood Handbook: Wood as an Engineering Material.*

6.6.2 Wood Products

6.6.2.1 Commercial lumber.
Lumber is graded to provide the buyer with an indication of the strength and appearance of the lumber. Because wood is not homogeneous like most engineering materials, the grading provides an averaging of the quality of boards in a given grade. Visual grading spots defects including knots, stains, bark pockets, decay, shakes (cracks), pitch pockets (resin buildup), and checks (splits). *Softwood lumber* is a vital material for construction and manufacturing and is graded as construction and remanufactured lumber based on the American Lumber

Standard. *Construction grade* includes stress-graded, nonstress-graded, and appearance lumber. *Stress-graded* provides an indication of engineering properties and covers pieces of 2 to 4 in. nominal size, larger timbers (5 or more in. nominal thickness), posts, beams, decking, stringers, and boards (less than 2 in. nominal thickness). *Nonstress-graded* lumber includes a combination of allowable properties and visual defects that is used for boards, lath, batten, and planks. *Appearance lumber* that is nonstress-graded is cut to patterns such as trim, siding, flooring, and finished boards. *Remanufactured* grade is lumber that will receive further processing or secondary manufacturing such as cabinet stock, molding, and pencil stock. Grading specifies conformity of the lumber to its end use. *Hardwood lumber* is graded as factory lumber (amount of usable lumber in a piece), dimensional parts, and finished market products (maple flooring, stair treads, trim and molding). The furniture industry is the prime user of most hardwood grading.

A system for designating the degree of dressing or finish *surfacing* specifies one side surfaced (S1S), two sides surfaced (S2S), one edge surfaced (S1E), two edges surfaced (S2E), and various combinations, such as one side and one edge surfaced (S1S1E) or all sides and edges surfaced (S4S).

6.6.2.2 Composite wood. To gain the maximum use of trees and to achieve properties not possible from solid wood, composite woods have been developed for use in construction and manufacture. Among composite woods are laminated timber, impreg-wood, plywoods and veneer, particle board, hardboard, sandwiched materials, and insulation board.

6.6.2.2.1 Laminated Timber. Laminated timber is a product of the adhesive joining technology. Through adhesive bonding of pieces of lumber so that the grain of all pieces is parallel to the length of the timber, it is possible to produce straight to curved structural wood members of large size and outstanding strength. Laminated beams offer the architect flowing lines and the warmth and beauty of wood. Custom-ordered laminated beams have the capability to span more than 100 m of unobstructed space for sports arenas, convention centers, entertainment halls, and churches (Figure 6-33). Seasoning of lumber for laminated beams prior to gluing provides both improved strength, dimensional stability, and elimination of unsightly surface cracks and shrinkage. The ablative nature of the large beams provides an advantage from the hazards of fire, since the thin layer of char protects the interior of the beam and the wood does not become plastic at high temperatures, as do metal and plastic members.

6.6.2.2.2 Impreg-wood. Impreg-wood achieves a very stable lumber by the bonding of phenolic resins to the cell wall microstructure. This is accomplished through saturation of thin veneers with phenolic resin that is polymerized before the veneers are stacked into thick laminates. Applications of impreg-wood include sculptured models for the huge metal dies used to stamp automobile sheet metal parts. *Compreg*-wood employs impregnation of veneers with phenolic resin, but the polymerization is accomplished when the veneers are stacked and compressed into the desired final shape. The dimensionally stable, high-density, hard compreg laminate

Figure 6-33 Custom laminated beams. (U.S. Forest Products Laboratory.)

yields glossy finish throughout the material. Knife handles, bowls, jigs for manufacturing, and parts for textile looms use the compreg laminates.

6.6.2.2.3 Plywood. Plywood is another form of laminate that is most common in building construction. It is produced by stacking veneer with the grain direction in each layer at right angles to the next, beginning with the grain running the length of the panel (Figure 6-34a). This cross-laminating takes advantage of the fact that wood is much stronger in the direction of the grain than across the grain. Thus, plywood yields equalized strength in the plane directions, while also being shear, puncture, and split resistant. Numerous grades of plywood are available; Figure 6-34b shows the marking applied to three grades of plywood using the American Plywood Association standards. Letters assigned to veneer grades specify the quality of exposed plys (Figure 6-34c). Typing of plywood divides it into *interior* plywood, which has its laminates glued with moisture-resistant glue, including phenolformaldehyde, urea, and melamine. Veneers in interior plywood may be of lower quality than exterior type. In contrast, *exterior* plywood is glued only with phenolformaldehyde to provide a waterproof bond. The waterproof glue can withstand aging and boiling water. Marine plywood is the superior-quality exterior plywood. The exterior and interior typing are applied to both appearance- and engineering-grade plywood.

Thin appearance plywood in the form of paneling is very popular for walls in homes, mobile homes, recreational vehicles, and boats. There is a nearly unlimited variety of hardwood faced paneling. An inexpensive grade of Philippine mahogany

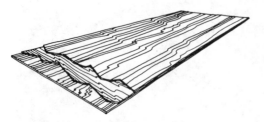

Figure 6-34 (a) Plywood. (b) Marking used by American Plywood Association. (c) Veneer Grades (American Plywood Association)

3 ply construction (3 layers of 1 ply each)

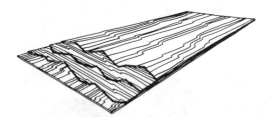

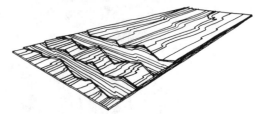

4 ply construction (3 layers: Plies 2 and 3 have grain parallel)

5 ply construction (5 layers of 1 ply each)

(a)

Typical APA Registered Trademarks

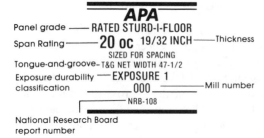

Panel grade —— RATED STURD-I-FLOOR
Span Rating —— **20 oc** 19/32 INCH —— Thickness
SIZED FOR SPACING
Tongue-and-groove– T&G NET WIDTH 47-1/2
Exposure durability —— EXPOSURE 1
classification ————— 000 —— Mill number
NRB-108
National Research Board
report number

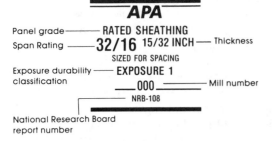

Panel grade —— RATED SHEATHING
Span Rating —— **32/16** 15/32 INCH —— Thickness
SIZED FOR SPACING
Exposure durability —— EXPOSURE 1
classification
————— 000 —— Mill number
NRB-108
National Research Board
report number

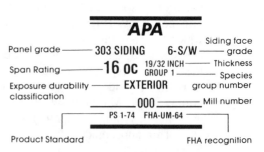

Panel grade —— 303 SIDING   6-S/W —— Siding face grade
Span Rating —— **16 oc** 19/32 INCH —— Thickness
GROUP 1 —— Species
Exposure durability —— EXTERIOR   group number
classification
————— 000 —— Mill number
PS 1-74   FHA-UM-64
Product Standard   FHA recognition

(b)

N	Smooth surface "natural finish" veneer. Select, all heartwood or all sapwood. Free of open defects. Allows not more than 6 repairs, wood only, per 4 × 8 panel, made parallel to grain and well matched for grain and color.
A	Smooth, paintable. Not more than 18 neatly made repairs, boat, sled, or router type, and parallel to grain, permitted. May be used for natural finish in less demanding applications.
B	Solid surface. Shims, circular repair plugs and tight knots to 1 inch across grain permitted. Some minor splits permitted.
C Plugged	Improved C veneer with splits limited to 1/8-inch width and knotholes and borer holes limited to 1/4 × 1/2 inch. Admits some broken grain. Synthetic repairs permitted.
C	Tight knots to 1-1/2 inch. Knotholes to 1 inch across grain and some to 1-1/2 inch if total width of knots and knotholes is within specified limits. Synthetic or wood repairs. Discoloration and sanding defects that do not impair strength permitted. Limited splits allowed. Stitching permitted.
D	Knots and knotholes to 2-1/2 inch width across grain and 1/2 inch larger within specified limits. Limited splits are permitted. Stitching permitted. Limited to interior (Exposure 1 or 2) panels.

(c)

Figure 6-34 (continued)

known as lauans receives many types of coatings and textures for decorative paneling. Thicker appearance-grade plywood is used for furniture and cabinets. Engineering-grade plywood serves as forms for casting concrete, underlayment for flooring and carpets, structural panel, roofing, decking, walls, and cabinets. Plywood siding has gained wide acceptance in contemporary-style buildings. Siding offers warmth, low maintenance, and good thermal and acoustical insulation, among other properties, and is available in various species with textures including smooth, deep grooves, brushed, rough sawn, and overlays. *Veneer* in thin sheet form is used in single thickness for baskets, boxes, and as ornamental inlays.

6.6.2.2.4 Wood-based Fiber and Particle Panel Materials. Wood-based fiber and particle panel materials include a variety of panels and boards used in building construction, packaging, furniture, and other manufactured products. *Particle board* panels, also known as reconstituted panels, are produced through the use of thermosetting resins such as urea formaldehyde and phenolformaldehyde, which serve as a matrix to bind together wood residues or shavings in the form of small wood flakes, wood flour, and additives (Figure 6-35). Water resistance is improved through the addition of wax. The resins account for 5% of the dry panel's weight and polymerize as the particles are pressed in flat presses or extruded through thin die presses. A variety of particle sizes and shapes is manufactured to yield panels with specific properties. Particle board serves as backing for plastic and wood laminates or as decorated panel that is normally painted. *Hardboard* is produced by processing wood chips with steam and/or pressure, as the lignin bonds the fibers together. As medium-density, high-density, or special densified hardboard, this material offers high strength, wear resistance, moisture resistance, and resistance to

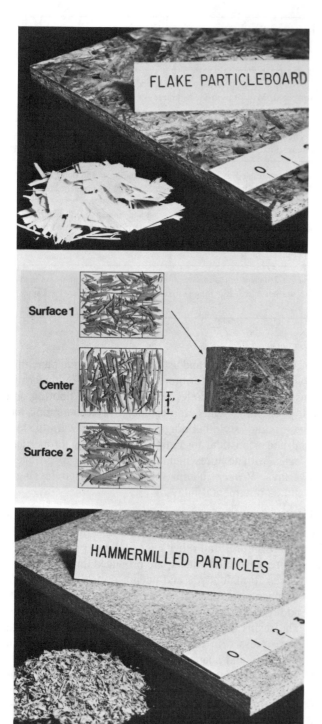

Figure 6-35 Particle board, also called reconstituted panel, products consist of a variety of compositions and organizations: flakeboard uses relatively square chips, oriented strandboard uses long, narrow chips that can be aligned for maximal directional strength. (U.S. Forest Products Laboratory.)

cracking and splitting and has good working qualities. Tempered hardboard has oils added to increase water resistance, hardness, and strength. Hardboard finds use as painted house siding, diestock, and electrical panels, with plastic and other laminates, and as components in furniture such as drawer bottoms, mirror backs, and television sides and backs.

6.6.2.2.5 Sandwiched Materials. Sandwiched materials include treated paper honeycomb faced with laminating veneers. Doors and panels employ these materials (Figure 8-23 page 379). Fiberboards made of wood fibers bonded with rosin, asphalt, alum, paraffin, oils, fire-resistant chemicals, and plastic resins serve as insulation panels on walls and roofing onto which some other material is added, such as siding or tar and gravel.

6.6.3 Wood By-products

The most common by-product of lumber and plywood processing is *chips* for the production of pulp used for a wide variety of paper products. Pulp is obtained through chemically defibrating the chips by dissolving the lignin to obtain fibers. Other by-products include naval stores (turpentine and rosin), bark for mulching plants, charcoal briquettes, and various resins from the tree.

6.6.4 Physical Properties

6.6.4.1 Appearance. When selected for furniture, house trim, doors, floors, wood turning, or other applications in which wood is exposed, appearance is paramount as it lends warmth and beauty. Color, figure (patterns produced by grain, texture, and machining), luster, and the manner in which finishes affect wood all become important factors in selection.

6.6.4.2 Sawing effects. The Forest Products Laboratory of the U.S. Department of Agriculture defines grain and texture as follows:

> *Grain*—The direction, size, arrangement, appearance, or quality of fibers in wood or lumber.
> *Texture*—Finer structure of wood rather than annual rings.

Twenty grain specifications cited in the *Wood Handbook* provide precise meaning to various grain patterns.

The angle at which a log is cut is one factor that determines grain pattern. Figure 6-36 shows the difference in quarter-sawed and plain-sawed boards.

While lumber is used in many stages from logs with the bark removed or rough-sawn timbers, finished lumber is the most common form encountered. Standard dimensions include 2×4, 1×12, 4×4 which are nominal measurements. Nominal measurements indicate the approximate size of the lumber when it is in the rough state. A finished 2×4 is actually $1\frac{1}{2} \times 3\frac{1}{2}$ in. Following rough sawing, lumber is *seasoned* to remove the moisture and reduce warping and cracking, while also reducing weight and increasing strength and many other properties. There are a variety

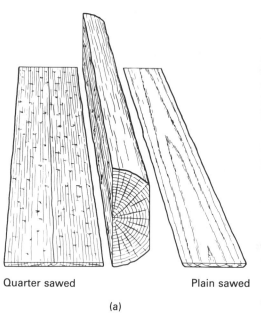

Quarter sawed   Plain sawed

(a)

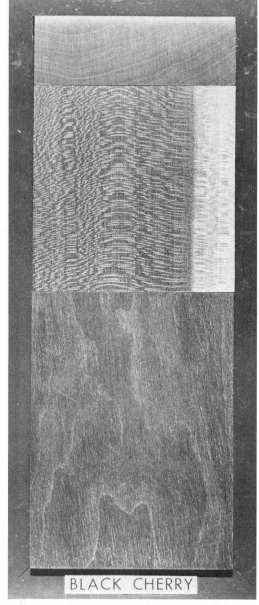

(b)

Figure 6-36 (a) Sawed boards. (b) Black cherry. Top segment illustrates end grain. Middle segment is quarter sawed. Bottom segment is plain sawed.

of seasoning methods that include air drying and kiln drying. Air drying is a slow process in which rough timber is stacked in a specified manner and allowed to dry naturally. Kiln drying employs ovens in which the rough lumber is stacked and hot air is forced over the wood. Research into solar kilns shows promise for low cost and quick seasoning.

Beyond seasoning, many methods are employed to improve the physical, chemical, and mechanical properties of wood. *Chemical modification* involves a chemical reaction (covalent bonding) between the cellulose, lignin, and hemicellulose and a chemical that acts as a preservative. Woods are also impregnated with polymers, oils, salts, and other solutions. Coatings of a wide variety from natural clear finishes, to stains and paints provide protective and decorative value.

6.6.4.3 Moisture content.
Moisture content is a measure of the water in wood. Normally expressed in a percentage of the amount of water weight to the weight of oven-dry wood, moisture content is an extremely important factor in the mechanical and other physical properties of wood. Trees have moisture content ranging from 30% to 200%. Drying of freshly cut wood removes moisture, but the hygroscopicity of wood substance allows it to pick up moisture from the surrounding atmosphere. For this reason, seasonal changes and environmental changes in relative humidity cause a constant shift in moisture content. You may recall during humid and rainy seasons that doors and drawers become hard to open, while during dry winter days the same doors or drawers become easy to use. Treatment of wood must include steps to minimize moisture content changes, weathering, chemicals, and biological attack. Expanding and shrinkage or improperly finished wood cause *warping, splitting, checking* (lengthwise separation across the annual rings), and other problems. Damp wood also provides opportunity for bacteria and fungi to *decay* or *rot* wood. Seasoned lumber has been dried in varying degrees. To ship lumber and avoid molding, moisture is reduced to 25% or below. Thoroughly dried lumber has from 12% to 15% moisture. The key to seasoning lumber is to attain a moisture content corresponding to the average atmospheric conditions in which the lumber will serve.

6.6.4.4 Weatherability and decay resistance.
Similar to the reaction of plastics, woods change color and structure due to heat, light, moisture, wind, and chemicals in the environment. Ultraviolet (UV) light causes chemical degradation and graying of color, while metal fasteners (screws, nails, etc.) and hardware (hinges, etc.) also cause color changes. The UV light breaks down cellulose in the fibers. However, the degradation of the surface fibers, which is rapid when wood is first exposed, slows to a negligible rate of about 1 mm per 100 years. Other effects of weathering include warping, checking, abrasion, and surface roughening. While the even gray color is attractive, the appearance of dark gray and blotchy colors indicates attack by biological organisms and probability of rotting. See Table 6-7 for the varying resistances of woods to decaying. The cut of boards aids weatherability, such as using vertical grain rather than flat grain, but special treatment with preservatives is normally required.

TABLE 6-7 PHYSICAL AND MECHANICAL PROPERTIES OF IMPORTANT COMMERCIAL WOODS GROWN IN THE UNITED STATES (AVERAGE VALUES OF SMALL, CLEAR, STRAIGHT GRAINED SPECIMENS)

Common species	Specific gravity (12% moisture)	Hardness[1] (newtons)	Modulus of[2] elasticity (mega pascals)	Decay resistance[3]
Hardwoods				
Ash, White	0.60	5,900	12,000	S
Basswood	0.37	1,800	10,000	S
Beech, American	0.64	5,800	11,900	S
Birch, Yellow	0.62	5,600	13,900	S
Cherry, Black	0.50	4,200	10,300	V
Elm, American	0.50	3,700	9,200	S
Maple, Sugar	0.63	6,400	12,600	S
Oak, Northern Red	0.63	5,700	12,500	S
Oak, White	0.68	6,000	12,300	V
Walnut, Black	0.55	4,500	11,600	V
Yellow, Poplar	0.42	2,400	10,900	S
Softwoods				
Bald Cypress	0.46	2,300	9,900	V
Cedar, Eastern Red Cedar	0.47	4,000	6,100	V
Northern White	0.31	1,400	5,500	V
Western Red Cedar	0.32	1,600	7,700	V
Douglas Fir—				
Coast	0.48	2,200	13,400	M
Interior West	0.50	2,300	10,300	M
Fir—				
Eastern Species	0.38	2,200	10,300	S
Western Species	0.36	1,800	8,500	S
Pine—				
Eastern White	0.35	1,700	8,500	M
Western White	0.38	1,900	10,100	S
Sugar	0.36	1,700	8,200	S
Redwood—				
Old-growth	0.40	2,100	9,200	V
New-growth	0.35	1,900	7,600	V

(Taken from *Wood Handbook: Wood As An Engineering Material*, by Forest Service, U.S. Department of Agriculture)

[1]Newtons required to imbed a 11.26 mm ball one-half its diameter in a direction perpendicular to the grain.

[2]Measured from a simple supported, center loaded beam, on a span-depth ratio of 14 to 1.

[3]V—very resistant or resistant, M—moderately resistant, S—slight or nonresistant.

Generally, heartwood provides more resistance to decay than does sapwood. The naturally occurring wood extractives determine the resistance of a particular species of heartwood to attacking fungi. The southern and eastern pines formerly were slower growing and older and thus had more heartwood, which made them more resistant to decay. New forestry methods produce faster growth but more sapwood, which reduces these species' decay resistance. Bald cypress has outstanding natural durability.

6.6.4.5 Density. The variation in a wood's moisture content depends on the environment and preparation of the wood and will greatly affect the density, weight, and specific gravity. The range of density for most woods falls between 320 and 720 kg/m³. Calculations of specific gravity are determined on either kiln-dried wood or wood with a specific percentage of moisture. Twelve percent is frequently used. At a 12% moisture level, most species of trees yield specific gravities in the range of 0.32 to 0.67, as seen in Table 6-7.

6.6.4.6 Working qualities. In hand working, normally the lower-specific-gravity wood works best. Machinability of wood depends on such characteristics as *interlocked grain, hard deposits* of such minerals as calcium carbonates and silica, *tension wood,* and *compression wood,* in addition to density. High density and hard deposits will dull tools. Reaction wood is tension wood and compression wood. Reaction wood and interlocked grain cause binding of boards as they are fed through saws and planers. Other considerations in the workability of wood include nail splitting, screw splitting, and surface results of sanding.

6.6.4.7 Ablation. An advantage offered by the use of wood timbers and boards in building is their ability to char when burned. This ablation property allows woods to retain much of their strength since charring slows the burning. Conversely, metals act as good thermal conductors and uniformly transmit heat of a fire throughout the metal. This can cause metal supports to lose much of their strength and bend

Figure 6-37 Ablative nature of wood: steel beams became plastic and draped over wooden beams in a plant fire. (U.S. Forest Products Laboratory.)

under their loads. Figure 6-37 shows the ablative value of wood; the steel beams failed (softened) under the heat, while the wood structural members held some of the load.

6.6.5 Mechanical Properties

Mechanical properties of woods can vary widely because of the lack of homogeneity in wood structure. Knots, cross grain, checks, and growth rings provide varying properties within boards and in boards from the same tree. The properties shown in Table 6-7 reflect values obtained from small, clear, and straight-grained specimens. They provide average values for making comparisons of common woods and for comparing wood with other materials on the basis of modulus of elasticity.

6.7 ADHESIVES

Until the mid 1940s, when a designer wished to join metallic materials together, the selection normally involved either mechanical fasteners, such as bolts, rivets, and pins, or a thermal bonding method, such as welding, brazing, or soldering. Beyond that period up to today, the knowledgeable designer learns that adhesive bonding must receive equal consideration over the traditional methods of joining parts together. The technician and craftsperson involved in maintenance and repairs also can choose adhesives as an alternate technique for fastening broken or unjoined components. Generally, welding, riveting, and adhesive joining are classified as permanent joining methods, while mechanical fasteners such as bolts, screws, nuts, and pins allow disassembly.

Adhesives often hold the advantage over fasteners and thermal bonding in situations requiring the joining of dissimilar materials, light weight, joint sealing, sound and vibration dampening, thermal and electrical insulation, uniform strength, and low-cost, low-skill techniques. These advantages mean adhesives find use in joining many parts in building construction, including the replacement of nails with mastic adhesives to bond plywood or particle board to flooring joists and drywall panels to wall studs, elastomeric adhesives for joining wooden furring strips to concrete, and various other adhesives used to fasten vinyl and ceramic tiles (see Figure 6-38).

The wide use of adhesives by the aerospace industry results from their need to join thin sheets of aluminum, plastics, or composites to frame work and still maintain smooth aerodynamic surfaces, light weight, uniform strength, and good fatigue resistance. Epoxy adhesives stick together structural members formerly joined by rivets, which caused problems due to their nonuniform stress distribution that created stress concentrations; also, rivets are heavy and prone to corrosion.

To help the automotive industry achieve more fuel-efficient vehicles through reduction of weight, adhesive joining technology provides many unique joining methods. Certain structural acrylic and plastisol adhesives offer the auto industry and other

Figure 6-38 Adhesives in construction. Low cost, light weight panel resulting from strong honeycomb paper core adhesively bonded between plywood sheeting. (U.S. Forest Products Laboratory.)

such assembly lines an advantage because of their compatibility with oily metal surfaces (Figure 6-39a), and they bond many dissimilar metals and plastics in addition to wood, glass, hardboard, and asbestos board. These acrylics resist moderate temperatures, have high impact and peel strength, plus they cure rapidly with minimal shrinkage. Urethane adhesives bond together the steel and glass sun roofs, while elastomeric and plastic bumper systems and plastic body side molding use adhesives that dissolve the contaminating oils and films found on production lines, thus providing a dependable joint. The successful use of cyanoacrylates or "super glues" on electronic, electrical, appliance, and instrument assembly lines represents the widespread acceptance of these instant bonding adhesives (Figure 6-39b).

Among the major reasons for rejecting adhesives bonding over mechanical or thermal joining is heat sensitivity, with their normal service temperature ranging from 160° to over 500°C. However, most common adhesives withstand a lower range from −90° to 290°C. The Space Shuttle vehicle employs an adhesive bonding system for its ceramic heat shield tiles with design temperatures up to 1260°C. Epoxy adhesives in the form of microcapsules serve as locking systems for mechanical fasteners. With the shearing force of a nut on an epoxy-coated bolt, the adhesive mixes and cures quickly.

6.7.1 Adhesives Systems and the Bond Joint

An adhesively bonded joint results from the adhesive's ability to flow, wet, and set. ASTM defines *adhesive* as a substance capable of holding materials together by surface attachment. The materials held together are *adherends* or *substrates*. Substrate is the broader term that defines a material upon the surface of which an adhesive-containing substance is spread for any purpose such as bonding or coating. As a liquid of varying viscosity that ranges from low viscosities below that of water to a semisolid (viscoelastic) or syrup consistency, including hot melted solids, adhesives must wet the surfaces to which they bond. *Wetting* involves flowing 1) into large openings such

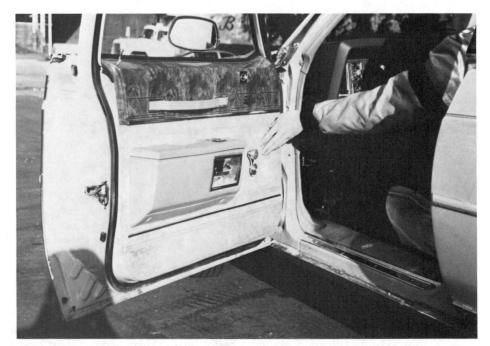

(a)

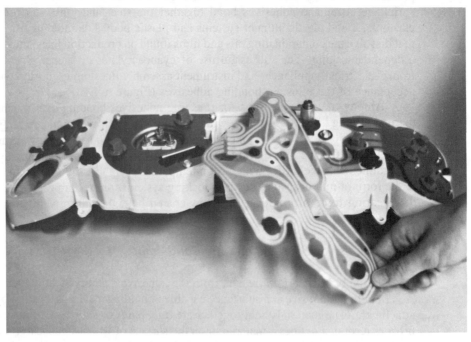

(b)

Figure 6-39 Adhesives in auto and electronic industries (a) The compatibility that acrylics and plastisol adhesives have with oil makes them candidates in auto manufacture. (b) Cyanoacrylics set rapidly for electronic component assembly. (General Motors.)

as pores of woods or cells of foam plastics or 2) into the microscopic hills and valleys that exist on the smoothest of surfaces, such as polished metals or glass. A key factor in adhesive selection is its ability to wet all or most of the available bonding sites of the *substrates* (surfaces making the bond joint) (Figure 6-40). Refer to Figure 6-29 for a microscopic view of wood to contrast with smoother surfaces. The free spreading of the adhesive on the substrate surface may be interfered with due to such conditions as air entrapment, moisture or oxide buildups, and contaminants such as oil and dirt. Figure 6-40 illustrates a nonwetting condition, such as water beading up on a waxed surface, and a wetting condition, such as water absorbed in tissue paper. To achieve wetting of an adherend, the adhesive must not have *cohesive forces* (attraction of molecules inside the adhesive) that provide a very strong *surface tension* (force contracting the liquid into a droplet). The surface tension must be less than the adhesive forces between the adherend and adhesive. A high-surface-energy liquid such as epoxy adhesive will not wet a low-energy polyethylene solid. But the polyethylene liquid with lower surface energy will wet and form good adhesion to an epoxy solid.

Once the wetting has occurred, the adhesive must *cure* into a solid (occasionally a viscoelastic) or *set* so that it develops adhesion, thus bonding materials *(adherends)* together into an adhesive joint.

Four theories exist that attempt to explain the forces involved in the adhesion of an adhesive to its substrates. Both chemical and physical bonds are involved in adhesive bonding, but the exact nature of all forces is not fully known. Covalent bonds and physical attraction through van der Waals forces seem to operate in adhesives as they do in other polymers, but scientists differ on how the adhesion develops. On porous materials such as wood, *mechanical interlock* (penetration of adhesives into openings) provides added surface area for the adhesive to attach, but again some scientists feel mechanical interlock has a minor role in adhesion. Testing of the adhesive bonding systems must deal with the *interfacial adhesion* (attraction between adhesives and substrates) and *practical adhesion* (breaking strength of joint).

Synthetic resins and natural adhesives cure through the various polymerization processes covered earlier in this module. Addition and condensation polymerization develop in the thermosetting adhesives that form permanent bonds; thermoplastic adhesives can be softened to reposition parts or for disassembly.

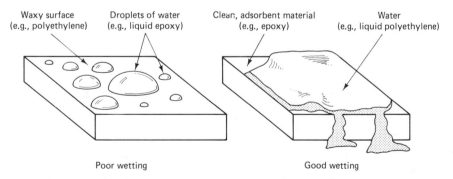

Poor wetting   Good wetting

Figure 6-40 Wetting of substrate. Poor wetting: liquid beads up on surface. Good wetting: liquid flows into micro hills and valleys on surface and porous materials absorb liquid.

6.7.2 Adhesive Bond Chain Links

To grasp the complexities of adhesive bonding, a model developed by Alan A. Marra (Figure 6-41) illustrates the nine links in the adhesive bond chain. The chain-link model reveals the interrelations of factors in the bonded product that include (1) adhesive composition, (2) adhesive application, (3) the bonded materials' properties and (4) their preparation, plus (5) the stress and environment that the product will encounter in actual service. The physical and chemical theories and principles associated with the chain links become quite involved and require adhesive specialists to deal with them. The user of adhesives must recognize that proper adhesive bonding depends on the choice of proper adhesive for the intended environment, which must be applied according to specifications. The best adhesives will fail if not used properly. Link 1: the adhesive film must have proper *cohesion* to stick together when in service. Links 2 and 3: the intra-adhesive boundary layers are strongly influenced by the chemical and physical properties of the adherend; for example, extractives in wood may have an undesirable chemical interaction with the adhesive and change the composition and/or action of the adhesive as it is curing. Links 4 and 5 are critical links in the systems because this is where the adherend and adhesive engage. Such problems as surface contamination can prevent the anchoring of the adhesives at the necessary bonding sites. The compatibility of adhesive to adherend is required here so that proper wetting takes place; several theories attempt to explain the actual forces operating in these two links. Links 6 and 7: the adherend/substrate is critical in a material such as wood or ceramic where a damaged surface must be repaired through

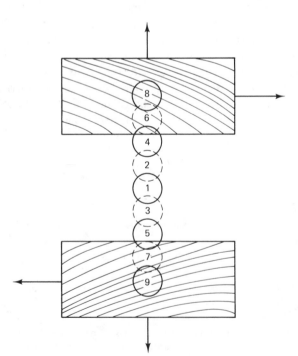

Figure 6-41 Marra's adhesive joint model. Links of an adhesive bond: *link 1*—the adhesive film; *links 2 and 3*—intra-adhesive boundary layer, strongly influenced by the adherend; *links 4 and 5*—adhesive-adherend interface, site of adhesion forces; *links 6 and 7*—adherend subsurface, partially fractured in preparation; *links 8 and 9*: adherend proper. (Dr. Alan A. Marra, University of Massachusetts)

penetration of the adhesive into the pores. Mechanical interlocking may help with the repair. Finally, the adherend material's links 8 and 9 must possess sufficient cohesive strength and compatibility with the bonded joint. For example, a porous material such as a cellular plastic or elastomer may absorb and transport harmful liquids to the bonded joint where the chemicals could degrade the adhesive.

These problems lead to four main factors in adhesive selection:

1. Materials to be bonded and type of joint.
2. Conditions under which the bonding must be done.
3. Durability and strength levels required during service of the product.
4. Cost of the bonding process, including joint preparation, material cleaning, cost of adhesives, and so on.

These four factors will vary considerably and will require close attention to specifications by those making adhesive selections. Technical sales representatives, handbooks, and other technical information guide the user, but as with any industrial production system, the manufacturer of the product must evaluate the adhesively bonded product throughout its development and maintain quality control over the adhesives they purchase and the application of those adhesives to their products.

Advances in microprocessors and computers resulted from improvements in semiconductors and ceramic substrates. Many of those advances were possible because of continually improved *interface technology* which focused on joining materials. Multilayered ceramics (MLC), such as shown in the TCM (Figure 4-1), are an example of improved integration, one product of interface technology. Very large scale integration (VLSI) of semiconductors also presents interface challenges as solder balls only .002 inches join microthin metal circuitry which has been deposited on ceramics.

6.7.3 Types of Adhesives and Their Properties

The wide variety of adhesives fits into various classifications depending on such factors as the *substrate,* physical form (solid, liquid, viscoelastic), and bonding characteristics. Among the characteristics is the determination of whether the adhesive is structural or nonstructural. *Structural adhesives,* by ASTM definition, are those used to transfer required loads between adherends exposed to service environments typical for the structure involved. Some authorities classify structural adhesives as those whose failure could result in a threat to public safety, loss of life, or substantial property damage. These adhesives were formulated to replace nuts, bolts, rivets, and welds. A recent U.S. government study labeled PABST (Primary Aircraft Bond Structure Test) demonstrated that adhesive bonding is feasible and economical for replacing mechanical fasteners and welds for the main structural components on aircraft. Adhesives must possess *serviceability* (withstand design conditions such as heat, cold, stress, moisture, or chemical attack) and *durability* (maintain serviceability of the bond joint over its expected life).

Structural adhesives for wood include *phenol-resorcinol-formaldehyde* polymers, which are thermosetting all-weather exterior adhesives used on laminated structures; *casein* adhesives are structural water-resistant adhesives (moisture content of wood less than 16%) commonly used on softwood laminated beams and doors. Nonstructural adhesives include *urea-formaldehyde,* which is used on particle board and furniture, *polyvinyl acetate* (Elmer's™) for furniture and interior woodwork, and *starch* and *dextrins* for paper products. *Protein* adhesives include animal glues made from bone, hide, and fish (they are used for woodworking and gummed labels); casein from skimmed milk; and *blood* and *soybean flours* formulated with phenolic and formaldehyde resins, which crosslink to serve as water-resistant adhesives on interior plywood.

Thermoplastic adhesive generally has serviceability at temperatures below the thermosets. Among the thermoplastics, *cyanocrylates* (Superglue™) has a service range of $-5°$ to $80°C$; *polyamides,* $15°$ to $100°C$; *polyvinyl acetate* (up to $45°C$); and *butadiene styrene* ($5°$ to $70°C$). These compare with such thermosets as *epoxy,* $-10°$ to $140°C$; *phenolic epoxy,* $-25°$ to $260°C$; and *phenolic neoprene,* $-20°$ to $95°C$. Some synthetic adhesives come as thermosets and thermoplastics such as *polyimides* (TS up to $370°C$ and TP $-160°$ to $350°C$); *silicones* (TS $-25°$ to $250°C$ and TP $-75°C$); and polyurethanes (TS and TP $-60°$ to $175°C$).

Elastomeric adhesives are formulated with natural or synthetic rubber or blended with other resins to provide impact resistance and withstand wide temperature variations. These adhesives include *neoprene, polysulfides, butyls,* and *nitriles.* A phenolic and rubber blend adhesive bonds brake linings that withstand severe shock, temperature, and saltwater exposure.

Except for pressure-sensitive tapes that remain tacky, adhesives change from the liquid state to the solid state upon curing. Tapes such as adhesive tape, duct tape, and transparent tape do not develop the strength found in solid adhesives; however, double-coated pressure-sensitive tapes serve as hold-down tapes for use on machine tables of milling machines and surface grinders.

Curing of adhesives involves the cooling of hot melts, loss of solvents, anaerobic environments, and two-part mixtures. *Hot-melt* adhesives usually begin as a solid that becomes fluid in heat guns, furnaces, or with other heat sources that develop around $150°C$ temperatures; most cool and set within seconds after removal of the heat source. *Loss-of-solvent* adhesives employ volatile liquids, including water and organic solvents, that dissolve the adhesive base material to form the fluid; upon application to the joint, the adhesives harden as the solvents evaporate or penetrate porous adherends. With solvent adhesive, especially, and other adhesives, one must determine whether the adhesive will attack the substrate so severely that it may produce weakness in the material being joined. *Anaerobic adhesives* retain their fluid state when exposed to oxygen, but when squeezed into thin joints that block the oxygen, they set up into hard, strong adhesives. *Two-part* mix adhesives are stored with the resin and catalyst (or hardener) separated; when mixed the catalyst causes crosslinking of the resin.

Loss-of-solvent adhesives often have long setting times; and except for water-solvent types, the solvents present health and fire safety problems. Also, they rely on our short supply of petroleum, which is causing a trend toward water-based adhesives. Anaerobics come in one part, cure quickly, develop high shear and impact strength, and some have gap-filling abilities. Other adhesives such as cyanoacrylates cure as a result of exposure to the moist alkali environment found on most surfaces. Contact cements are usually rubber based structural adhesives that form an instant bond when substrates coated with the cured adhesives are brought into contact, thus eliminating the need for clamping. Most other adhesives require pressure long enough for the adhesive joint to cure.

Table 6-8 lists some common adhesives and their typical applications. The table provides an overview of selected adhesives, adherends, and applications. Adhesives have been formulated to meet the most varied of circumstances. In medicine, methylmethacrylate resins join artificial hip sockets to natural hip bones and repair broken, damaged bones. Refer to Applications and Alternatives on page 16. Cyanoacrylates bond eye tissue and other skin tissue. Thermoplastic polyester resins bond copper foil to flexible backing on printed circuit boards and along with epoxy and phenolic adhesives withstand etchants and the heat of soldering operations. *Weld bonding* combines spot welding and adhesives. The weld immobilizes the adherends to allow the adhesives to cure. The increased use of resin-based composites promises even wider use of adhesives.

TABLE 6-8 COMMON ADHESIVES

Adhesive	Typical adherends and applications
Acrylics (2 part)	Plastics (ABS, phenolic-melamine, polyester, polystyrene, polysulfone, PVC) to metal, rubber to metal, and plastics to plastics
Casein	Interior wood components
Cyanoacrylates (anaerobic)	Thread locking, gear and bushing mounting, aluminum, copper, steel, ceramics, glass, rubber, polycarbonate, phenolic, acrylics
Epoxy	Most adherends except some plastics such as acetals, polypropylenes, polyurethanes, and silicones; many modifications including nylon epoxies and epoxy phenolics
Melamine formaldehyde	Paper, textiles, hardwood, plywood for interior use
Natural rubber	Cork, foam rubber, leather, paper-building materials, pressure-sensitive tapes
Neoprene rubber	Many plastics (ABS, fluorocarbons, polystyrene), aluminum, ceramics, copper gasketing and laminating
Nitrile rubber and phenolic	Many plastics (ABS, cellulose, fluorocarbons, PVC, polyester), aluminum, ceramics, glass, magnesium, microwave isolators
Phenol-formaldehyde	Cork, cardboard, softwood, plywood for exterior use
Polyvinyl acetate	Textiles, polystyrene, wood, other porous and semiporous materials, interior furniture and construction
Resorcinol-formaldehyde	Asbestos, cork, paper, rubber, wood furniture, laminated exterior wooden beams, wooden boats
Silicone	Aluminum, ceramics, glass, magnesium, phenolics, polyester, acrylics, rubber, steel, textiles, integrated circuits, sealants, gasketing
Urethane elastomer	Many plastics (ABS, polyester, acrylics, PVC), aluminum, ceramics, copper, glass, magnesium, steel-bond solid propellants, cryogenic applications, insulation

APPLICATIONS AND ALTERNATIVES

Protective Coatings

Because polymers include synthetics, an endless variety of them can be synthesized. Protective coatings cover another category of polymers not discussed in this module. Protective coatings include natural polymers such as shellac, oil based paint, and many synthetic polymers including the same general types used for plastics, elastomers, and adhesives; for example epoxies, silicones, latex, urethanes, and acrylics. These coatings develop their protective skins in a number of ways ranging from chemical reaction (e.g. polymerization), cooling of hot melts, and evaporation of solvents or water. Evaporation leaves behind a film of the polymer which was dissolved in solvent or suspended in water. Much of current development in protective coatings focuses on eliminating solvents since they create health hazards to workers and harm the atmosphere. With water as the vehicle to carry the polymer dispersions, it evaporates with few harmful effects. Much of the paint being used today is water based.

A new polymerization process known as group transfer polymerization (GTP) allows for unlimited polymer chain growth in acrylics with the potential for high performance auto finishes which will cure at lower temperatures (180°F versus 250–300°F) and uses lower concentrations of polluting solvents.

Polymer Conductors

A major disadvantage to lead-acid and other metallic batteries is their weight. The development of conductive polymers such as polyacetylene electrodes could open the way to long sought electric automobiles. We remember that polymers do not normally serve as conductors because their electrons are locked in covalent bonding versus the free electrons in metals. Oxidation, resulting from doping polyacetylene with iodine, allows free electron movement. Reduction in the same polymer, achieved by reaction with the metal sodium, turns polyacetylene into an anion while the sodium becomes a cation. The polymeric form of sulfur nitride becomes conductive at low temperatures without doping. Poly(para-phenylene) or PPP has a make up of single and double bonds that can achieve electron mobility through doping. Polypyrrole also has conductor potential because of bonding similar to PPP. Both have capabilities to be switched back and forth between conductors and insulators. Each of these polymers offers potential as conductors but major problems exist in converting them to reliable engineering materials.

Oils, Lubricants, and Fluids

Polymers play an important role in many types of lubricating, cutting and hydraulic applications as solids, and additives or suspensions in liquids. Engine oil viscosity for multigrade oils such as SAE 10W 40 are able to change viscosity when the ambient and engine temperatures change because of polymers which control how thick or thin the oil is. New long lasting engine oils are the result of synthetic polymers.

Electro-rheological (ER) fluids congeal from a liquid into a solid mass in a few milliseconds when a current is passed through them. When the current is removed the solids return to liquids. ER fluid has promise as a hydraulic fluid for robots, automotive clutches, and artificial human limbs. Experimentation with combinations of selected oils, with particles (microscopic solids of polymethacrylic acid and water in tangled long chains) will continue in the quest for the ideal ER fluid. What possibilities might this type of solid lubricant hold for the problems encountered with high temperature ceramic engines or automotive brakes.?

Advances in polymer materials development, much as those discussed here, occur daily. With the basic concepts covered in this module you possess the ability to understand these developments. To stay current with polymer innovations, make it a habit to regularly read some of the periodicals listed at the end of this module.

6.8 SELF-ASSESSMENT

6-1 Define the term polymer, list the major groups of polymers, and cite examples of materials in each group.

6-2 Discuss the sources of raw materials used in the production of synthetic polymers. Explain how the depletion of fossil fuels can affect synthetic polymers and what alternatives may be possible to meet the needs served by present-day synthetic polymers.

6-3 Explain the necessity of unsaturated hydrocarbons and bifunctional monomers for polymerization. Describe the two main polymerization processes and typical products of each.

6-4 Discuss the effect of crystallinity on polymers and explain methods of achieving crystallinity.

6-5 Calculate the molecular weight for polyvinyl chloride, polystyrene, and nylon 6.

6-6 Calculate the average molecular weight of a polyethylene compound with 10,000 monomers. Would this be classified as a low, medium, or high molecular weight polymer?

6-7 The Pause and Ponder section suggested that you consider containers as you read this module. For each of the following types of containers, explain why crystallinity may or may not be a factor in the selection of a polymer; use Figure 6-8 to classify the molecular weight and degree of crystallinity for each application: pocket calculator, personal computer cabinet, outdoor public telephone, shampoo tube, chlorine bleach bottle, candy box, lawn-mower engine shroud, egg carton, automobile body, camera, disposable pop bottle, electrical box or receptacle, and automobile tire. For example, consider which would be more flexible: a low molecular weight, low crystalline polymer, or a low molecular weight, highly crystalline polymer.

6-8 What mechanism produces network polymers? Do thermoplastic polymers form networks?

6-9 Does condensation polymerization usually produce thermosetting or thermoplastic polymers? What are the exceptions? What terms describe crosslinking in rubber? What chemical produces this crosslinking?

6-10 What provides for greater crosslinking in polymers? What is the nature of a polymer with a netting index of 50?

Plastics

6-11 What advantages do most plastics offer over linear plastics?

6-12 Describe the plastic classifications of (a) general purpose, (b) engineering, and (c) cellular.

6-13 Why are cellular plastics in ever increasing use? List typical applications of foam plastics.

6-14 What is FRP?

6-15 What element serves as the backbone for most plastics? Name one plastic with an inorganic backbone.

6-16 What does ASTM stand for?

6-17 Referring again to our Pause and Ponder problem on the use of polymers as containers, select a plastic from Appendix 10.5.1 that would be suitable for the following containers: (a) candy wrapper, (b) thermos bottle, (c) medicine bottle, (d) television cabinet, (e) suitcase, and (f) vacuum cleaner housing.

6-18 What additive would be valuable in a plastic that would be frequently exposed to sunlight? How does it chemically alter the plastic?

6-19 List additives for plastics to deal with the following service conditions: (a) aircraft storage compartments subject to fire, (b) coating for wrench handle, (c) vegetable storage drawers in refrigerated bins, (d) flexible gloves, (e) lightweight, stiff automobile interior door panels.

6-20 Determine whether thermosetting or thermoplastic plastics are *generally* best for the following type of service conditions: (a) high impact, (b) creep, (c) chemical exposure.

6-21 While a certain plastic may have lower specific gravity, better toughness, and equal optical properties, why may glass still be preferred for a service condition such as storm doors, eye glasses, or lenses on instrument panels? What plastic competes strongly with optical glasses? List some of the competing factors.

6-22 What electrical property is of key consideration in selecting an electrical insulator? How do plastics compare as electrical insulators?

6-23 When would it be undesirable to have a plastic with low water absorption resistance? Name three plastics with good water absorption resistance.

6-24 What are the combined stresses that act on a material subjected to weathering? Select three plastics most capable of service conditions that involve weathering. How would the three compare to (a) plain carbon steel, (b) alumina, and (c) glass? Give the coefficient of thermal expansion and rate of thermal conductivity for each of the three materials.

6-25 List five properties essential for the rope used to pull water skiers. Select two plastics suitable for this service and explain the selections. Use the ASTM abbreviations.

6-26 Given the choice between PS, PMMA, ABS, and PF as possibilities for motorcycle helmets, select one and explain your choice.

6-27 Select two plastics that would withstand the service conditions under the hood of an automobile. Use the ASTM abbreviation. Explain the selection.

6-28 Give some reasons that viscoelasticity offers the designer certain advantages and disadvantages when using polymers in product design.

6-29 From the tables in this module and current prices listed in such reference issues as *Machine Design, Modern Plastics,* or *Materials Engineering,* calculate the specific strength and cost per unit volume for the following materials: acrylic, polycarbonate, plain carbon, steel, wood, epoxy, and aluminum.

Elastomers

6-30 What is the major source of natural rubber? What are other possible sources?

6-31 Define the term elastomer and diagram a polymer of natural thermoplastic rubber and a polymer of vulcanized rubber.

6-32 Diagram the polymer chain structure of rubber at rest, and then with a tensile stress applied to it.

6-33 What are the major advantages and disadvantages of natural rubber? Give examples of applications of NR to show how the advantages caused selection and the disadvantages would prevent its selection.

6-34 For the following containers use Appendix 10.5.2 to choose a suitable elastomer: (a) hydraulic hose, (b) hot water bottle, (c) skin diver's wet suit, (d) bicycle tire inner tube.

6-35 What is the major raw material for synthetic rubber?

6-36 At what stage in tire manufacture does vulcanizing occur?

6-37 Why is natural rubber used in automobile tires? Why are synthetic rubbers also used?

6-38 What additives to synthetic polymers accomplish the following: (a) retard aging from ultraviolet rays, (b) soften raw rubber, (c) increase rigidity, (d) strengthen?

6-39 Explain TR in terms of structure and applications.

6-40 Explain the following elastomer designation: ASTM BH 530.

6-41 What is the importance of hardness in rubber? Compare hardness of elastomers to other polymers.

6-42 What is the term to describe loss of energy through heating in elastomers? When is it an advantage for this property to be (a) high or (b) low?

6-43 Cite examples of an application where the cost of an elastomer may be so high that another material would be selected rather than the elastomer.

6-44 Select elastomers for the following applications; specify their ASTM abbreviation and their favorable properties: (a) fire hose, (b) lawn-mower wheels, (c) stopper on chemical test tubes, (d) golf ball winding, (e) racquet ball, (f) tennis shoe soles, (g) rubber band, (h) seal on auto brake master cylinder, (i) carpet backing, (j) work shoe soles, (k) pad for bicycle caliper brakes, (l) wrestling mat.

Woods

6-45 In light of diminishing supplies of many raw materials, what makes wood such a promising engineering material for future generations?

6-46 What is the relation of lignin to cellulose in a tree? When is lignin removed?

6-47 Explain how anisotropy develops in wood both at the microstructure (cellular) and macro-structure level?

6-48 How are hardwoods different from softwoods? Cite examples to illustrate these differences.

6-49 Describe how the various layers in a cross section of a tree affect the properties of the lumber produced from the tree.

6-50 Name typical applications for the following grades of lumber: (a) nonstressed graded, (b) construction appearance, (c) remanufactured, (d) stress.

6-51 Name four wood products that are not solid wood and cite typical uses of each.

6-52 Explain the following APA designation:
B-D
Group 3
Interior

6-53 What properties make wood a desirable material for furniture and paneling?

6-54 What properties can limit the use of wood as an engineering material?

6-55 What effects can moisture have on wood? How can they be altered?

6-56 Name a wood to fit into the following categories: (a) high hardness, high decay resistance, and high specific gravity, (b) very low hardness, low specific gravity, good modulus of elasticity.

Adhesives

6-57 List three advantages of adhesive bonding over other joining methods. Name three applications.

6-58 (a) In the adhesive system, what three events must occur to obtain a bonded joint? (b) What types of forces are involved in adhesion?

6-59 What is the difference in adhesion and cohesion?

6-60 (a) What is the difference in practical adhesion and interfacial adhesion? (b) What can prevent wetting?

6-61 (a) Describe structural adhesives and list two examples. (b) What has PABST proved?

6-62 (a) Name the five integrated factors that can affect the bonded product. (b) What must be considered in adhesive selection? (c.) What contributions have interface technology made to new computers?

6-63 (a) Why would a thermoplastic adhesive be selected over a thermosetting adhesive? (b) Name two protein adhesives and give an application for each.

6-64 What adhesive could be used to bond structural components that might meet service conditions in the southern deserts or the Arctic regions?

6-65 Describe three methods in which adhesives develop their cured attraction between the substrate and the adherend.

6-66 Select an adhesive for the following joining applications: (a) aluminum electronic component to fiber-glass printed circuit board, (b) PUR foam bumper facia to steel bumper, (c) ABS control mount to porcelain enamel dishwasher panel, (d) plywood veneers, (e) aluminum identification label to PC plastic electric drill housing, (f) aluminum towel rack to ceramic tile wall, (g) glass walls to chrome-plated steel aquarium, (h) PS foam pad to wooden seat, (i) exterior laminated beam, (j) copper foil to flexible printed circuit board, (k) brake liners to steel brake shoe.

6-67 What is a major thrust in development of protective coatings? List three broad groups of protective coatings.

6.9 REFERENCES AND RELATED READING ON PLASTICS

ALLCORK, HARRY R., and FREDERICK W. TAMPE. *Contemporary Polymer Chemistry*. Englewood Cliffs, N.J.: Prentice-Hall, 1981.

Annual Book of Standards Part 35—Plastics—General Test Methods; Nomenclature. Philadelphia: American Society for Testing and Materials, 1983 or current edition.

Annual Book of Standards Part 36—Plastics—Materials, Film, Reinforced and Cellular Plastics; High Modulus Fibers and Their Composites. Philadelphia: American Society for Testing and Materials, 1983 or current edition.

ARNOLD, LIONEL K. *Introduction to Plastics*. Ames, Iowa: Iowa State University Press, 1968.

BOSICH, JOSEPH F. *Corrosion Prevention for Practicing Engineers*. New York: Barnes and Noble, 1970.

BUDINSKY, KENNETH. *Engineering Materials: Properties and Selection*, 2nd ed. Reston, Va.: Reston, 1983.

DE YOUNG, H. GARRETT. "Conductive Polymers Imitate Metals," *High Technology*, January, 1984, pp 65–70.

FERADOS, JOEL, ed. *Plastics Engineering Handbook of the Society of the Plastics Industry*, 4th ed. New York: Van Nostrand Reinhold, 1976.

FONTANA, MARS G., and NORBERT D. GREENE. *Corrosion Engineering*, 2nd ed. New York: McGraw-Hill, 1978.

Fundamentals of Plastics (programmed text). 1974 ed. Cleveland: Penton Publishing.

HARPER, CHARLES A., editor-in-chief. *Handbook of Plastics and Elastomers*. New York: McGraw-Hill, 1975.

HESS, HARRY L. *Plastics Laboratory Procedures*. Indianapolis: Bobbs-Merrill, 1980.

MILBY, ROBERT V. *Plastics Technology*. New York: McGraw-Hill, 1973.

MITTAL, K. L. ed. *Polyimides: Synthesis, Characteristics, and Applications, Vol 1 & 2*. New York: Plenum Press, 1984.

PATTON, W. J. *Materials in Industry*, 2nd ed. Englewood Cliffs, N.J.: Prentice-Hall, 1976.

Plastics: Programmed Instruction Text. Detroit: Society of Manufacturing Engineers, 1970.

RICHARDSON, TERRY A. *Industrial Plastics: Theory and Application*. Cincinnati: South-Western, 1983.

SCOTT, DAVID. "Amazing Hardening Fluid," *Popular Science*, April, 1984 pp 82–85.

SEYMOUR, WILLIAM B. *Modern Plastics Technology*. Reston, Va.: Reston, 1975.

TAUBES, GARY. "Electrifying Plastics," *Discover*, June 1984, pp 46–49

Periodicals

Materials Engineering.
Modern Plastics.
Modern Plastics Encyclopedia, annual periodical.

Plastic Design Forum.
Plastics Engineering.
Plastics Technology.
Plastics World.

6.10 REFERENCES ON RUBBER

Annual Book of Standards Part 35—Rubber, Natural and Synthetic—General Test Methods, Carbon Black. Philadelphia: American Society for Testing and Materials, 1983 or current edition.

Annual Book of Standards Part 38—Rubber Products, Industrial—Specifications and Related Test Methods: Gaskets: Tires. Philadelphia: American Society for Testing and Materials, 1983 or current edition.

HARPER, CHARLES A., ed. *Handbook of Plastics and Elastomers.* New York: McGraw-Hill, 1975.

Kraton Thermoplastic Rubber. Shell Chemical Company, Polymer Division, Houston, 1972.

Synthetic Rubber: The Story of an Industry. New York: International Institute of Synthetic Rubber Products, 1973.

Periodicals

Automotive Engineering.

Industrial Research.

Materials Engineering.

"Materials Selector" issues, *Materials Engineering.*

NASA Tech Briefs.

Product Engineering.

Rubber Age.

Rubber World.

6.11 REFERENCES ON WOOD

BAKER, GLENN E., and L. DOYLE YEAGER. *Wood Technology.* Indianapolis: Howard W. Sams, 1974.

CHAMPION, F. J. *Products of American Forests.* Washington, D.C.: Forest Products Laboratory, U.S. Department of Agriculture (G.P.O. Stock Number 0100-2592), 1973.

DIETZ, A. G. H., E. L. SCHAFFER, and D. S. GROMALA, eds. *Wood as a Structural Material,* Educational Modules for Materials Science and Engineering. University Park, Pa.: Pennsylvania State University, 1982.

FEIRER, JOHN L. *Woodworking for Industry.* Peoria, Ill.: Charles A. Bennett, 1971.

SAEMAN, JEROME F. "Wood as an Engineering Material—An Urgent Issue for Educators," a paper presented to American Society for Engineering Education, June 14–17, 1976, Knoxville, Tenn.

THOMAS, RICHARD J. "Wood Anatomy and Ultrastructure." *Journal of Educational Modules for Materials Science and Engineering,* vol. 1, no. 3, Fall 1979, pp. 451–532.

WANGAARD, F. F., ed. *Wood: Its Structure and Properties.* Educational Modules for Materials Science and Engineering. University Park, Pa.: Pennsylvania State University, 1981.

Wood—A Modern Structural Material. American Institute of Timber Construction, American Plywood Association, Southern Forest Products Association, and Southern Pressure Treaters Association. Tacoma, WA.

Wood Handbook: Wood as an Engineering Material. Washington, D.C.: Forest Products Laboratory, U.S. Department of Agriculture (G.P.O. Stock Number 0100-103200), 1974.

6.12 REFERENCES AND RELATED READING ON ADHESIVES

Adhesive Bonding of Wood and Other Structural Materials, Educational Modules for Materials Science and Wood. University Park, Pa.: Pennsylvania State University, 1984.

"Adhesives," *Technology,* Jan./Feb., 1982, vol. 2, no. 1.

Adhesives in Building Construction. Washington, D.C.: U.S. Department of Agriculture, Forest Products Laboratory, 1978.

ALNER, D. J. ed. *Aspects of Adhesion.* Cleveland: Chemical Rubber Company, 1973.

Annual Book of ASTM Standards Part 22: Woods; Adhesives. Philadelphia: American Society for Testing and Materials, current edition.

BLOMQUIST, R. F., et al. *Adhesive Bonding of Wood.* Washington, D.C.: U.S. Department of Agriculture, Forest Products Laboratory, 1983.

CAGLE, CHARLES V. *Adhesive Bonding.* New York: McGraw-Hill, 1963.

DELOLLIS, N. J. *Adhesives for Metals.* New York: Industrial Press, 1970.

MITTAL, K. L. ed. *Adhesive Joint Formation, Characteristics, and Testing,* New York: Plenum Press, June, 1984.

MITTAL, K. L. ed. *Adhesive Aspects of Polymeric Coatings* New York: Plenum Press, 1983

MITTAL, K. L. ed. *Physicochemical Aspects of Polymeric Surfaces, Vol 1 & 2,* New York: Plenum Press, 1983.

Modules on "Adhesive Bonding of Wood and Other Structural Materials." *Journal of Educational Modules for Materials Science and Engineering.* University Park, Pa.: Pennsylvania State University, 1981–82.

RIDER, D. K. *Principles and Applications of Adhesives,* Educational Modules for Materials Science and Engineering. University Park, Pa.: Pennsylvania State University, 1979.

SKIST, I., ed. *Handbook of Adhesives,* 2nd ed. New York: Van Nostrand Reinhold, 1977.

Periodicals

Adhesive Age.   *Materials Engineering.*

High Technology.   *Popular Science.*

Iron Age.

7

CERAMICS

7.1 PAUSE AND PONDER

In designing an engine a major objective is to obtain maximum energy output for the energy input. The conventional cast-iron, gasoline engine is quite inefficient. At a speed of 30 miles per hour on level terrain, the typical heat-powered engine loses up to 30% of the 100% energy input due to cooling and lubrication. Water, air, and oil circulating through the engine system are necessary to keep it cool to prevent damage to the metal's structure. Of course, the cooling robs the valuable heat energy necessary to power the vehicle. Cast iron is also very heavy.

Realizing the inefficiency of cast-iron engines, designers have long sought alternative materials. Currently, ceramics are under investigation by several major engine manufacturers (Figure 7-1). One example is the use of sintered ceramic partially stabilized zirconia (PSZ) for gas turbine engines. Why ceramics? As you will learn in this module, many ceramics perform well under high heat conditions (operation at 1000°C with low linear coefficients of thermal expansion), yet cast iron and most other metals must operate at much lower temperatures while exhibiting high linear coefficients of thermal expansion. High expansion rates create problems with dimensional stability and affect the close fits required of engine parts.

Ceramics offer advantages for engine materials but pose some serious challenges because of their strength, ductility, and fracture-resistant properties and the available processing technology. As you study this module on ceramics, consider their possibilities in relation to use in an

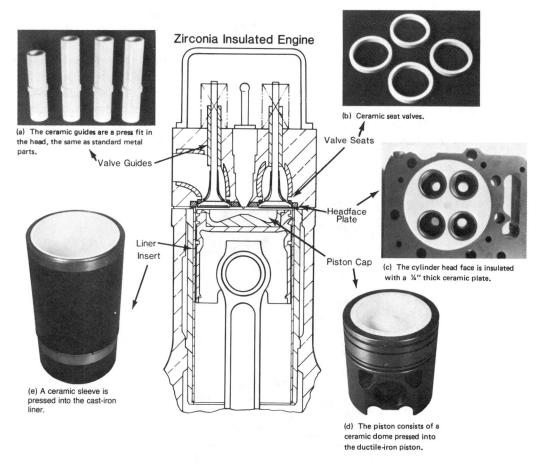

(a) The ceramic guides are a press fit in the head, the same as standard metal parts.

Valve Guides

Zirconia Insulated Engine

(b) Ceramic seat valves.

Valve Seats

Headface Plate

Liner Insert

Piston Cap

(c) The cylinder head face is insulated with a ¼" thick ceramic plate.

(e) A ceramic sleeve is pressed into the cast-iron liner.

(d) The piston consists of a ceramic dome pressed into the ductile-iron piston.

Figure 7-1 Ceramic engine. Cut-away view shows how Cummins Engine Company has used zirconia to insulate cast iron in a heat engine in order to eliminate the cooling system and make it an adiabatic engine. Photos of actual ceramic parts: (a) Valve guides are press fit into head. (b) Valve seats. (c) Cylinder head face insulated with ¼" thick plate. (d) Zirconia dome pressed into ductile-iron piston. (e) Sleeve pressed into cast iron liner. (Cummins Engine Company.)

automobile engine. Also, view their properties to see what advantages or limitations they may offer as alternates to other materials. Bear in mind that raw material sources of many ceramics are quite plentiful when contrasted to raw materials for plastics and metals, which makes them even more attractive.

7.2 INDUSTRIAL CERAMICS

Ceramics are usually crystalline compounds made up of metals and nonmetals (frequently oxygen). Glass is a notable exception with its amorphous structure. Normally, they are very hard and brittle and withstand high temperatures. Ceramics in the form of stones were one of the first materials used for tools and as pottery for cooking and eating implements, dating back to 2000 B.C. Only recently has industry started to make greater use of this group of materials in engineering. Because of the obscure location of many ceramics, the average person is not aware of them. Rock wool, one of several unseen ceramics, is produced by melting stone and using high-pressure steam to blow it into fibers. This excellent insulation material is hidden behind walls and in attics. Also, hidden silicon chips and ceramic substrates allow computers to continually increase in computing power as their sizes shrink dramatically.

The furnaces used to melt metals have walls constructed of ceramic *refractory bricks* that can withstand very high temperatures. The porcelain insulator on a spark plug hides under the hood of an automobile where it resists high temperatures as a

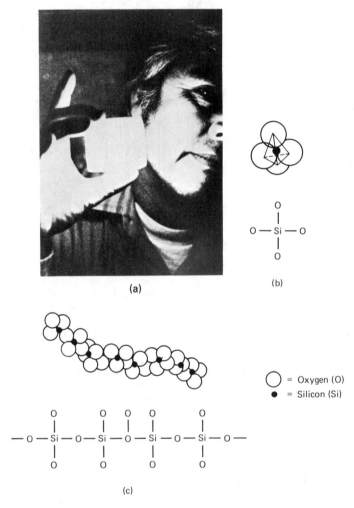

Figure 7-2 (a) Silica insulation tiles for NASA Space Shuttle shed heat so effectively that white hot block just out of 1260°C furnace can be held on edges with bare hand. (Corning Glassworks.) Silica structure: (b) Silicate tetrahedral. (c) Single chain of silicate.

part of the ignition system, while refrigeration applications subject porcelain to subfreezing temperatures. NASA's use of silica tiles as a part of the Space Shuttle thermal protection system marks a significant advance in ceramic high technology (Figure 7-2a).

On the other hand, many ceramics are quite visible and familiar to us, including the porcelain-coated slip cast sinks and toilets that provide scratch and corrosion-resistant sanitary surfaces. Many forms of glass serve our needs, ranging from the windows in our homes, cars, and offices to optical lenses used in eyeglasses and microscopes. Bricks and other building blocks of nearly unlimited colors, sizes, and shapes serve as both structural and decorative material for construction of shelters. Many other ceramics are used in conjunction with other ceramics or additional materials, so they are seen but not often recognized as ceramics. For example, limestone, clay, and certain elements are used to produce Portland cement, which is mixed with rock, sand, and often steel reinforcing rods to form sidewalks, roadways, and bridges. Masonry cement also serves as the cement adhesive to bond together bricks and other building blocks.

7.3 BASIC STRUCTURE AND PROPERTIES

Silicates, which make up many ceramics, are the most common mineral on earth. They include sand, clay, feldspars ($KAlSi_3O_8$),* quartz (SiO_2), and the semiprecious stone, garnet. The silica structure is the basic one for most ceramics. It has an internal arrangement consisting of pyramid (tetrahedron or four sided) units, as seen in Figure 7-2b. Chains, the same as in polymers, are formed as the silicate tetrahedrals combine in the chemical formula SiO_2 (Figure 7-2c). Four larger oxygen (O) atoms surround each smaller silicon (Si) atom. The silicon (Si) atoms occupy the openings (interstitials) between the oxygen atoms and share four valence electrons with the four oxygen atoms through covalent bonding. Ionic bonds hold together the long covalently bonded chains. The most familiar ceramic chains resemble organic polymers with their carbon backbone, except the backbone of ceramics consists of alternate silicon and oxygen atoms.

The unique structure of ceramics leads to the properties possessed by most ceramics. Recall that ductility in metals allows plastic deformation and necking, which results from the dense packing and slip planes as discussed in Module 5. However, ceramics are generally brittle and fracture without plastic deformation. This is because ceramics lack the uninterrupted slip planes that are found in metals. Also, metals have equal bonding between atoms and grains, whereas ceramics have covalent, ionic, and van der Waals bonds of varying strength. Fracture in ceramics is promoted by stress concentrations, which develop from the atomically sharp corners, porosity from trapped air, and microcracks that develop as the ceramic cures from its green state. Tensile strength is hard to ascertain on ceramics because of these flaws. Figure 7-3 provides a comparison of the strength of ceramics with metals.

*References most often call out the ceramic's group structure, such as $Al_2O_3 \cdot 2SiO_2$, instead of exact chemical formulas.

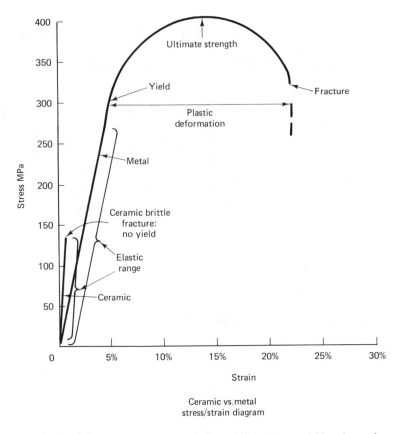

Ceramic vs. metal
stress/strain diagram

Figure 7-3 Ceramics brittle fracture results from their inability to yield as do metals. This diagram shows a large percentage of plastic deformation in metals that occurs because localized stresses are relieved by localized plastic deformation. No yielding point shows on the ceramic curve so localized stress eventually causes catastrophic failure when the elastic range is exceeded. Percentage of strain before failure on ceramics can be measured in hundredths of a percent compared to tenths of percents for metals.

The covalent and ionic bonding in ceramics tie up the electrons. The lack of free electrons makes ceramics good electrical and thermal insulators (dielectrics), whereas metals are good conductors due to their free electrons resulting from metallic bonds. Heat conduction in ceramics results from vibrations of the atomic lattice (phonons). The bonding of ceramics also contributes to their stability for good chemical resistance.

The combination of silica with aluminum, magnesium, oxygen, and other elements yields a broad range of ceramic compounds. One example is kaolin clay, known as kaolinite, which consists of aluminum, silicon, oxygen, and hydrogen, with the formula $Al_2(Si_2O_5)(OH)_4$. Kaolinite has a complex unit cell consisting of alternate layers of aluminum ions and silicon ions with SiO tetrahedra existing in the cell. Kaolinite crystals consist of parallel sheets bonded together by van der Waals forces. These sheets slip easily over each other in cleavage, which makes clay easy to mold. Ionic and covalent bonds provide strength within the sheets of clay. The processing used on this white clay typifies the manner in which many ceramic products are made.

The wet kaolin clay is shaped by hand or machine while in the *plastic* or moldable state. This plastic or rheological nature is important in the processing of clay. *Rheology* concerns itself with the deformation and flow of matter. It is necessary to determine how well wet ceramics will flow into formed plastic shapes without further deforming from the material's own weight. Some ceramics are cast while in a slurry or *colloidal* state in which the ceramic particles float in a liquid (usually water). These slurries, such as concrete or slip clay, are poured into a plaster of Paris mold, and most of the water is absorbed into the pores in the mold, leaving the "plastic" casting in the mold as seen in Figure 7-4a. With concrete, the solids set up and become very hard through hydration (components chemically combine with the water). During hydration, heat is given off. Sufficient curing time (days or weeks) must be provided in the case of concrete to ensure complete combining. But slip clays and other clays must be *fired* to achieve maximum strength and hardness. These methods of processing do not lend themselves to close tolerances such as one becomes accustomed to in metals and plastics. Some newer improved processes are discussed later.

Firing (sintering) is a process that uses high temperatures to create ample atomic movement of the solids so surface tension will bring the material together and reduce the openings (pores) between the constituents. The high temperature may cause melting of some elements, which forms a glassy substance (vitreous) that seals the pores and cements the crystals together (Figure 7-4b) in a ceramic bond. This process is known as *vitrification*; vitreous materials are glasslike. During the vitrification process, polymorphic (change in internal lattice) structural changes occur, just as allotropic changes occur in iron as it is heated. Like iron, three phases occur in the ceramic silica: (1) *quartz* exists at room temperature, (2) *tridymite* forms at 867°C (1592°F), and (3) *cristobalite* forms at 1470°C (2678°F). New technology uses a slurry composed of ceramic powder mixed with meltable organic material for injection molding. The mixture can form complex shapes and set quickly, much like plastic injection molding. Powder metal processes (Figure 7-4c), as discussed in Module 5, can produce close tolerance ceramics.

Typically, ceramics are crystalline compounds. However, glassy (vitreous) materials have an amorphous structure. In other words, glasses are solids that retain their unordered lattice structure of the liquid state when they cool. Common silica sand (silicon dioxide, SiO_2) is the source of most glass. An exception to amorphous glass materials is a group of technical glasses or glass ceramics that attain a fine crystalline structure as a result of heat treating.

7.4 CERAMIC CLASSIFICATIONS AND PROPERTIES

It is difficult to place ceramics into distinct groups because of the considerable blending of organic and inorganic elements that occurs in ceramic compounds. The majority of ceramics are made up of compounds of oxygen, carbon, and nitrogen along with other elements. These compounds include *silicates* (silicon and oxygen), *oxides* (ox-

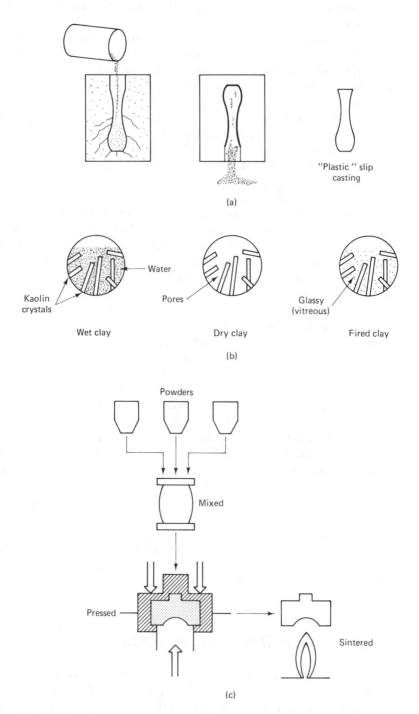

"Plastic" slip
casting

(a)

Kaolin
crystals

Water

Wet clay

Pores

Dry clay

Glassy
(vitreous)

Fired clay

(b)

Powders

Mixed

Pressed

Sintered

(c)

Figure 7-4 Basic processes for ceramics. (a) Casting. (b) Sintering (firing). (c) Powder metal.

ygen that combines with *electropositive elements,* those that easily give up electrons), *carbides* (carbon and metals), *nitrides* (nitrogen and metals), *aluminates* (alumina with metallic oxides), *borides* (boron combined with electropositive elements), and *hydrides* (hydrogen and electropositive elements).

The many forms of carbon serve to illustrate the wide range of ceramic properties. Carbon exists as diamond, the hardest material known, and graphite, an effective lubricant. Diamond has a cubic symmetry (Figure 7-5a) with perfect covalent bonding that joins all atoms into a single molecule. Graphite develops into sheets with good covalent bonding, but between sheets semimetallic bonds are weak and easily shear, thus allowing the sheets to slide past each other (Figure 7-5b). Module

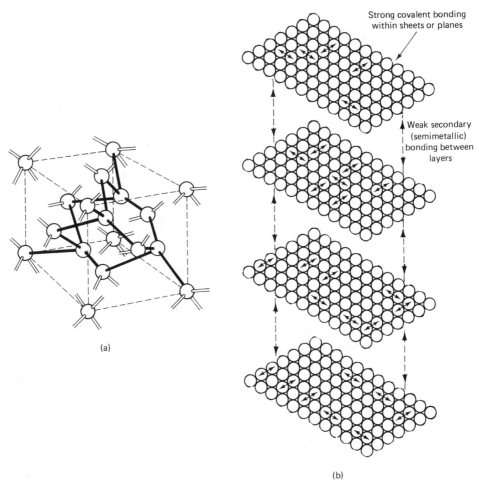

Figure 7-5 Polymorphism of carbon. (a) Diamond's cubic structure: each carbon atom forms strong covalent bonds with 4 other carbon atoms to develop a tetrahedron in same manner as silicon and germanium; diamond cutters use sharp instrument and a sharp blow to split crystal along cleavage planes (111) to produce perfect smaller jewels. (b) Layered structure in graphite: semimetallic bonding between layers allows easy cleavage into sheets thus providing good lubrication properties.

8 describes methods of producing reinforcing carbon fibers through carbonization and graphitization.

Ceramics are used in many areas of manufacturing and building construction, including building materials, electrical and thermal insulators, optical equipment, corrosion-resistant containers, tooling, electronic circuitry, and protective coatings. They are also common in such everyday uses as dishes, eyeglasses, and appliance coatings and components. Ceramics are generally classified under the following headings:

Engineering and technical ceramics	Refractories
Structural clay (brick, tile)	Whiteware and stoneware
Abrasives	Cement and concrete
Insulation	Protective coatings
Porcelain enamels	Fluxes
Glasses	Ceramic glasses
Carbon, diamond and graphite	Electronic ceramics
Ceramic magnets	Ceramic lubricants
Nuclear ceramics	Lime, plaster, and other minerals

As usual, headings do not provide separate and distinct groups. There is an overlap because some groups reflect properties, while others emphasize applications. This discussion will deal with several of the more common and important ceramics. The books cited at the end of this module deal only with ceramics and provide complete coverage of most ceramics.

7.4.1 Engineering and Technical Ceramics

Those ceramics that a designer might select to replace other engineering materials, such as metals, woods, and plastics, are termed *engineering ceramics*. Ceramics have a brittle nature and present problems in holding close tolerances during processing, so in the past they had little engineering use. Plastic deformation in ductile metals spreads stress over the material, thus allowing localized stress to be redistributed, but the lack of plasticity in ceramics results in rupture with relatively small localized stresses. However, development of new technologies has allowed us to take advantage of the high resistance to heat and thermal corrosion, low creep rate, the resistance to wear, and the light weight (compared to high temperature metals) of ceramics, thereby prompting designers to make adjustments to permit their use.

Engineering ceramics commonly include oxides of aluminum, zirconium, magnesium, and titanium along with compositions of carbides, nitrides, and borides of silicon. Covalent bonding of rigid atomic lattices, which are further strengthened through processing, provide strong microcrystals whose electrons are locked up and not easily dislodged by heat, chemicals, and electricity. While ceramics such as silicon carbide only possess tensile strength around 15,000 psi (103.4 MPa) at room temperatures compared to 100,000 psi (689.5 MPa) for certain metal alloys, they often have superior compressive strength. Then when service conditions rise to 1500°F,

(816°C) metals give up their superiority since many ceramics retain strength beyond 3000°F (1650°C). Zirconia (ZrO_2) seems to offer the best combination of properties for many heat engine applications: thermal expansion near that of steel and cast iron, low thermal conductivity, high strength and toughness, good wear resistance, and high thermal shock resistance. See Figure 7.1.

7.4.1.1 Technical ceramics. These include the engineering ceramics and other ceramics used by industry, such as semiconductors, spark plug insulators, laser crystals, and lamp bulbs. *Alumina,* Al_2O_3 (aluminum oxide), is one of the most commonly used engineering ceramics because of the ease of and versatility in its processing. Alumina is less costly than many other ceramics (Figure 7-6 shows a photomicrograph of alumina). Other engineering ceramics of the metal oxide group include BeO, *beryllia* (beryllium oxide), and ZrO_2, *zirconia* (zirconium oxide) (Figure 7-7). Silicon carbide, SiC, and silicon nitride, SiN, provide high strength, thermal shock resistance, and relatively low thermal expansion.

In heat engine development zirconia is joined by silicon carbide and silicon nitride for high temperature (near 2500°F) (1370°C) uses such as power turbines, gasifiers, combustion liners and other insulators, plus rotors. Aluminum silicate and lithium aluminum silicate are designated for engine components where higher temperatures can offer a 30% increase in fuel efficiency. In 1975 a ceramic material of magnesium and partially-stabilized zirconia (PSZ) was discovered that possesses both an increased fracture toughness over brittle ceramics (about $30 \times$) and the capability of being heat treated in various ways to fit specific applications. Commercial production includes two grades, MS (Maximum strength) and TS (thermal shock resistant) that are being evaluated by various engine companies.

Ceramics and *cermets* (ceramics-metal composites) have served industry well

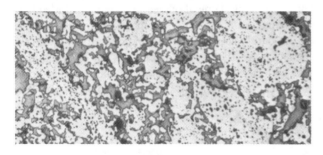

Figure 7-6 Bonded alumina refractory, $125 \times$. (Buehler Ltd.)

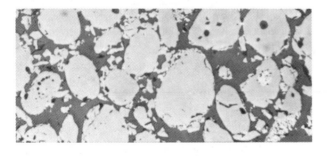

Figure 7-7 Bonded zircon refractory, $125 \times$. (Buehler Ltd.)

as *cutting tools*. Industrial-quality natural diamonds and synthetic diamonds serve as cutting tools as well as abrasives. Aluminum oxide and carbide PM tools allow cutting speeds hundred of times faster than high speed steel (HSS). Newer ceramic cutting tools use silicon nitride (Si_3N_4) formulated with yttrium oxide, silicon oxide, and other elements to provide speeds up to 2000 surface feet per minute (sfm) on rough cuts of cast iron compared to typical speeds of 75 sfm for HSS and 600 sfm for sintered carbide. Sialon (silicon-aluminum-oxygen-nitrogen) is a hot-pressed cermet that works well on nickel-based superalloys. Another composite ceramic achieves improved fracture toughness by dispersing zirconia particles in a Al_2O_3 matrix. This adds strength at grain boundaries and imparts compressive forces which overcome tensile stress that causes failure in conventional Al_2O_3 cutting tools. See Section 7.6.1 for more detail on these prestressing techniques.

7.4.1.2 Glass ceramics. This group includes beta cordierite ($2MgO \cdot 2Al_2O_3 \cdot 5SO_2$), and beta spodumene ($Li_2O \cdot Al_2O_3 \cdot 4SiO$), which are specific phases of glass that possess very low expansion coefficients and superior resistance to oxidation when subjected to high heat. As discussed earlier, glass will crystallize under slower cooling processes than are normally used in glass manufacture. Crystals form on the surfaces of glass with impurities. These surface crystals will cause the failure of the glass part. However, controlled heat treatment can produce a fully crystalized material known as *glass ceramics*.

Glass ceramics are *polycrystalline glass* that has four to five times the strength of glass, with high mechanical hardness about equal to tool steel. This polycrystalline glass is capable of developing structures that resist extreme thermal shock (rapid change from cold to hot, or vice versa). Cookware is a familiar form of glass ceramics. The fine crystalline structure of glass ceramics is achieved by the introduction of a nucleating agent into the ceramic compound. Typical nucleating agents include phosphorus oxide (P_2O_5), fluorides, titanium oxide (TiO_2), platinum, and zirconium oxide (ZrO_2). These nucleating agents are barely soluble in glass and remain in solution at high temperatures. At lower temperatures the agents precipitate out of solution to become seeds or nuclei around which crystals grow. Therefore, the process shown in Figure 7-8a involves allowing a ceramic composition to cool down (b–c), raise

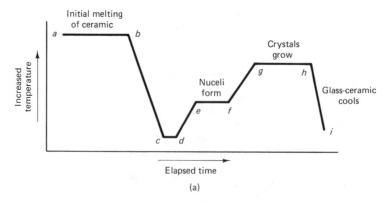

Figure 7-8 (a) Heating curve for crystal formation in glass ceramics.

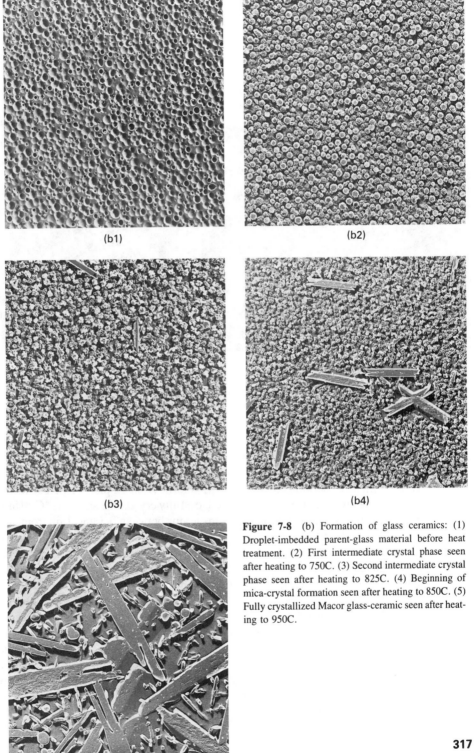

(b1)

(b2)

(b3)

(b4)

(b5)

Figure 7-8 (b) Formation of glass ceramics: (1) Droplet-imbedded parent-glass material before heat treatment. (2) First intermediate crystal phase seen after heating to 750C. (3) Second intermediate crystal phase seen after heating to 825C. (4) Beginning of mica-crystal formation seen after heating to 850C. (5) Fully crystallized Macor glass-ceramic seen after heating to 950C.

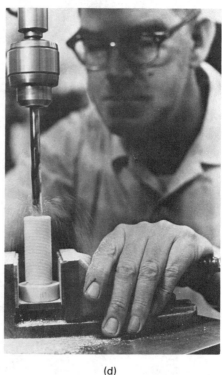

(c)   (d)

Figure 7-8 Machinable glass ceramic: (c) turning Macor™ and (d) drilling previously machined Macor.™ A machinability index allows comparison of Macor™ with graphite and metal going from easiest (1) graphite, to machinable glass ceramic (25), free machining brass (36), 2024 T-4 aluminum (50), no. 10 copper alloy (97), less steel at (229). (Corning Glass Works.)

slightly to nucleation temperature and hold (d–e–f), and then raise and hold the temperature for crystal growth (f–g–h). Figure 7-8b shows the formation of a glass crystal from an amorphous opal glass with fluorine-rich droplets dispersed throughout. Through heat treatment, it is transformed to a fully crystalline structure. Crystallization of glass ceramics exceeds 95% and forms grains between 0.1 and 1 micrometer (μm), which are smaller than in typical ceramics. Formation of glass ceramic parts follows the processes used for glass making, including pressing, blowing, drawing, and centrifugal casting. One polycrystalline glass ceramic consists of mica crystals in an opal glass matrix. Known by the trade name of Macor, it can be machined with standard metalworking tools (Figure 7-8c). Randomly oriented crystals about 20 μm in length and width keep fracture localized instead of propagating (spreading), as would a crack in an amorphous glass or plastic. Table 7-1a provides a comparison of the glass ceramic Macor™ with boron nitride, alumina, and thermoplastic poly-ester. Rather than tensile strength, Table 7-1a gives flexural strength for Macor™. This is a common ceramic test conducted with the setup shown with the modulus of rupture curve. Note that this glass ceramic retains full strength of 15,000 psi to beyond 600°C, but thereafter the rupture strength falls off sharply. Table 7-1b shows the relative machinability of Macor™ in terms of the horsepower required in drilling.

TABLE 7-1a MACOR PROPERTIES

Property	Units	Macor™ machinable glass ceramic	Boron nitride 96% BN	Alumina nominally 94% AL_2O_3	Valox® 310 thermoplastic polyester
Density	gm/cc	2.52	2.08	3.62	1.31
Porosity	%	0	1.1	0	.34
Knoop hardness	NA	250	<32	2000	NA
Maximum use temp. (no load)	°F °C	1832 1000	5027 2775	3092 1700	204° 140°
Coefficient of thermal expansion	in/in °F in/in °C	52×10^{-7} 94×10^{-7}	23×10^{-7} 41×10^{-7}	39×10^{-7} 71×10^{-7}	530×10^{-7} 934×10^{-7}
Compressive strength	psi	50,000	45,000	305,000	13,000
Flexural strength	psi	15,000	11,700	51,000	12,800
Dielectric strength (A.C.)	Volts-mil	1,000	950	719	590
Volume resistivity	ohm-cm	$>10^{14}$	$>10^{14}$	$>10^{14}$	$>10^{14}$

®Registered trademark General Electric

TABLE 7-1b

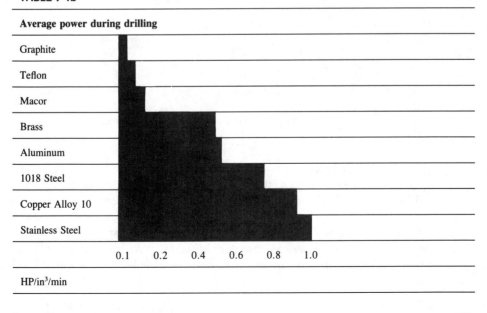

Average power during drilling

Graphite
Teflon
Macor
Brass
Aluminum
1018 Steel
Copper Alloy 10
Stainless Steel

0.1   0.2   0.4   0.6   0.8   1.0

HP/in³/min

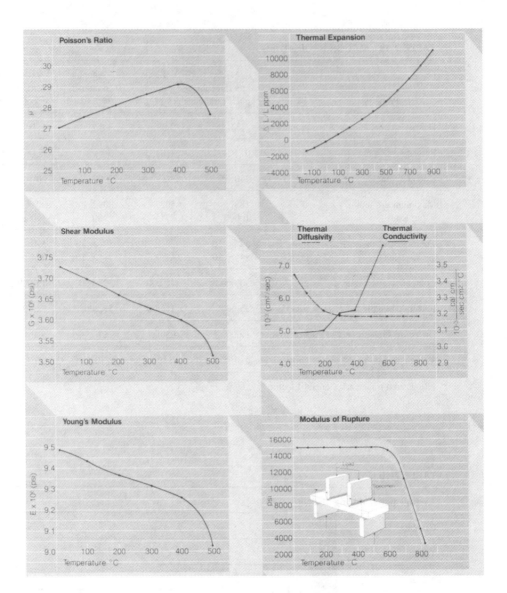

Effect of temperature on mechanical properties.

Another glass ceramic, ceramic monolith, is produced with parallel cellular openings that can be used to direct the flow of gas in heaters or heat exchangers for turbine engines. This material withstands operating temperatures close to 1000°C.

7.4.1.3 Abrasives. Abrasives are very hard particles used for grinding, sanding, and polishing. Abrasive materials include particles of flint, garnet, diamond, aluminum oxide, silicon carbide, emery, pumice (pulverized volcanic lava), rotten

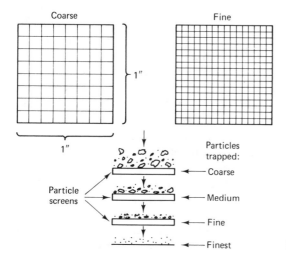

Figure 7-9 Abrasive screening.

stone (shale rock), and corundum. Aluminum oxide and silicon carbide find the greatest industrial application, although garnet is often used as *sandpaper,* an abrasive paper or belt used for sanding wood. *Abrasive papers* include paper and cloth sheets, belts, discs, and drums onto which an abrasive is cemented. *Wet* or *dry* abrasive paper employs waterproof adhesives. Grinding wheels consist of abrasive particles held together by a bonding material that is tough, such as rubber or organic resins. The particles constantly fracture on the wheel to expose new sharp cutting edges, or they break off as the wheel becomes smaller in size. Abrasive particles are also used in the loose form for sand blasting, hand rubbing, or abrasive drilling. Selection of the grit (particle size) is based on a screening method that determines the number of particles per inch (Figure 7-9), such as 60 grit or 600 grit. A 120 grit would allow approximately 100 particles to cover a square inch of screening. Each hole and grit would be about 0.01 inch.

7.4.1.4 Protective coatings.

These ceramic particles are used to coat and decorate pottery and metals when they are melted into a glassy state. *Glazes* and *enamels* impart color, hardness, and corrosion resistance when they are suspended in a slurry and fired to a vitreous state; as the *substrate* (base material) cools, the coating bonds tightly to the substrate surface. *Flame spraying* uses a vaporizing gun, which melts the ceramic oxide powder then sprays or blasts the coating onto the surface of a metal. The oxide material cools and bonds to increase its surface hardness while offering insulation against heat and oxidation.

7.4.1.5 Electronic ceramics.

Typical electronic ceramics include the ferrites (ZrFeO),* silicon (SiO_2),* zirconia (ZrO_2),* steatite (SiO_2MgO),* porcelains (Al_6SiO_{13}),* and alumina (Al_2O_3).* Module 9 deals with some of these and other electronic materials.

*Representative compounds.

7.4.1.6 Ceramic magnets. Those ceramics that contain sufficient *dipoles* (atoms with electron spin in the same direction, discussed in Section 4.6.5) are ferromagnetic. Ceramic ferromagnetics are both soft and hard magnets. High electrical resistivity of ceramic magnets gives them an advantage over metal magnets for use in high-frequency devices. Their permanent magnetic behavior finds application in microwave devices. The group known as *ferrites* are iron oxides consisting of ions of such elements as zinc (Zn^{2+}), iron (Fe^{2+}), magnesium (Mg^{2+}), and nickel (Ni^{2+}) that combine with oxygen. Ferrites are commonly found in the computer industry as memory cores. Newer computer memory materials rely on magnetic bubble technology. *Magnetic bubbles* are tiny magnetic domains on ceramic magnets that can be quickly rearranged along precise paths to serve as digital memories. These *thin-film* devices consist of crystalline materials of magnetic rare-earth garnet over a nonmagnetic substrate. Module 9 gives more detail on magnetic bubbles.

7.4.1.7 Thick- and thin-film ceramic devices. Of major significance to the computer industry is the group of technical ceramic substrates (mostly alumina) onto which semiconductor integrated circuits are mounted. These substrates generally fit into categories of single layer and multilayer ceramic (MLC) devices which employ thick- or thin-film metal layers to serve as circuitry. *Substrate* usually means a base material, but, in these devices, it also includes the composite of ceramic with the metal layer. Figure 4-1 shows an application of such devices for a computer.

For *thick-film* devices, the Al_2O_3 substrate receives a layer of metal paste such as gold, platinum, palladium, or silver. The paste is printed by silk screen in a circuitry pattern (designed to carry the electron signal) onto either greenware or bisque. With the bisque substrate, the paste gets a sintering around 850°C which changes the paste structure into resistors and conductors; greenware and paste receive cofiring. Computer-controlled lasers may trim resistors to exact dimensions before the chip is soldered onto the composite substrate. The MLC shown as a part of Figure 4-1 employs thick-film printing to produce multilayer substrates. The layers begin as continuous slip-cast sheets of greenware which are blanked to size, then silk screened. The ceramic substrates consist of 92% alumina and frit made up of alumina, calcia, magnesia, and silica. Frit is a form of glass which has been melted, cooled, then broken into small particles. The substrates with metal paste are stacked into multilayers (5 to 35 layers), then sintered around 1600°C.

Thin-film circuits composed of micro thin layers (about 150 nm) of material such as gold or nickel-chromium alloys are produced through vacuum deposition onto such ceramic substrates as alumina, glass or beryllia. Further study in the references listed at the end of the module can focus on specific techniques such as sputtering, evaporation and electron beam deposition. Also see Section 9.3.2. for more detail.

Engineering ceramics are normally cast, extruded, press molded, and sometimes injection molded. Alumina, the most versatile of the group, is used for bearing seals, spark plugs, oxygen sensors in auto exhaust manifolds, terminals, bushings, and high-frequency insulators. Beryllia finds use for transistors and resistors. Zirconia changes from an electrical conductor at higher temperatures to an electrical insulator at lower temperatures. Silicon carbide is the most common abrasive used to grind cast iron and nonferrous and nonmetallic materials. The push for higher temperatures and

TABLE 7-2 TYPICAL PROPERTIES OF SELECTED ENGINEERING AND TECHNICAL CERAMICS

Properties	Ceramics				Comparison materials		
	Alumina	Beryllia	Silicon carbide	Zirconia	Mild steel	Aluminum	Nylon
Melting point (approximate °C)	2050	2550	2800	2660	1370	660.2	215
Coefficient of thermal expansion (m/m/K) 10^{-6}	8.1	10.4	4.3	6.6	14.9	24	90
Specific gravity	3.8	–	3.2	–	7.9	2.7	1.15
Density (kg/cu. m)	3875	2989	3210	9965	7833	2923	1163
Dielectric strength (v/m) 10^6	11.8	–	–	9.8	–	–	18.5
Modulus of elasticity (MPa) 10^4	34.5	39.9	65.5	24.1	17.2	6.9	.33
Hardness (Mohs)	9	9	9	8	5	3	2
Maximum service temperature (K)	2222	2672	2589	2672	–	–	422

lighter weight makes engineering ceramics excellent candidates. While many concepts have been advanced for their use in internal combustion engines, such problems as the bonding of ceramics (interface technology—see references on adhesives) to metal and their lack of ductility will require newer technologies before they gain wider use.

The key to broader acceptance lies in processing that will insure highly reliable parts to fit into the materials systems. Among advanced techniques are multilayer ceramics (MLC) sintered reaction-bonding, hot isostatic pressing (HIPS), sol-gel processing, and slip casting using wax molds to replace plaster molds. These innovations couple with advanced design like finite element analysis and improved X-ray and ultrasonic NDT to detect flaws in grain structure and thereby increase reliability. Also, composite development can use ceramic fibers in metal matrices or polymer matrices. Silicon carbide fibers suspended in aluminum give a composite with high fracture toughness suitable as structural members in aircraft, ablative heat shields for spacecraft and rocket nozzles. Other composites involving silicon carbide fibers are candidates for turbines in advanced auto engines.

Table 7-2 provides a comparison of typical properties of some technical ceramics. Moh's hardness scale is not precise, but allows a comparison of the hardness of materials measured with different industrial hardness tests.

7.4.1.8 Advances in technical ceramics.

The evolution of the spark plug used in internal combustion engines represents the advances made in technical ceramic technology. This development began with triaxial porcelain (flint–feldspar–clay), which replaced mica as an insulator. Triaxial porcelain could not withstand the thermal shock nor did it possess the high-temperature electrical resistivity needed for World War I aircraft engines. Mullite ($3Al_2O_3 \cdot 2SiO_3$) overcame some of the shortcomings

CERAMIC DIVISION
body making

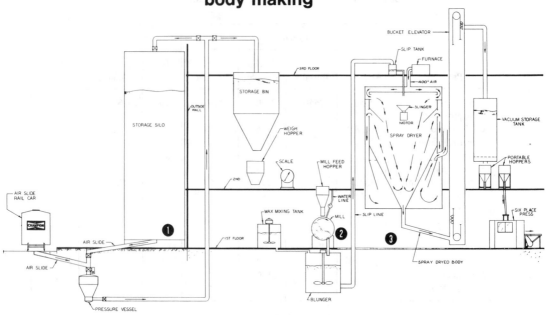

CERAMIC DIVISION
insulator manufacturing

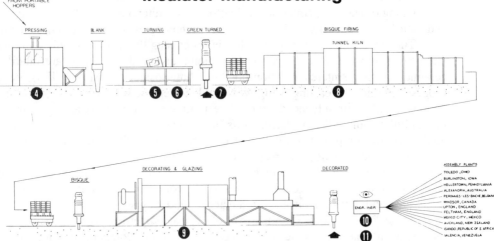

Figure 7-10 Spark plug insulator body manufacturing. (1) Raw materials. (2) Wet grinding of ceramics. (3) Spray drying to obtain definite size and shape of ceramic pellets. (4) Pressing in rubber molds to give "green" unfired insulator blank. (5) Turning of "green" blank to closer tolerances. (6) Comparator inspecting for tolerances. (7) Loaded into refractory trays. (8) Firing at 2700°F to produce bisque. (9) Decorating with glass glaze and refired at 2000°F to produce high gloss finish. (10) Production inspection as done at every other stage. (11) Final engineering inspection statistically used on each batch. (Champion Spark Plug Company.)

TABLE 7-3   TYPICAL PROPERTIES OF HIGH Al_2O_3 INSULATORS

Property	Measurement
Al_2O_3 content	87 to 95%
Specific gravity	3.48 to 3.73
Fired grain size	5 to 10 μm
Flexural strength	40,000 to 60,000 psi
Compressive strength	300,000 to 400,000 psi
Modulus of elasticity	37 to 46 $\times$ 10^6 psi
Thermal expansion (72°–1800°F)	4.5 $\times$ 10^{-6} in.$^{-2}$ °F^{-1}
Dielectric strength	380 to 420 V/mil (0.050-in. thickness)
T_e value*	1550° to 1750°F

*Temperature at which resistivity = 1 MΩ · cm. Courtesy Champion Spark Plug Company.

of triaxial porcelain because it contains alkaline earth silicate rather than feldspar with alkali, which had low electrical resistance at high temperatures. Also, calcined (cooked) kaolin clay replaced the quartz to give the mullite improved insulator properties; this was further improved through use of sillimanite. Mullite, as did triaxial porcelain, possessed too much SiO_2. The SiO_2 mixed with lead antiknock additives in gasoline to form low melting eutectics and reduce lead corrosion resistance. The development of high alumina (Al_2O_3) insulators provided the necessary properties, seen in Table 7-3, required by high-compression engines, that is, high strength, high electrical resistance, high dielectric strength, refractiveness, high thermal conductivity, chemical inertness, and good thermal shock resistance.

Typical of the processing of technical ceramics is the processing of the spark plug shown in the flow chart of Figure 7-10. Processing through step 3 yields high alumina pellets in the range of 75 to 100 μm. The pellets are compacted in rubber molds (Figure 7-11) at 20,000 to over 41,000 psi with isostatic pressure (equal pressure on all sides) at stage 4. Next they are turned to the final shape (Figure 7-12) through stages 5, 6, and 7. Firing in stage 8 is done in a mullen kiln at over 1700°C to produce the bisque insulator, which is glazed, decorated, and then assembled with the metal portions for the completed spark plug assembly.

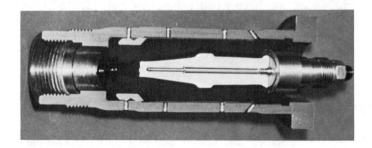

Figure 7-11   Isostatic pressing mold in cross section showing insulator blank. (American Ceramic Society.)

Sec. 7.4   Ceramic Classifications and Properties

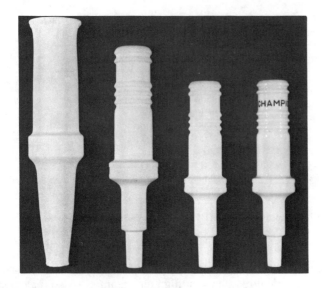

Figure 7-12 Stages of insulator body development (left to right) pressed "green" blank, turned "green" insulator, bisque fired insulator, glass glazed and decorated refired to finished insulator. (American Ceramic Society.)

7.5 REFRACTORIES

Refractory materials include a broad range of ceramics with very high melting points above 1500°C and usually serve as structural materials in high-temperature environments. They are products of sintered clay consisting of crystals in a glassy matrix; these materials find use in ovens, kilns, furnaces, and melting pots, in welding and cutting, and as engine parts. Some common refractory ceramics are alumina, alumina-silica, chromite, bauxite, zirconia, kaolin, silicon carbide, magnesite, and graphite. In addition to heat resistance, chemical properties bear heavily on the selection of refractories that come into contact with molten metals and glasses. The acidic or basic nature of a refractory will determine how it reacts with a given molten material. The reaction between the materials may change the properties of the metal or glass being processed.

Module 5 deals with refractory metals, but *refractory hard metals* (RHM) fits between metals and ceramics. RHMs consist of metal carbides sintered or cemented into a metal matrix. They possess high hardness and wear resistance, but withstand thermal shock and impact better than most ceramics. Serving in high-temperature uses as valves, seals, thermal metal-working tools, and rolls and dies, some RHMs include tungsten carbide in a cobalt matrix, titanium carbide in a molybdenum or nickel matrix, and tantalum carbide in a chromium matrix.

7.6 GLASS

The American Society for Testing and Materials (ASTM) classifies glass as those inorganic products of fusion that have cooled to a rigid condition without crystallizing. They differ from glass ceramics in their lack of a polycrystalline structure. Some authorities prefer not to consider glass as a ceramic because of its structure, and thus give it a separate classification.

7.6.1 Nature of Glass

The base of glass is the very pure white silica sand found in abundant supply in the central United States and other parts of the world. While there are approximately 750 different glasses and glass ceramics, most can fit into six groups: soda-lime, lead-alkali, borosilicate, aluminosilicate, 96% silica, and fused silica.

The vitreous state of glass is mechanically rigid, like crystalline materials, yet has an atomic space lattice that is amorphous like a liquid. Even though such raw materials as quartz sand (SiO_2) have a naturally crystalline structure (Figure 7-13a), when they melt, the lattice breaks up (Figure 7-13b). Slow cooling will allow some crystals to form, but normal cooling produces the amorphous vitreous or glassy structure (Figure 7-13c), such as the SiO_4^{4-} tetrahedron (silicon oxide ion with a net charge of negative four or 4^-).

As discussed under phase diagrams, the amorphous materials do not have clearly defined freezing or melting points, as do crystalline materials; rather, they harden or

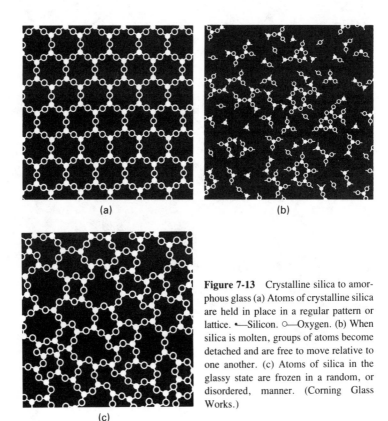

(a)

(b)

(c)

Figure 7-13 Crystalline silica to amorphous glass (a) Atoms of crystalline silica are held in place in a regular pattern or lattice. ●—Silicon. ○—Oxygen. (b) When silica is molten, groups of atoms become detached and are free to move relative to one another. (c) Atoms of silica in the glassy state are frozen in a random, or disordered, manner. (Corning Glass Works.)

liquify over a range of temperatures. By cooling molten glass, it becomes more viscous (Figure 7-14) and thickens to a *working* or *softening* point. Figure 7-15 plots the heating of crystalline raw materials. The curve *a–b* shows an increase in volume until it becomes molten (*liquidus* temperature); increased temperature yields a corresponding increase in volume as the atoms expand. Glass is formed as the raw materials cool through curve *d–c–e*, but the volume remains higher at point *e* than when the raw materials were in a closely packed crystalline state. This curve represents a normally cooled glass.

Slower cooling permits some crystallization in the process known as *devitrification* (Figure 7-8). Devitrification can only occur below the liquidus. As with metals,

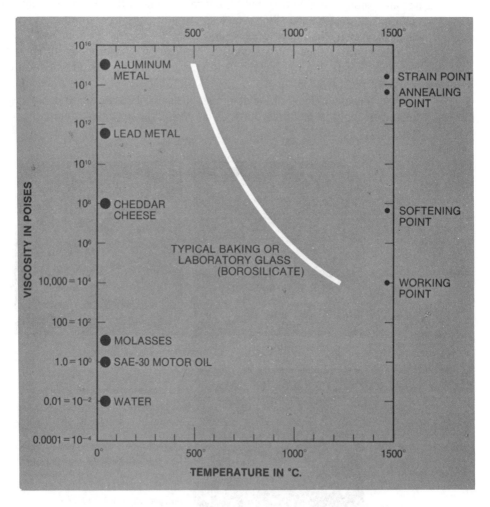

Figure 7-14 Glass viscosity. As glass is heated it becomes gradually less viscous. (Poise—absolute viscosity in Pascals seconds [Pa · s]. The measure of force required to overcome resistance to flow.) (Corning Glass Works.)

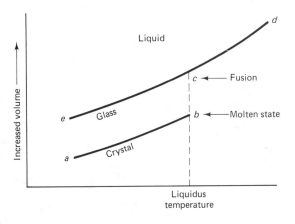

Figure 7-15 Glass formation.

it is important to know the liquidus point to control the glass structure. High resistance to devitrification is desirable and prominent in commercial glasses.

A variety of raw materials yields glass compositions of assorted oxides. Oxides such as SiO_2 (silicon dioxide), B_2O_3 (boron trioxide), GeO_2 (germanium dioxide), and V_2O_5 (vanadium pentoxide) are glass formers. *Glass formers* or *network formers* are those oxides that promote the ionic linking or *polymerization* of oxide molecules in the glass compound. *Network modifiers* or *fluxes* such as lead ions, zinc ions, and alkali ions lower the liquidus temperature, improve workability, and change thermal and optical properties. Other *stabilizers* such as CaO (calcium oxide) improve chemical properties.

Heat treatment of glass, as with metals, can reduce internal stress or create high internal stress. These stresses could lead to breakage from minor force. *Annealing* of glass through slow cooling provides homogeneous structure by reducing internal stresses to give isotropic (equal in all directions) properties. *Tempering* involves rapid cooling of the outer surface of glass while still in the plastic state. The tempering results in compressive stress on the surface and tensile stresses in the core (Figure 7-16). This condition occurs because the slower cooling core tries to contract. The nonequilibrium condition causes tempered glass to fracture into small pieces rather than large splinters. Tempered glass is now required on doors in buildings because of this safety feature.

Glass is nearly a perfect elastic material. At any stress under breaking stress, glass will return to its original shape when the stress is removed. Some plastic flow can be achieved, but the amount of stress must nearly equal its ultimate strength. Except in the case of internal flaws, glass will fail in tension on the surface opposite the compressive stress. Whereas crystalline materials fracture along planes of slippage that may not be normal with the tensile stress, the amorphous structure of glass causes fracture to be normal to the stress. Tempering glass provides a compressive stress throughout the material, which must first be overcome before the tensile stresses can act on the surfaces (Figure 7-16). This is similar to prestressing concrete, which is also weak in tension.

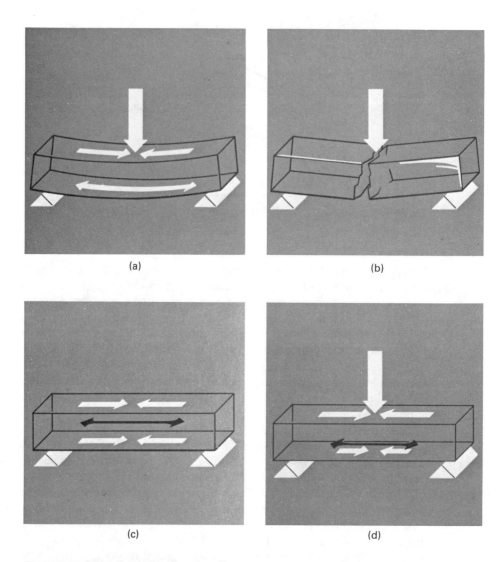

(a)

(b)

(c)

(d)

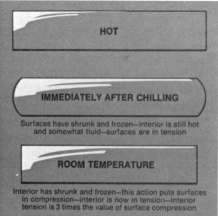

HOT

IMMEDIATELY AFTER CHILLING

Surfaces have shrunk and frozen—interior is still hot
and somewhat fluid—surfaces are in tension

ROOM TEMPERATURE

Interior has shrunk and frozen—this action puts surfaces
in compression—interior is now in tension—interior
tension is 3 times the value of surface compression

(e)

Figure 7-16 (a) Bending load produces tension in the lower surface of test bar and compression in the upper surface. (b) The bar breaks when the tension on the lower surface exceeds the ultimate strength of the glass. (c) Pre-stressed bar shows compression in all surfaces. Reactive tension is buried within the bar. (d) Bending load applied to pre-stressed bar must first overcome built-in compression before surface can be put in tension. (e) Pre-stressing by physical tempering consists of heating glass until it begins to soften. Then an abrupt chill shrinks and freezes surfaces. When interior of glass cools and shrinks, surfaces are compressed. (Corning Glass Works.)

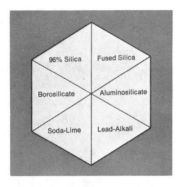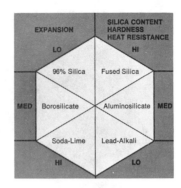

Figure 7-17 Categories of Glass. *Hard* glasses are lower in thermal expansion and higher in both heat resistance and silica content than *soft* glasses. (Corning Glass Works.)

7.6.2 Types and Properties of Glass

The six glass categories are diagrammed in Figure 7-17 and broken into soft glasses and hard glasses. The classification of hard or soft does not denote mechanical hardness but, rather, ability to resist heat. *Soft glasses* are those that soften or fuse at relatively low temperatures and include soda-lime and lead-alkali. Soft glasses have lower heat resistance and a higher coefficient of thermal expansion than the hard glasses. *Hard glasses* are borosilicate and aluminosilicate, and the hardest or high silica glasses are 96% silica and fused silica.

Soda-lime silica glass is the oldest glass; it dates back 4000 years and is most familiar to you in the form of window panes and bottles. It accounts for more than 90% of the glass produced. The typical composition is 74% silica (SiO_2), 15% soda (Na_2O), 10% lime (CaO), and 1% alumina (Al_2O_3). The least expensive of glasses, soda-lime glass has a composition of oxides of silicon (from silica sand), calcium, and sodium. While easy to form and cut into many designs, soda-lime glass has fair chemical resistance but cannot endure high temperatures or rapid thermal changes (thermal shock). Because it shatters so easily into dangerous splinters, it is not recommended for doors or other glazings where people might accidentally hit it. Thinner soda-lime glass as used for small window panes is *single-strength* glass. Increased thickness is referred to by strength, for example, *double-strength* glass.

Lead-alkali silica glass is slightly more expensive than soda-lime but both are considered soft glass; neither will endure high temperatures or thermal shock. *Borosilicate glass* is considered a hard (thermally) glass; it is the oldest type of heat-resistant glass. Pyrex is the familiar Corning trade name for this glass. *Aluminosilicate glass*, while three times as costly, can sustain higher service temperatures than the borosilicates and is similar in its ability to handle thermal shock. The *96% silica glass*, a valuable industrial glass, is capable of setting on a block of ice and having molten metal poured over it without breaking, attesting to its superior thermal hardness (Figure 7-18). *Fused silica glass* is composed of at least 99% silicon dioxide (SiO_2); this pure glass is the most costly and most resistant to heat. It is capable of service

Figure 7-18 Bowl of 96% silica glass sits on block of ice as molten bronze (2000°F) is poured. Failure to break attests to ability of this *hard* glass to withstand thermal shock. (Corning Glass Works.)

temperatures from 900° to 1200°C and also has the highest corrosion resistance. It is superior for the transmission of ultraviolet rays.

Colors are achieved in soda-lime, lead, borosilicate, and 96% silica glasses by the addition of elements that change the glasses' structure to absorb certain bands of light spectra. For example, manganese ions (Mg^+) can produce purple glass and iron yields green and brown bottle glass.

Speciality Glasses. Using the six basic groups just covered, it is possible to develop glasses with unique properties to meet special needs. Some speciality glasses include tempered, optical, colored, sintered, glazing, fibrous, laminated, cellular, photosensitive, and light-sensitive glasses. Advances in *optical glass* technology have resulted from a number of societal and economic demands, including the need for greater transmittance (light transmission), safety (impact resistance), styling and convenience, and competition from the plastics industry. The hard, stable nature of glass permits grinding

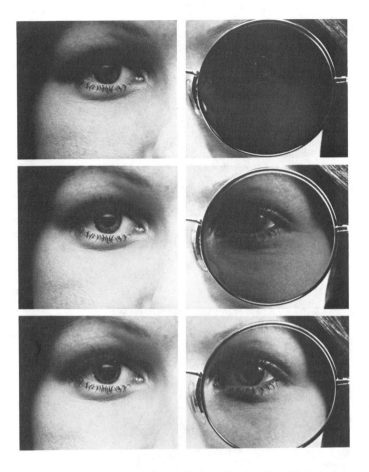

Figure 7-19 Photochromic optical glass. (Corning Glass Works.)

and polishing of optical lenses for telescopes to 1/6000 of 1 percent. For styling and convenience, photochromic lenses provide eyeglasses that darken in bright light but lighten up with subdued light (Figure 7-19). This ability to change the transmittance of light results from the use of silver ions. When photons of light strike these silver particles, they change from silver ions to metallic silver and absorb more light. When the light source is removed, they revert back to silver ions (Ag^+), and the lenses lighten in color.

Chemically Toughened Glass. This type of tempered glass is produced by using a molten salt bath in which eyeglass lenses are soaked for hours to achieve appropriate impact strength. During the long-term soaking, larger potassium ions from the salt bath replace sodium ions, as seen in Figures 7-20 and 7-21. This crowds the surface and causes compressive stresses. The impact strength of the tempered glass is then a product of its untempered strength and the amount of surface compression developed in chemical toughening. A chemically tempered glass is stronger than a heat tempered glass.

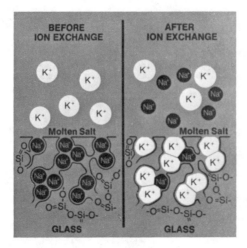

Figure 7-20 Chemically strengthened glass. Ion exchange causes relatively large potassium ions (K+) to replace smaller sodium ions (Na+), thus crowding surface and producing compression (prestressing).

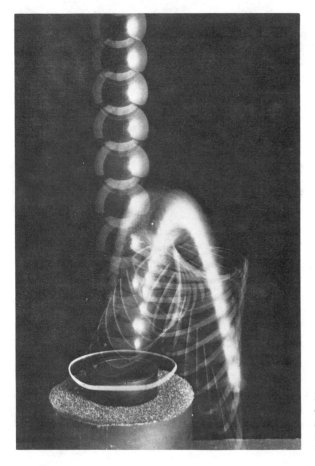

Figure 7-21 A steel ball drops on chemically-strengthened *(chemtempered)* glass from a height of 15 feet (4.5m) and bounces away leaving the lens undamaged. (Corning Glass Works.)

Glazing. Glass is an important architectural material that adds beauty, and it can improve energy efficiency if properly used in design. A variety of types of glass serve as *glazing,* or glass windows. Most older buildings utilize single soda-lime sheet glass.

Glass Fibers. These threads of glass are produced through various techniques, including drawing and blowing operations. Fibers may be produced as thin continuous filaments or discontinuous fiber segments. These thin fibers serve as reinforcers in standard, woven, and other forms for plastic resins such as fiber glass. (Fiber glass is discussed extensively in Module 8.) Segmented or discontinuous glass fiber is also used as a very effective thermal insulation in buildings, on refrigeration and heating units, and in land and air vehicles. Its light weight and good insulation properties coupled with low cost make glass fiber very popular. It is used in sheet or loose form. Glass fiber is also woven and matted into sheets for filters in heating and air-conditioning equipment.

Optical fibers, used to transmit light, are gaining wide acceptance in the communications field because they can be produced from the plentiful silica sand to replace heavier, bulkier, and more expensive copper and aluminum conductors. Optical fibers of glass and plastic have been used for a number of years in the medical profession and by engineers as inspection tubes. Flexible fibers on probes or endoscopes are inserted into the human body or into a motor and attached to a television system or magnifying lens to allow viewing of these otherwise inaccessible places. A single optical fiber the size of a hair has the potential for transmitting several

Figure 7-22 Optical fibers. Fiber-making process uses highly purified glass layers which form as chemical vapors react within the silica tube; tube and contents are heated until they collapse into solid glass rod, from which nearly 10 miles of fiber are drawn in the Modified Chemical Vapor Deposition process. (Bell Laboratories.) Single-strand, optical waveguide has the potential, through fiber-optic technology, to transmit 10,000 simultaneous telephone conversations when used in pairs. (Corning Glass Works.)

TABLE 7-4 COMPARISON OF ELECTRICAL PROPERTIES OF INSULATING MATERIALS AT ROOM TEMPERATURE (CORNING GLASS WORKS)

Thickness Material	Intrinsic dielectric strength				Dielectric constant	Volume resistivity (ohm–cm)
	mm	in. $\times 10^{-3}$	kV/cm	kV/in.		
Cellulose acetate	0.025—0.12	0.98—4.7	2300**	5840	5.5	10^{12}
Glass						
Borosilicate code 7740	0.10	3.9	4800*	12200	4.8	10^{16}
Soda lime	0.10	3.9	4500*	11400	7.0	10^{12}
Soda lead	0.10	3.9	3100*	7880	8.2	10^{14}
Mica, muscovite clear ruby	0.020—0.10	0.79—3.9	3000—8200**	7620—20850	7.3	10^{17}
Phenolic resin	0.012—0.04	0.47—1.6	2600—3300**	6600—8380	7.5	10^{11}
Porcelain, electrical			380**	965	4.4—6.8	10^{14}
Silica, fused			5000*	12700	3.5	10^{18}
Rubber, hard	0.10—0.30	3.9—11.8	2150**	5460	2.8	10^{13}
Porcelain, steatite—low loss			500**	1270	6.0-6.5	10^{15}

Intrinsic dielectric strength can be realized only under special test conditions and is very much higher than the working dielectric strength attainable in ordinary service. These data are listed for purposes of comparison.

*Values of P. H. Moon and A. S. Norcross, Trans. AIEE 49, 755, (1930).

**Values of S. Whitehead, "World Power," p. 72, Sept. 1936.

Table from "Glass, the Miracle Maker," by C. J. Phillips (Pitman Publishing Co., New York, NY)

thousand voice signals, compared to less than 50 voice signals carried on a copper wire of the same size. For this reason, telephone, TV, computer and other communication systems are moving to the smaller coaxial fiberoptics cable to replace copper stranded cables. Several forms of optical fibers exist, including glass cores clad with silica and coated with plastic (Figure 7-22). This high-purity glass fiber can transmit for over 1 km, more than 95% of the light beamed into it, and has a tensile strength over 4137 MPa (6×10^5 psi) and good flexural strength.

Table 7-4 provides a comparison of the electrical properties of various insulating materials. Most of the glasses and ceramics have higher dielectric constants than plastics and rubber. Except for soda-lime glass, they offer greater volume resistivity than the organic polymers. Table 10.5.5 provides some comparison of the mechanical, physical, and chemical properties of selected glasses. With petroleum getting scarcer, it is conceivable that glass and other ceramics with their plentiful raw materials will replace plastics in many applications.

7.7 CEMENT AND CONCRETE

Cement can include any material that acts as an adhesive to bind components together. *Cementitious materials* include pozzolans, lime, gypsum, asphalt, tar, synthetic plastics, and cement. Ceramic cements are most common to us in the form of *concrete* and *mortar*. Concrete and mortar are similar in that both use cement to bind together the *aggregate* (rocks or sand, which combine only through adhesion with cement) that adds bulk to the material. Concrete is more correctly classified as a composite, especially when rocks, reinforcing rods, and wires are added to it.

7.7.1 Portland Cement

Portland cement is the primary cement used in making concrete for building construction. The ingredients of Portland cement include lime (CaO), silica (SiO_2), alumina (Al_2O), and iron oxides (FeO). These ingredients are obtained from such raw materials as limestone, clay, shale, oyster shells, silica sand, iron ore, blast furnace slags, marl, and chalk. Lime comes from dolomite limestone ($CaMg2CO_3$) and consists of calcium carbonate and magnesium carbonate (Figure 7-23). The production of Portland cement comes about through the *calcination* (heating to remove impurities) of the raw materials to yield a fine powder. Control of the specific type of Portland cement depends on the mixture of raw materials. The raw materials yield compositions that are abbreviated as C_3S, C_2S, C_3A, and C_4AF. C_3S is tricalcium silicate ($3CaOSiO_2$), C_2S is dicalcium silicate ($2CaOSiO_2$), C_3A is tricalcium aluminate ($3CaOAlO_3$), and C_4AF is tetracalcium alumino ferrite ($4CaOAl_2O_3Fe_2O_3$). These compositions are varied to produce five types of Portland cement, as seen in Table 7-5.

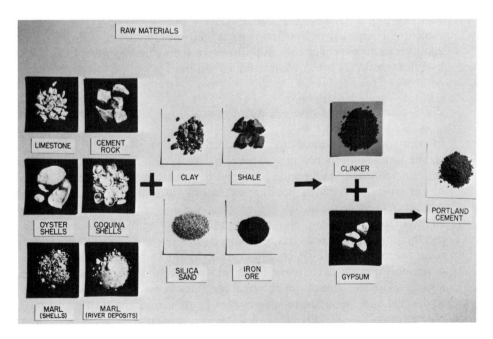

Figure 7-23 Raw materials to make Portland cement. [From Design and Control of Concrete Mixtures (EB001.11T), Portland Cement Association, Skokie, IL.]

Sec. 7.7   Cement and Concrete

TABLE 7-5 PORTLAND CEMENT

ASTM Type	Name	Composition				Nature and applications
		C_3S	C_2S	C_3A	C_4AF	
I	Normal	50%	24%	11%	8%	General construction, e.g., sidewalks, roadways, bridges, normal soil
II	Modified	42%	33%	5%	13%	Better resistance to sulfates in ground water than type 1; drainage systems, foundations, floor slabs
III	High early strength	60%	13%	9%	8%	High strength soon after pouring (week or less), about 190% strength of type 1 in 3 days, good when freezing protection required
IV	Low heat	26%	50%	5%	12%	Minimum heat, slow setting, massive structures, e.g., large dams
V	Sulfate	40%	40%	4%	9%	Highly sulfate and sea water resistance, slow hardening, bulkheads, sea dams

7.7.2 Masonry Mortar

Masonry mortar consists of masonry cement plus a fine aggregate sand. Masonry cement is comprised of Portland cement with hydrated lime, silica, slag, and other additives to improve plasticity and slow down setting time. Masonry mortar is applied by brick masons between blocks, brick, and tile.

7.7.3 Cement and Concrete Additives

In efforts to produce more attractive concrete structures and mortar joints, decrease setting time, and produce more complex shapes, *additives* change the properties of cements, concrete, and mortar. The ratio of water to cement is critical in providing proper workability during pouring and shaping to achieve maximum strength after the concrete is in place. Techniques that vacuum out water after placement and additives to provide extra plasticity are now being employed. Mortar used to cement together bricks and blocks has been improved with color additives and additives that reduce corrosive efflorescence. *Efflorescence* normally results in a white discoloration of masonry and is especially noticeable with dark-colored brick. Corrosion in concrete also develops because of the chemical reactions between high-alkali cement and an acid aggregate. An additive of pozzolan helps to prevent such corrosion. Just as billions of dollars annually are wasted due to corrosion of metals in the United States, corrosion of concrete is highly wasteful. Bridges, roadways, and other concrete structures are deteriorating from corrosion tens of years before they fulfill their designed service life.

7.7.4 Plaster of Paris

Plaster of Paris, another hydraulic (water combined) cement, is used for wall coverings as well as molds for the casting of precision low-temperature metal. It is also used in slip casting of ceramics. The calcining of gypsum rock provides a pure white plaster of Paris [$Ca_2S_2O_7(OH)_2$]. The powder is then mixed with water for placing on walls. When set, the plaster is highly crystalline with poor strength and weak cementing power. To reduce labor cost and improve strength, dry wall board has rapidly replaced plaster. The wallboard of gypsum has improved strength since it becomes a composite with paper sides containing the gypsum.

7.8 CLAY

Clay is broadly grouped as either residual or sedimentary. Residual clay results from the wearing down of rock from the mechanical and chemical action of wind, water, earth movements, and chemicals in the soil. Residual clays are found where they have formed. Sedimentary clays also form through mechanical and chemical erosion of rock, but they are moved by wind or water to places other than where they were formed. Common clays include kaolin ($Al_2O_3 \cdot 2SiO_2 \cdot 2H_2O$), ball clay, fire clays, montmorillonites ($Al_2O_3 \cdot 4SiO_2 \cdot H_2O$), stoneware clay, aluminous clays, slip clays, and flint clays. In various forms, clay finds many applications such as the ingredients for most of the ceramics previously mentioned, as structural products of brick and tile, as raw materials for sculptured art, as china dishes, and as electronic components. Silicon and aluminum are prominent in clay. This is due to the fact that silicon represents nearly 28% of the earth's upper layer and aluminum accounts for 8%. Most other elements are only in small percentages, such as carbon with nine thousandths of a percent.

APPLICATIONS AND ALTERNATIVES

Ceramics in transportation. Because the design of this "space truck" (Figure 7-24) required that it be reusable instead of being a one-shot vehicle, new materials were needed. Earlier space vehicles, such as those that carried men to the moon and back, used polymer heat shields that served as ablatives (discussed in Module 6). Upon reentry into the earth's atmosphere, these heat shields were subjected to temperatures over 1350°C, which charred them beyond further use. Such materials are unsatisfactory for the Space Shuttle Orbiter because these vehicles must fly 100 missions through the atmosphere. Figure 7-24 shows the range of temperatures that the vehicle will encounter. The shuttle's external protection system (Figure 7-30) has reinforced carbon–carbon (RCC), consisting of *pyrolyzed* (chemical decomposition) carbon fabric and binder with an overcoat of silicon. As a result of firing operations, the silicon layer consists of ceramic silicon

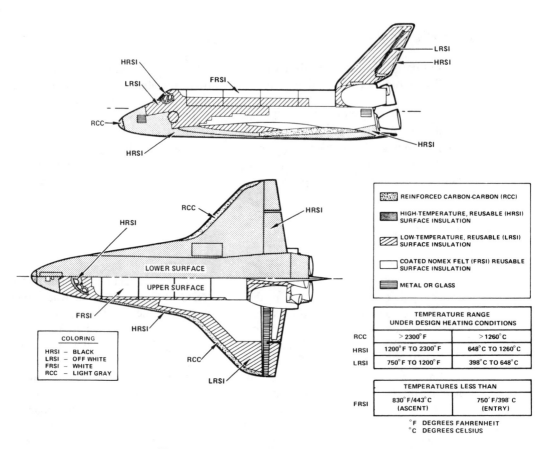

Figure 7-24 Space shuttle thermal protection system. (NASA.)

Within the figure:

LRSI
HRSI
FRSI
HRSI
LRSI
RCC
HRSI
HRSI

RCC
HRSI
HRSI
LOWER SURFACE
UPPER SURFACE
FRSI
HRSI
RCC
LRSI

	REINFORCED CARBON-CARBON (RCC)
	HIGH-TEMPERATURE, REUSABLE (HRSI) SURFACE INSULATION
	LOW-TEMPERATURE, REUSABLE (LRSI) SURFACE INSULATION
	COATED NOMEX FELT (FRSI) REUSABLE SURFACE INSULATION
	METAL OR GLASS

	TEMPERATURE RANGE UNDER DESIGN HEATING CONDITIONS	
RCC	> 2300° F	> 1260° C
HRSI	1200° F TO 2300° F	648° C TO 1260° C
LRSI	750° F TO 1200° F	398° C TO 648° C

	TEMPERATURES LESS THAN	
FRSI	830° F/443° C (ASCENT)	750° F/398° C (ENTRY)

°F DEGREES FAHRENHEIT
°C DEGREES CELSIUS

COLORING

HRSI	–	BLACK
LRSI	–	OFF WHITE
FRSI	–	WHITE
RCC	–	LIGHT GRAY

oxides on the outer surface and interfaces with the carbon through a silicon carbide layer. This silicon layer provides oxidation protection for the carbon network. The high- and low-temperature reusable surface insulation (HRSI and LRSI) consists of a low density (9 lb/ft^3) rigidized silica network (brick) overcoat about 0.015 in. thick with a black glass coating in the case of HRSI and a white glass coating for the LRSI. Both coatings are essentially borosilicate glasses (Figure 7-25).

Flexible reusable insulation (FRSI) consists of an Aramid™ polymer felt that has been heat treated to reduce shrinkage and overcoated with a layer of white paint (Figure 7-25). The HRSI and LRSI are applied to the vehicle in 6- to 8-in. squares of various thickness (0.3 to 4 in.) by use of nonheat-treated Aramid™ felt. One layer of Aramid pad is bonded to the HRSI tile base and the other layer is bonded to the aluminum structure. Room temperature vulcanizing (RTV) adhesives serve as the bonder in layers of about 0.007 in.

FRSI is bonded to the Orbiter vehicle in large sections with RTV

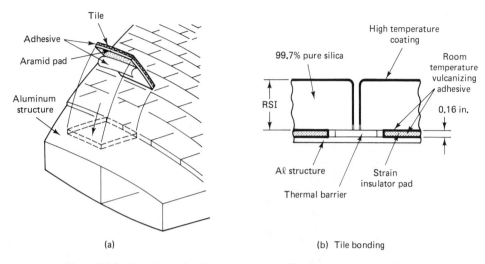

Figure 7-25 Reusable surface insulation and reusable external insulation. (NASA.)

adhesive. The HRSI and LRSI are installed in arrays of as many as 30 to 40 tiles. Filler tiles are individually applied between the arrays. Each Orbiter vehicle requires 30,000 to 40,000 individual tiles of HRSI and LRSI. These tile profiles are produced with computer-controlled machining, with each tile a different size. Areas between and under the tile joints are filled with thermal barrier made of aramid felt coated with RTV.

Ceramics in construction. Proper design of homes, offices, and factories can utilize the natural solar energy in cold weather or block it during hot weather to conserve valuable energy. Solar lighting is also free and pleasant to work with under the proper conditions. A well designed and well-managed building compensates for the location of the sun in winter and summer. Large glassed-in areas south of a building permit maximum sunlight to enter in the winter. Table 7-6 provides a comparison of the fossil fuel investment required in the production of typical building materials. Additionally, the table compares the weight of these materials. In addition to the fact that float glass is produced with good cost effectiveness, sand, its principal ingredient, is quite plentiful.

Other factors must also be considered in the final material cost in terms of energy. These include exploration, mining, refining, and shipment of the raw material and finished product. Beyond that are costs of installation, maintenance, and replacement. How do glass and other ceramics compare on these factors? What possibilities do these materials offer for recycling?

TABLE 7-6 FOSSIL FUEL FEED STOCK AND PRODUCTION FUEL CONTENT OF TYPICAL INSTALLED BUILDING ENVELOPE COMPONENTS

MATERIAL	PLYWOOD	ACRYLIC PLASTIC	FACE BRICK & MORTAR	CONCRETE	FLOAT GLASS	ALUMINUM (11 GAUGE)	STEEL (18 GAUGE)
Weight of Material (Lbs/Cubic Ft.)	40	72	120	148	156	173	490
Typical Envelope Thickness (Inches)	0.5	0.25	4	4	0.09 0.125 0.25 0.5	0.125	0.05
Weight of Envelope Material (Lbs/Sq. Ft.)	1.7	1.5	40	49	1.2 1.6 3.3 6.5	1.8	2
Envelope Feed Stock & Fuel Content (1000 BTU/Sq. Ft.)	1.2	84.7	308.3	128.5	7.3 14.7 29.3 58.6	92.6	27

7.9 SELF-ASSESSMENT

7-1 In Section 7.1, you were asked to consider the possible uses of ceramics in an automotive engine as you studied the module. List at least four such possibilities and explain the reason for the chosen ceramic in each application.

7-2 Explain the type of bond(s) found in ceramics.

7-3 Explain how bonding affects the following properties of ceramics: (a) electrical conductivity, (b) ductility, (c) chemical resistance.

7-4 Name three elements commonly found in ceramics.

7-5 In terms of microstructure, why do crystalline ceramics have different properties from metals that are also crystalline?

7-6 (a) Explain some reasons that silica tiles were chosen over ablative polymers or metals as the thermal protection system of the Space Shuttle Orbiter. (b) Why is it important that ZrO_2 have a thermal expansion rate near cast iron when ZrO_2 is a candidate for engine use?

7-7 Why do graphite and kaolin clay crystals slip easily, while diamond and other ceramics are quite hard?

7-8 What type of flaw might develop in the manufacture of a beer bottle that would cause it to break with only a slight jolt?

7-9 Match the name of the ceramic to its chemical formula or composition.

Portland cement	Al_2O_3
Soda-lime silica glass	SiO_2
Alumina	99% SiO_2
Quartz	$Al_2(Si_2O_5)(OH)_4$
Fused silica glass	74% SiO_2, 15% Na_2O, 10% CaO, 10% Al_2O_3
Kaolinite	C_3S, C_2S, C_3A, C_4AF

7-10 Explain (a) how crystals are formed in glass, (b) when crystals are undesirable, and (c) when desirable.

7-11 (a) What effect does tempering have on glass? (b) How does light cause photochromic lenses to darken? (c) How does toughened glass obtain improved impact strength?

7-12 State the approximate viscosity values in poise for the liquid state of the following materials at the designated temperatures: (a) borosilicate glass at $>500°C$; (b) water at $<100°C$; (c) motor oil SAE 30 at $<100°C$.

7-13 Match the ceramic to the application for which it is best suited:

a) Tempered glass	1) Grinding and polishing
b) Fused silica	2) Furnace windows
c) Ferrite	3) Protect metal from oxidation
d) Glaze (porcelain)	4) Computer memory discs
e) 96% silica glass	5) Storm door glazing
f) Diamond	6) Contain hot acids

7-14 Consider the characteristics of the general classes of ceramics, plastics, and metals. List the class that is generally superior in the following characteristics; then write down a specific material and value to illustrate its superiority: (a) stiffness, (b) chemical resistance, (c) abundance of raw materials, (d) tensile strength, (e) lowest coefficient of thermal expansion, (f) thermal insulation, (g) hardness, (h) impact strength, (i) flexural strength, (j) density.

7-15 Match the correct ceramic to its favorable characteristics. Some will have the same characteristics.

Boron nitride	Low cost
Graphite	Machinable
Al_2O_3	Best lubricity
Glass ceramic	Maximum hardness

7-16 Name three engineering ceramics, cite a typical application, and give the property that probably caused its selection for that use.

7-17 (a) How are the sizes of abrasive grit determined? (b) Name five types of materials used for abrasives. (c) List three industrial applications of abrasives.

7-18 (a) State three uses for glass fibers. (b) What advantages do optical fibers have over copper and aluminum conductors for the transmission of electrical signals?

7-19 List the density for each of the following: (a) aluminum, (b) steel, (c) Al_2O_3, (d) nylon, (e) 96% silica glass, (f) ZrO_2.

7-20 Provide the following property values that represent the ceramics listed: (a) Knoop or Mohs hardness, (b) dielectric strength, (c) Young's modulus, (d) refraction index, (f) thermal expansion.

(1) Soda-lime glass	(4) ZrO_2
(2) Glass ceramic	(5) Al_2O_3
(3) Borosilicate glass	(6) SiC

7-21 (a) What is the purpose of calcination? (b) Name two ceramics that undergo this process.

7-22 (a) Name four cementitious materials. (b) Which are not ceramics?

7-23 (a) What are the ingredients of Portland cement? (b) Why would additives be used with cement and concrete?

7-24 (a) What are the main ingredients in clay? (b) List two industrial applications of clay and explain why clay was selected as the material.

7-25 Select applications of plastics and metals, (a) list the specific material and its use, then (b) list a good ceramic substitute.

7.10 REFERENCES AND RELATED READING

All about Glass. Corning, N.Y.: Corning Glass Works, 1968.

Annual Book of ASTM Standards Part 17. Refractories, Glass and Other Ceramic Materials; Carbon and Graphite Products. Philadelphia: American Society of Testing and Materials, 1983 or current edition.

Automotive Engineering, February 1984, vol 92 number 2 pp. 68–72, "New Ceramics Advance Adiabatic Diesel Engine."

BABCOCK, C. L. *Silicate Glass Technology Methods*. New York: Wiley, 1977.

BLANKS, ROBERT F., and HENRY L. KENNEDY. *The Technology of Cement and Concrete*. New York: Wiley, 1955.

BORAIKO, ALLEN. "The Chip." *National Geographic*, October, 1982, Vol. 162, No. 4. pp. 420–476.

CHISSICK, S. S., and R. DERRICOTT. *Asbestos: Properties, Applications, and Hazards*, Vol. 2. New York: Wiley, 1983.

COLEMAN, JOHN R. "New-Generation Ceramics Cut Costs," *Tooling and Production,* May, 1984 pp. 46–49.

CORDON, WILLIAM A. *Properties, Evaluation, and Control of Engineering Materials.* New York: McGraw-Hill, 1979.

Design and Control of Concrete Mixtures. Engineering Bulletin of Portland Cement Association, Skokie, IL, 1968.

FLINN, RICHARD A., and PAUL K. TROJAN. *Engineering Materials and Their Applications,* 2d ed. Boston: Houghton-Mifflin, 1981.

FRENCH, WILLIAM G. *Materials for Fiber Optical Communication.* Educational Modules for Materials Science and Engineering. University Park, Pa.: Pennsylvania State University, 1976.

GARDINER, KEITH M. (ed) *Systems and Technology for Advanced Manufacturing,* Dearborn, MI: Society of Manufacturing Engineers, 1983.

GRAFF, GORDON. "Ceramics Take on Tough Tasks." *High Technology,* December, 1983, vol. 3, no. 12, pp. 68–73.

HANNANT, D. J. *Fibre Cements and Fibre Concrete.* New York: Wiley, 1978.

"IBM 3081 Systems Development Technology," *IBM Journal of Research and Development,* vol. 26, no. 1, Jan. 1982.

JONES, J. T., and M. F. BERARD. *Processing and Testing.* Ames, Iowa: Iowa State University Press, 1972.

KINGERY, W. D., H. K. BOWEN, and D. R. UHLMANN. *Introduction to Ceramics,* 2d ed. New York: Wiley, 1976.

Lectures and pass out material from IBM Manufacturing Technology Institute, Corporate Technical Institute, New York, April 22–June 29, 1984.

MCDERMOTT, JEANNE. "Ceramics: The Future Beyond Plastics," *Technology,* vol. 1, no. 1, Nov./Dec. 1981, pp. 18–30.

NORTON, F. H. *Elements of Ceramics,* 2d ed. Reading, Mass.: Addison-Wesley, 1974.

ONODA, GEORGE Y., and LARRY L. HENCH. *Ceramic Processing before Firing.* New York: Wiley, 1978.

PERUCHER, K. N., and CONRAD HEINS. *Materials for Civil and Highway Engineers.* Englewood Cliffs, N.J.: Prentice-Hall, 1981.

Polycrystalline and Amorphous Silicon Photovoltaic Cells. SERI/SP-281-1704. Washington, D.C.: Solar Energy Research Institute, U.S. Department of Energy, 1982.

Properties of Glasses and Glass-Ceramics. Corning, N.Y.: Corning Glass Works, 1973.

RAMACHANDRAN, V. S., R. F. FELDMAN, and J. J. BEAUDOIN. *Concrete Science: A Treatise on Current Research.* New York: Wiley, 1983.

TILL, WILLIAM C. and JAMES T. LUXON. *Integrated Circuits: Materials, Devices, and Fabrication,* Englewood Cliffs, N.J.: Prentice-Hall, Inc., 1982.

Use Solar Daylight and Heat. Pittsburgh: Pittsburgh Plate Glass.

VAN VLACK, LAWRENCE H. *Physical Ceramics for Engineers.* Reading, Mass.: Addison-Wesley, 1964.

Periodicals.

Ceramic Bulletin, American Ceramic Society.

Ceramic Monthly.

Corning Product News, Corning Glass Works.

8

COMPOSITE MATERIALS

8.1 PAUSE AND PONDER

How much do you know about the tires that support you when you ride in a car or bus, or land in a high-speed jet liner? If you have ever purchased tires for a car, you know they seem quite costly and the salesperson probably gave you a number of reasons to choose one brand of tire over another, but did he or you fully understand this marvel of engineering design and technological development?

Radial tires offer the most advanced stages of rubber and tire technological developments. Most, if not all, tires today are composites. They incorporate a variety of materials that act together to meet the demands of high speeds and a wide range of road conditions.

Steel-belted radial tires use a materials system of steel belts in rubber in which the steel belts move and generate heat. This very rough description of a tire is used to point out one of many problems that must be overcome in designing a materials system. In this case the incompatibility of the original constituents precludes their use in tires for high speeds and very hard cornering as experienced by highway police vehicles. So new constituents had to be employed to meet the demands.

The strength of a tire and its ability to minimize the heat generated by the flexing of the cord body, as well as the relative movement of the steel fibers, depends greatly on the type of fabric used—the basic material. Figure 8-1 identifies the main components of a tire such as the body cord piles, stabilizing belts, and bead, each made of different materials because each has different and specific requirements. Fabrics made from nylon,

Bias Ply Tire

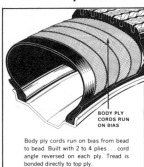

Body ply cords run on bias from bead to bead. Built with 2 to 4 plies . . . cord angle reversed on each ply. Tread is bonded directly to top ply.

BODY PLY
CORDS RUN
ON BIAS

Radial

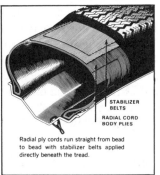

Radial ply cords run straight from bead to bead with stabilizer belts applied directly beneath the tread.

STABILIZER
BELTS

RADIAL CORD
BODY PLIES

Figure 8-1 What's inside a passenger car tire? (Firestone Tire and Rubber Co.)

rayon, polyester, fiber glass, aramid, and steel fibers make up to 12% of the weight of a tire but provide almost 100% of its strength. Steel belts are woven from ultrathin wire, with each strand containing up to 12 wires. Kevlar™ aramid fiber, developed especially for tires, is also used as a belt material for both radials and bias-ply tires. Its strength is so great that a $\frac{1}{2}$ lb of Kevlar can replace $2\frac{1}{2}$ lb of steel. Kevlar fibers in tires made especially for highway police cars permit sustained speeds of 200 km/h. Different types of rubber, both natural and synthetic, go into the tire; treads must be hard to withstand wear and puncture, while also being resistant to various chemicals; sidewalls must be flexible, reduce heat buildup, and resist attack by ozone. After molding, each tire is cured a prescribed amount of time to allow vulcanization to change the soft pliable rubber compounds into the tough elastic material required for good tire performance. A tire is an excellent example not only of a composite material but of the many advanced materials making up a materials system for the fabrication of today's products.

The tire, as an example of a composite, illustrates the manner in which materials from our various family groups can be brought together to meet a design objective. As you study this module, note that composite design strives to have one material complement the other. Ask why specific constituents are used. What applications can you see for the composites discussed? Why are more composites not used to replace traditional engineering materials such as steel and wood?

8.2 INTRODUCTION

A *composite material* or *composite* is a complex solid material composed of two or more materials that on a macroscopic scale form a useful material. The composite is designed to exhibit the best properties or qualities of its constituents or some properties possessed by neither. The combining of these two or more existing materials is done by physical means, as opposed to the chemical bonding taking place in the case of alloys of monolithic solid materials. A true composite might be considered to be one that has a matrix material completely surrounding its reinforcing material in which the two phases act together to produce characteristics not attainable by either constituent acting alone. It is to be noted that the insoluble phases or main constituents in a composite do not lose their identity. This characteristic is not true with solids such as metallic alloys or copolymers, whose phases are lost to the naked eye.

The broad definition of composite materials includes the naturally occurring composites such as wood, as well as the synthetic or man-made composites.

In answer to the question, why use composites? one can reply, in part, without attaching significance to their listing, as follows:

1. To increase stiffness, strength, or dimensional stability.
2. To increase toughness (impact strength).
3. To increase heat deflection temperature.
4. To increase mechanical damping.
5. To reduce permeability to gases and liquids.
6. To modify electrical properties (e.g., increase electrical resistivity).
7. To reduce costs.
8. To decrease water absorption.
9. To decrease thermal expansion.
10. To increase chemical and corrosion resistance.

The objective of this module is to present a fundamental knowledge of composites, the terms used in this specialized field, the nature, structure, and properties of composites, along with an arbitrary classification of the major types. Examples of the applications of some composites in a limited number of industries are included with a view toward providing the reader, many of whom occupy positions in industry that require a degree of expertise in the selection of materials, with a base of knowledge of composites upon which to make intelligent selection decisions.

8.2.1 Development of Composites

Composite materials have been used from the earliest of times. Mongol bows in the thirteenth century utilized a materials system consisting of animal tendons, wood, silk, and adhesives. The ancient Israelites used straw to reinforce mud bricks. The early Egyptians fabricated a type of plywood. Medieval swords and armor are examples of our present-day laminated metal materials. Nature has provided composite materials in such living things as seaweed, bamboo, wood, and even human bone.

Such materials were recognized for their strength and light weight when used in the construction of the first all-composite, radar proof airplane—the Mosquito—around 1940. Composite materials in the form of sandwich construction showed that primary aircraft structures could be fabricated of these materials. World War II saw the birth of glass fiber–polyester composites for radomes and secondary aircraft structures, such as doors and fairings which were designed and placed into production.

Glass fiber composites were recognized as valid space materials when they were picked for fabrication and production of Polaris submarine missile casings.

In the 1950s, fiber technology identified the need for fibers that could compete in strength and stiffness when the state-of-the-art development led to high-performance glass fibers such as the S-type. In the late 1950s, research efforts pointed to the lightweight elements in the search for even greater strength fibers that could successfully compete in the marketplace with aluminum and titanium. Boron fibers were the first result of this effort (1963), followed by carbon, beryllium oxide, and graphite. These filaments are surrounded by some material such as aluminum that serves as a matrix. Beryllium oxide fiber technology developed short fibers, whereas boron research developed continuous filaments that contained greater strength properties. These developments, made by the collective efforts of governments, NASA, industry, and universities, gave rise to *advanced composites* (now sometimes referred to as high-performance composites) that use fibers having high moduli, such as boron, carbon, and graphite (see Figure 8-2a). The other two major ingredients needed for successful production—engineering design and manufacturing—kept pace with these developments in materials and resulted in a constant introduction of new man-made composite materials for use throughout industry.

Due to the many advances in recent years in fiber composite systems, not only as spin-offs from the nation's space program but as a consequence of the growing energy consciousness and concern about the depletion of the world's nonrenewable material resources by a growing number of citizens, the emphasis placed on the group of composite materials in this module is deemed appropriate. This in no way is an attempt to downgrade the importance of the many other composite materials without which our economic well-being would suffer adverse consequences. In industries' new thrust for "design for manufacturability" and "design for automation," designers seek to reduce the number of components in a product, eliminate mechanical fasteners, and design to facilitate assembly by robots. Plastics and plastic based composites offer great promise to help meet certain of these goals.

8.3 CONSTITUENTS OF FIBER-REINFORCED COMPOSITES (FRP)

A fiber reinforced composite (FRP) is a material system primarily made of varying amounts of some particular fiber reinforcement embedded in a protective material called a matrix, with a coupling agent applied to the fiber to improve the adhesion of the fiber with the matrix material. Many of the terms in this modern materials technology field originated in the textile industry (Figure 8-2b). A few such terms are selected for brief definitions to facilitate an understanding of these highly important engineering materials.

8.3.1 Fiber

A *fiber* is the basic individual filament of raw material from which threads or fabrics are made. Fibers constitute one of the oldest engineering materials. Jute, hemp, flax, cotton, and animal fibers have been used from the earliest days of history. Today,

COMPOSITE MATERIALS

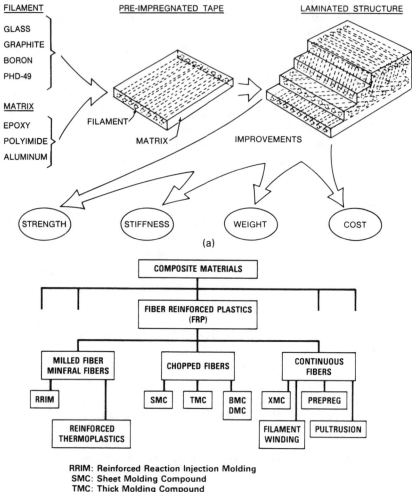

Figure 8-2 (a) Composite materials. (NASA.) (b) Major types of FRP composite materials. (General Motors.)

fibers can be organic (plant, animal, and mineral), synthetic (man-made from a polymer or ceramic material), or metallic. ASTM further defines a fiber as having a length at least 100 times its diameter, with a minimum length of at least 5 mm. A fiber can be a filament or a staple. *Filaments* are long, continuous fibers, whereas *staple* fibers are less than 150 mm in length. Most natural fibers are in staple form, whereas synthetic fibers may exist in both forms.

The *whisker* is a man-made nearly perfect single crystal with a diameter of about 1 micrometer (μm). Because it contains but few crystalline defects in the direction of the applied load, its strength approaches the theoretical value of over 6.9×10^3 MN/m^2 (meganewtons per square meter). Other materials have strengths well below this value. Man-made whiskers of many ceramic materials such as aluminum oxide (Al_2O_3 or sapphire), silicon carbide (SiC), and silicon nitride (Si_3N_4) with hardness values approaching that of diamond are being used as reinforcements in composite materials.

Continuous fibers are fibers that are essentially infinite in length and extend continuously throughout the matrix.

Discontinuous fibers are fibers less than 30 cm in length that tend to orient themselves in the direction of resin flow.

8.3.2 Matrix

The function of a *matrix,* the binder material, whether organic, ceramic, or metallic, is to support and protect the fibers, the principal load-carrying agent, and to provide a means of distributing the load among and transmitting it between the fibers without itself fracturing. When a fiber breaks, the load from one side of the broken fiber is first transferred to the matrix and subsequently to the other side of the broken fiber and to adjacent fibers. The load-transferring mechanism is the shearing stress in the matrix. Typically, the matrix has a considerably lower density, stiffness (modulus), and strength than the reinforcing fiber material, but the combination of the two main constituents (matrix and fiber) produces high strength and stiffness, while still possessing a relatively low density.

The matrix serves as the structure, while a filler provides the desired properties. In any case, there is no significant chemical reaction between the two phases except for the bonding action at their interface. If this were not the case, any reaction between the two materials would have a varying negative effect on the inherent properties of the filled composite.

The structure or matrix of a filled composite can take a porous or spongelike structure form. This network of open pores can be impregnated with a variety of materials, from plastics to metals to ceramics, and the final shape of the filled open pore composite is formed during the processing. Finally, the matrix may have a predetermined shape and size formed by an open honeycomb core made of metal impregnated with a ceramic filler for high-temperature applications.

Maximum operating temperatures for resins as matrices in fiber-reinforced composites are listed in Table 8-1 in order to give some idea of their limitations. No attempt is made to introduce the element of time in these figures other than to state that the time a material is subjected to these extremes in temperature is a critical factor that must be taken into consideration in any design. An example taken from the literature illustrates how this heat-resistant property is described for a series of polyimide molding compounds, Tribolo PI-600, available from Tribol Industries. Such compounds perform at 316°C continually and 427°C intermittently. A typical molding retains 56% of its mechanical strength after 100 hours at 316°C. Weight loss is 2.1% after 500 hours.

TABLE 8-1 OPERATING TEMPERATURES FOR SOME COMMON POLYMER RESINS

Resin	Maximum temperature, 0°C
Polyester	Room temperature (RT)
High-performance thermoplastic polyesters	150
Epoxies	200
Phenolics	260
Polyimides	300
Polybenimidazole	Above 300

A *coupling agent,* also known as bonding agent or binder, provides a flexible layer at the interface between fiber and matrix that will improve their adhesion and reduce the number of voids trapped in the material.

Voids are air pockets in the matrix, which are harmful because the fiber passing through the void is not supported by the matrix. The fiber under load may buckle and transfer the stresses to the weaker matrix, which could crack.

Interface and interphase: Composites, by definition, contain a combination of mutually insoluble components (constituents). With glass-fiber-reinforced plastics (GFRP), the glass fibers and the polymer resin are the main constituents. A wide variety of other material additives, such as fillers, pigments, catalysts, and fire retardants are also used to satisfy specific needs. Consequently, there exists an *interface(s),* a surface that forms a common area or boundary similar in many respects to grain boundaries in monolithic materials between any two constituents. An *inter-*

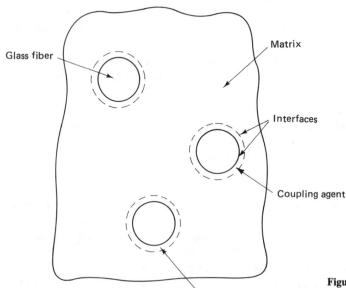

Glass fiber

Matrix

Interfaces

Coupling agent

Interphase

Figure 8-3 Cross section of a glass fiber reinforced plastic showing components.

Composite Materials   Module 8

phase is the region formed between two interfaces. It, therefore, is a distinct phase of itself with its own identity forming a region contiguous to two phases. Figure 8-3 is a sketch showing these main constituents of a GFRP.

8.3.3 Prepregs and Preforms

Fibers are obtainable saturated with resinous material through a process called preimpregnation. The resinous material is subsequently used as a matrix material. The preimpregnated fibers are called *prepregs*. Unidirectional preimpregnated fibers with a removable backing that prevents the fibers from sticking together in a roll are known as *prepreg tape* (see Figure 8-4). *Prepreg cloth* or *mats* are also available. Prepregs are ready-to-be-molded resin and fiber laminates. The fibers impregnated with resin are partially cured (B-staged) and ready to be softened and formed into a permanent shape with the application of heat and/or pressure. Some important advantages of using prepregs are the elimination of the handling of liquid resins, the simplification of manufacturing of reinforced plastic forms and shapes, and the savings in cost with small production lots. A few disadvantages are the elimination of room-temperature cures, the need for refrigerated storage for the prepregs, and the additional cost of the prepregs over the dry blended or undispersed materials. This latter disadvantage must be weighed against the use of higher mold pressures and increased molding cycle times for undispersed material to affect the required dispersion of the reinforcement throughout the matrix material for the attainment of the optimum properties of the composite.

Preforms are custom-shaped resin-bonded mats for reinforcement of molded parts with complex shapes.

A *lamina* is a flat arrangement of unidirectional fibers or woven fibers in a matrix.

A *laminate* is a stack of lamina with various orientations of principal materials directions in the lamina. Laminates can be built up with plates or plies of different materials or of the same material such as glass fibers. In any case, shear stresses are present between the layers of a laminate because each layer tends to deform independently of its neighboring layers due to each layer having different properties. These shear stresses, including the transverse normal stresses, are a cause of delamination.

Bulk molding compounds (BMC) are a premixed material of short fibers (chopped glass strands) preimpregnated with resin and various additives.

Sheet molding compounds (SMC) are impregnated continuous sheets of composite material. SMCs cut to proper size and stacked to provide the required thickness before heat curing in matched metal molds are used in such applications as automotive bodies and large structural parts (see Figure 8-4b). They come in three types, depending on the length of the glass fibers used in the reinforcement. Fibers may be only 25 mm in length and arranged in random fashion in the resin. Longer fibers (200 to 300 mm) may be oriented in one direction (directional fibers). Continuous fibers laid in only one direction make up another designation. Various combinations of these SMC types are used with fibrous glass reinforcements reaching 65% by

MAGNAMITE GRAPHITE PREPREG
X-AS4/1904

Magnamite graphite prepreg X-AS4/1904 is a 250°F curable epoxy resin reinforced with unidirectional graphite fibers. The reinforcement is Magnamite continuous AS4 graphite fiber that has been surface-treated to increase the composite-shear and transverse-tensile strength. The 1904 resin matrix was developed to cure under tape-wrap or vacuum-bag conditions at 250°F and to operate at temperatures up to 180°F.

	At 77°F	
Typical Composite Properties	Without Glass Scrim	With Glass Scrim
0° flexural strength, psi $\times$ 10^3	225	215
0° flexural modulus, psi $\times$ 10^6	18.4	16.1
0° tensile strength, psi $\times$ 10^3	240	—
0° tensile modulus, psi $\times$ 10^6	20.1	—
Shear strength, psi $\times$ 10^3	14.5	15.0
Cured ply thickness, mils	5.6	6.4
Fiber volume, %	58	58
Typical Prepreg Characteristics		
Tape width, in.	12	12
Resin content, %	40 ± 3	40 ± 3
Fiber content, [a] g/m^2	145 ± 4	145 ± 4
Glass scrim[b]	—	104
Yield, ft/lb	19	18

(a) Various fiber contents are available between 95 and 195 g/m^2.

(b) 104, 108, and 120 glass scrim have been used.

(a)

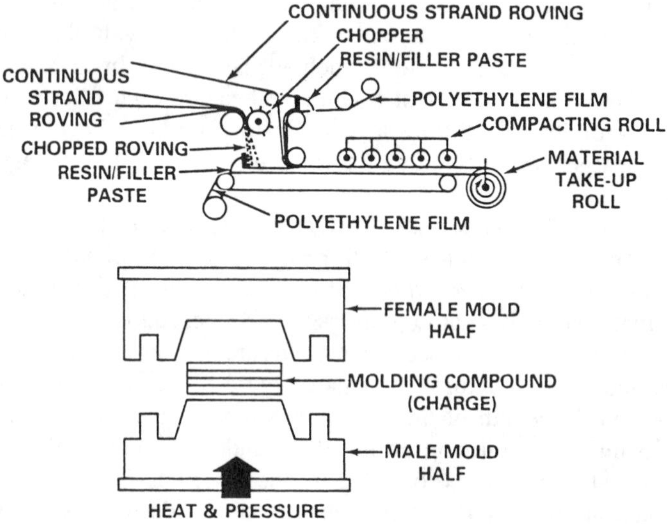

(b)

Figure 8-4 (a) Hercules prepegs. (Hercules, Inc.) (b) Schematic of machines used to make SMC sheet (top) and mold SMC plaques (bottom).

weight of the composite. A recent molding compound designated XMC may contain up to 80% glass fiber by weight and uses continuous fibers running in an X pattern (see Figure 8-15).

The more common fibers used in composites are described below. As research continues, more materials (organic, polymeric, ceramic, and metallic) are becoming sources of fibers for different composites designed for ever-increasing applications in an energy-conscious age.

8.4 REINFORCING FIBERS

8.4.1 Glass Fiber

Glass fibers are made by letting molten glass drop through minute orifices and then attenuating (lengthening) by air jet. The standard glass fiber used in glass-reinforced composite materials is E-glass, a borosilicate type of glass. The glass fibers produced with diameters from 5 to 25 μm are formed into strands having a tensile strength of 5 GPa. Chopped glass used as a filler material in polymeric resins for molding consists of glass fibers chopped into very short lengths.

E-glass is the first glass developed for use as continuous fibers. It is composed of 55% silica, 20% calcium oxide, 15% aluminum oxide, and 10% boron oxide and is the standard grade of glass used in fiber glass with a tensile strength of about 3.45 GPa.

S-glass was developed for high-tensile-strength applications in the aerospace industry. It is about one-third stronger than E-glass and is composed of 65% silicon dioxide, 25% aluminum oxide, and 10% magnesium oxide.

8.4.2 Boron Fiber

Advanced composites with boron fibers were developed in the early 1960s by the U.S. Air Force. 1963 marks the birth of boron technology. Boron fibers are composites with tungsten, silica coated with graphite, or carbon filaments as a substrate upon which boron is deposited by a vapor deposition process. The final boron fiber has a specific gravity of about 2.6, a diameter between 0.01 and 0.15 mm, a tensile strength of about 3.45 GPa, and a modulus around 413 GPa. Figure 8-5 is a representation of a boron fiber.

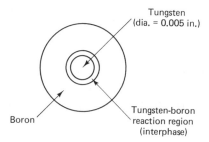

Figure 8-5 Structure of boron fibers.

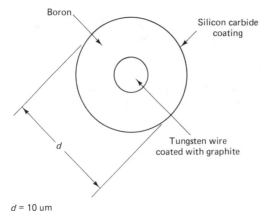

Figure 8-6 BORSIC fiber.

d = 10 um

Another boron fiber composite with the addition of a silicon carbide coating deposited over the boron surface provides a fiber that is more compatible with metal matrix materials, particularly at high operating temperatures. Figure 8-6 shows a sketch of a *Borsic* (trade name) filament and an aluminum matrix in which boron is deposited on a graphite-coated tungsten wire, which makes the fiber itself a composite.

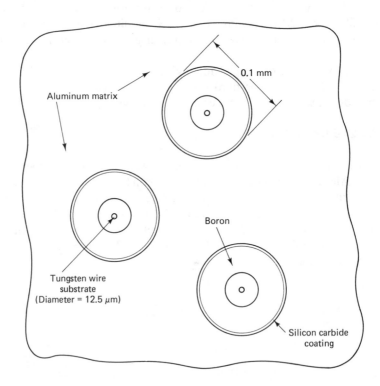

Figure 8-7 BORSIC fibers in aluminum (BORSIC/aluminum composite) B/SIC-AL.

Boron is more expensive than graphite and requires expensive equipment to place the fibers in a resin matrix with a high degree of precision. Problems with chemical reactions between boron and other metals continue to be a source of concern to materials technologists who are exploring ways to overcome this limitation. One approach to this problem was the coating of the *Borsic* fiber with silicon carbide (see Figure 8-7).

8.4.3 Carbon and Graphite Fibers

Carbon is a nonmetallic element. Black crystalline carbon, known as graphite, has a specific gravity of 2.25. Transparent carbon, diamond, has a specific gravity of 3.5. Graphite and diamond are allotropes of carbon, as are amorphous forms of carbon such as coke and charcoal. Both graphite and diamond have very high melting points of over 3732 K, which is explained upon examining the crystal structure and seeing the nearly infinite network of carbon–carbon covalent bonds that must be broken to melt these materials.

As with composite materials, graphite fibers are not a new development, having been first produced in small quantities in the nineteenth century for use in incandescent lamp filaments.

Several methods are used today to make carbon fibers of varying length and diameter with the versatility of glass fibers. The oldest method of producing these fibers in the late 1960s used *graphitization* of organic fibers, such as a rayon at temperatures up to 3250 K. An acrylic fiber polyacrylonitrile, abbreviated PAN, is also used as a source of such fibers, which the RAF in England produced in 1961. Heated (pyrolyzed) under tension to stabilize the molecular structure at temperatures between 920 to 1140 K, the noncarbon elements are driven off, leaving a fiber high in carbon content. A more recent process for producing high modulus carbon fibers uses low-cost pitch; the pitch is converted to a liquid crystal or mesophase state before it is spun into fibers.

Twenty different carbon fibers are now available from six manufacturers with cross-sectional configurations varying from circular to kidney shaped, in diameters of 0.008 to 0.01 mm, with strengths ranging from 1.72 to 3.1 GPa, and modulus values from 193 to 517 GPa.

One carbon yarn comes in plies from 2 to 30, with each ply composed of 720 continuous filaments of about 8 μm. This fiber is 99.5% carbon and maintains dimensional stability to 3420 K. Another grade of fiber consists of 2000 filaments gathered together into a uniform strand that provides a modulus of 379 GPa and a tensile strength of 2.0 GPa. Thornel™ 300 carbon fiber produced by Union Carbide consists of 6000 filaments in a one-ply construction. The surface of the fiber has been treated with sizing to increase the interlaminar shear strength in a resin matrix composite in excess of 89.6 MPa. The tensile strength is listed as 2.69 GPa, with an accompanying modulus (tensile) of 229 GPa. Carbon and graphite fibers and yarns are used to produce various fabrics to meet the ever-increasing demands of industry.

Carbon fibers first introduced in 1959 at a price of over $500 per pound have steadily dropped in price such that in 1977 the price was about $30 per pound. By

1984, the price declined to \$13–15 for lower grade graphite fiber and about \$19 a pound for aerospace grade fibers. The new process of making carbon fibers using ordinary pitch, a by-product of the coal-coking industry, promises a price of about \$10 per pound with multimillion pound usage in the 1980s.

The terms carbon and graphite are often used interchangeably. However, a line of demarcation has been established in terms of modulus and carbon content. Carbon fiber usually has a modulus of less than 344 GPa and a carbon content between 92% and 99%. Graphite fibers consequently have a modulus of over 344 GPa and carbon contents of 99% or greater.

Graphite fibers are linear elastic and like other carbon materials are anisotropic. An unusual feature of graphite is that its strength increases with temperature. Graphite possesses excellent creep and fatigue resistance, as well as dimensional stability at high temperatures. Graphite is one of the few truly efficient structural reinforcements, with both high specific modulus and strength. Thermal properties also must be considered unusual. The generally negative coefficient of expansion in the fiber direction coupled with the ability to tailor the orientation of the fibers permits the designer to adjust the thermal coefficient over a wide range of values. Table 8-2 is a comparison of the thermal stability of graphite with some selected materials.

One disadvantage of graphite is its poor oxygen resistance. It begins to oxidize in air at about 700 K. Unlike metals, graphite does not form a protective film. Instead, the graphite oxide is volatile. To protect graphite from oxidation, several coatings have been developed, one of which, silicon carbide, protects graphite for a limited time at temperatures approaching 1920 K (see Figure 8-7). Another limitation of carbon fiber is that it is a brittle, strain-sensitive material that cannot be depended on to offer much impact resistance.

Graphite fibers are available to the fabricator in the following forms:

1. Continuous fiber.
2. Unidirectional prepreg as tow or tape in widths from 3 to 36 in. to form laminations.
3. Chopped fiber.

TABLE 8-2 THERMAL (DIMENSIONAL) STABILITY

Material	Thermal Expansion (in./in./°F) $\times 10^{-6}$	
	$0°$	$0°, \pm 45°, 90°$
Graphite AS	-0.20	1.01
Graphite HTS	-0.25	0.80
Graphite HMS	-0.30	0.45
Fiber glass	3.50	6.00
Aluminum	13.0	—
Steel	7.0	—
Nylon (6/6)	45.0	—

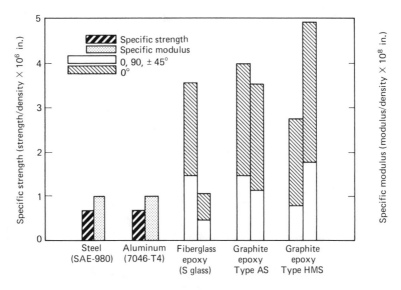

Figure 8-8 Comparison of strength and modulus of commonly-used materials. (Dupont)

4. Pultruded shapes.
5. Woven fabrics as dry cloth or prepreg for laminations.

8.4.4 Kevlar Fiber

Kevlar, a trademark for Du Pont's family of aromatic polyamide or aramid fibers introduced in 1972, has specific tensile strengths (Kevlar 49) greater than any commercially available reinforcing fibers and specific tensile modulus greater than glass and aluminum. However, they are not as stiff as graphite or boron fibers. These overall properties are the key to their successful use in weight-conscious industries such as space and aircraft. Figure 8-9 shows the position of these aramid fibers in relation to other materials from which reinforcing fibers are being made. Figure 8-9 is a plot of bending stress versus bending strain and compares Kevlar 49 in an epoxy resin matrix with its competitors. Another member of this aramid family of fibers is Kevlar 29, with a tensile strength of 3 GPa and modulus of 80 GPa. Compared to Kevlar 49 with the same tensile strength and modulus of 120 GPa, Kevlar 29 is especially suited for industrial applications in ropes, cables, protective clothing, and coated fabrics where resistance to stretch and puncture are also important.

8.5 FIBER PROPERTIES

8.5.1 Fiber Strength

A material's strength is directly related to its brittleness. Brittle materials have little resistance to the propagation of cracks. Pure metals possess varying degrees of ductility (opposite of brittleness), which allows for some yielding in the face of

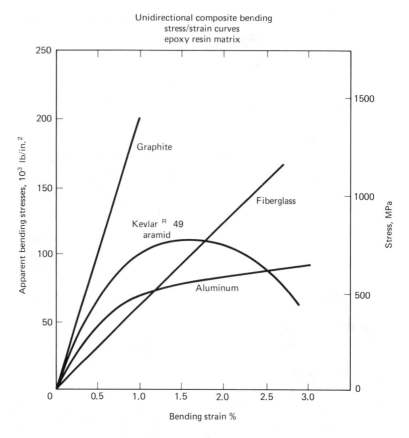

Figure 8-9 Bending strain. (Dupont)

longitudinal stress concentrations, which prevents fracture. Brittle fractures, on the other hand, are usually catastrophic and come without any advance warning. Alloying of metals and thermal treatments are two ways of improving the resistance of a material to plastic deformation (strength) by providing for the dispersion of harder particles within a matrix of softer material, which tends to limit the motion of dislocations through the matrix. Another way to increase the strength of a material is to add another material that has greater load-carrying capacity. The most prevalent form of this added material is fiber. How well fibers strengthen a material depends on the efficiency at which the relatively soft matrix material interacts with the fibers to transfer the load between them. The strength of an individual fiber is dependent on the absence of surface defects. The presence of microcracks anywhere in the small fiber causes localized stress and load concentrations and eventual failure of the individual fiber. The properties of a fiber are mainly dependent on the fiber's length, diameter, and orientation. The optimum reinforcing fiber for a composite material has a ratio of length (L) to diameter (D) of about 150. As this ratio, known as the *fiber-aspect ratio*, increases, the strength of the composite increases (see Figure 8-10).

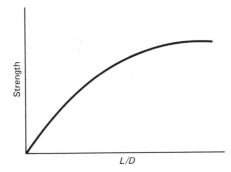

Figure 8-10 Strength versus fiber aspect ratio (*L/D*).

The shape of the reinforcing phase, rodlike in the case of fibers, plays an important role in determining the performance of a composite, as does the shape of the unit cell in a homogeneous solid. Their size and distribution control the texture of the material and determine, in part, the interfacial area between the fibers and the surrounding matrix. The topology of the fibers (i.e., their spatial relation to each other) is important because certain properties of the composite may be affected by the amount that the individual fibers or filaments touch each other (see Figure 8-11).

8.5.2 Specific Strength

Fiber composites find many weight-sensitive applications in space exploration, general aviation, automobile and sporting goods industries. One indicator used to portray the effectiveness of the strength of a fiber is specific strength.

Specific strength, the ratio of the tensile strength of a fiber material to its weight density, is an indicator of structural efficiency, giving the relative load-carrying ability of equal weights of material. Typical units of specific strength are millimeters (mm) or kilometers (km). Table 8-3 and Figure 8-12 list the specific strength for some typical materials from which fibers are formed for use in composite materials.

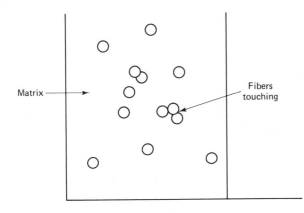

Figure 8-11 Cross-sectional sketch of a fiber reinforced material showing spatial arrangement of fibers within a matrix material.

TABLE 8-3 SPECIFIC STRENGTH

Materials	Weight Density, ρ (kN/m³)	Tensile Strength, S (GN/m²)	Specific Strength, S/ρ (km)
S-glass	24.4	4.8	197
E-glass	25.0	3.4	137
Boron	25.2	3.4	137
Carbon and graphite	13.8	1.7	123
Beryllium	18.2	1.7	93
Steel	77	4.1	54
Titanium	46	1.9	40
Aluminum	26.2	0.62	24

8.5.3 Specific Stiffness

Another indicator of the special properties of fiber composites, in particular, the effectiveness of a fiber, is *specific stiffness*. It is a ratio of the modulus of elasticity (or tensile stiffness) to the weight density of the fiber. Table 8-4, using the same materials as in Table 8-3, and Figure 8-13 show the average specific stiffness of such fiber materials, starting with materials with the highest values. Note that graphite, boron, and carbon have values almost six times that of steel. Graphite is known to have six times the strength of steel with six times less weight. These values of both strength and stiffness of materials in fiber form over the same materials in bulk form are most significant. As an example, the strength of glass in the bulk form, such as plate glass, is but a few MPa, yet in the fiber form this figure rises to around 3.5

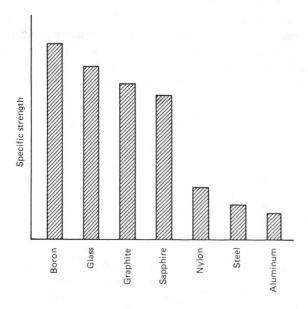

Figure 8-12 Relative specific strengths of typical materials used in composites.

TABLE 8-4 SPECIFIC STIFFNESS

Materials	Weight Density, ρ (kN/m^3)	Modulus of Elasticity, E (GN/m^2)	Specific Stiffness, E/ρ (Mm)
Graphite	13.8	250	18
Beryllium	18.2	300	16
Boron	25.2	400	16
Carbon	13.8	190	14
S-glass	24.4	86	3.5
E-glass	25.0	72	3.0
Steel	77	207	2.7
Titanium	46	115	2.6
Aluminum	26.2	73	2.8

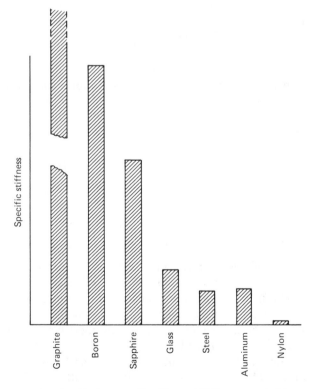

Note: Graphite has a specific stiffness of 5000 versus steel with 25 in units of 10^{6}N 'm/kg (using mass density in kg/m^3)

Figure 8-13 Relative specific stiffness of typical materials used in composites.

GPa. Structural steel (plain carbon steel) in bulk form has a tensile strength around 0.5 GPa, but in fiber form the value is 4 GPa. This contrast is due to the fact that the structure of the material differs between the two forms of the material. In fiber form the microstructure approaches the nearly perfect structure, with the crystals aligning themselves along the fiber axis in ordered fashion. As a result there are fewer internal defects or dislocations than would be present in the bulk form. As we recall from our previous discussions, the presence of dislocations explains many properties of a material. In this case the movement of dislocations permits the material to yield, changing the internal structure accordingly and accounting for ductility and accompanying decreases in strength and stiffness. Figure 4-5, page 93, shows the increase in stiffness gained by reinforcing materials (aluminum and titanium) with boron fibers.

Specific stiffness or specific strength may be expressed in terms of the mass density rather than the weight density. Such a practice is more compatible with the SI. Typically, units of mass density are expressed in kg m^{-3}. Modulus of elasticity and tensile strength can be expressed either in Nm^{-2} or Pa. The specific ratio will have units of m $\cdot$ N $\cdot$ kg^{-1}. For example a wood's longitudinal modulus (E) is 1.2×10^{10}N $\cdot$ m^{-2} and its density measured at 12% moisture content is 0.68×10^3kg $\cdot$ m^{-3}. The specific modulus of this wood is:

$$\frac{E}{\rho} = \frac{\dfrac{1.2 \times 10^{10} \text{ N}}{\text{m}^2}}{\dfrac{0.68 \times 10^3 \text{ kg}}{\text{m}^3}} = 17.6 \times 10^6 \text{ N} \cdot \text{m} \cdot \text{kg}^{-1}$$

Values of the stiffness-mass density parameter (stiffness modulus) for some other materials are:

graphite	5000×10^6N $\cdot$ m $\cdot$ kg^{-1}
boron	190×10^6
steel	25×10^6
nylon	3×10^6

8.5.4 Fiber Loading

Fiber loading refers to the amount of reinforcement in a composite material. The strength of the composite is directly proportional to the volume of fiber (*volume fraction*) present. In addition to the fiber loading, the arrangement or orientation of the fibers plays a major role in the strength of the given product. Using the analogy of filling a shoe box with pipe cleaners, the maximum number of fibers that can be placed in a given volume (shoe box) is determined by the arrangement of the fibers. If all fibers are placed parallel to each other, a maximum number can be attained. In fiber loadings a maximum of about 85% can be achieved. Between 50% and 75% load range can be reached if half the fibers are arranged at right angles to each other (fabrics), while a random arrangement (chopped strands) permits a range of only 15%–50% (see Fiber Orientation).

8.5.5 Fiber Orientation

As the direction of the applied load moves 90° to the fiber orientation, the strength of a directionally oriented FRP decreases to about 20% to 30% of the longitudinal direction. Several techniques are employed to orient the fibers in a FRP. If continuous fibers are oriented such that their length is in the direction (longitudinal) of the loading, this type of arrangement is known as directionally oriented. If the applied loads and/ or their directions are not known, a random oriented arrangement with continuous fibers running at various directions may be resorted to. Alternating layers of continuous fibers at various angles provide full strength at these various directions. However, this calls for additional lay-up time with its increased costs. Discontinuous fibers, less than 30 cm long, which tend to orient themselves in the direction of the resin matrix flow, provide another partial solution to providing a balanced strength in several directions. This technique is a compromise between the continuous and random orientation of continuous fibers. By sandwiching random short fibers between continuous fibers, a multiple configuration FRP results that is superior to a fully randomly reinforced material. The effect of adding some continuous fibers to a random, chopped-fiber reinforced SMC is highlighted in Figure 8-14a, which compares, among other

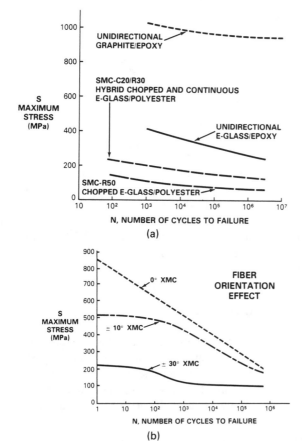

Figure 8-14 (a) Range of FRP composite tensile fatigue performance. (General Motors.) (b) Effect of fiber orientation on fatigue of XMC. (General Motors.)

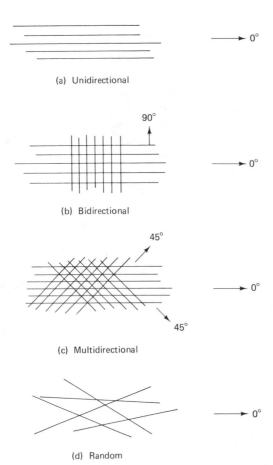

(a) Unidirectional

(b) Bidirectional

(c) Multidirectional

(d) Random

Figure 8-15 Fiber orientation.

things, SMC-R50, a 50% (by weight) of 25-mm chopped E-glass fibers, 16% calcium carbonate filler, and 34% polyester resin matrix formulation to SMC-C20/R30, a hybrid chopped (30%) and continuous (20%) E-glass fibers in a polyester resin matrix. Figure 8-14b shows the effect of a change in the fiber orientation on the fatigue failure of XMC, an X-pattern molding compound produced by PPG Industries. All the preceding arrangements are available in SMC form. Figure 8-15 contains sketches of some of these arrangements of fibers in a matrix.

8.6 STRUCTURE AND PROPERTIES OF COMPOSITES

Throughout this book we have stressed that the properties of a material are dependent on the material's structure. In monolithic materials such as steel or polymers, the structure commonly referred to is that of the crystalline or molecular material microstructure. Many times we peer beyond this atomic or molecular level to the subatomic level for further understanding of the behavior of such materials. With composite materials, we can identify with the naked eye (macroscopic) the major ingredients or constituents such as metal particles or glass fibers and different matrix materials involved in understanding the collective performance of these components in the composite material. Just as in the case of steel alloys where we can change the microstructure of these multiphase materials to obtain different properties by varying the amount of the basic ingredients (elements) or by changing the number, size, and shape of the ingredients through appropriate thermal processing, we can vary the properties of composite materials by varying their composition and their structure. We know also that the final behavior of materials such as steel depends on how well the various phases interact with each other. This is vitally important with composites. Being man-made materials, composites can be designed to have isotropic properties in all directions, in which case they probably would be homogeneous materials with their components distributed and arranged in a uniform pattern. Or the components may be distributed in a nonuniform manner, in which case the composites would have anisotropic properties. Third, similar to forged metals, a composite material may have its components distributed and arranged in a particular orientation that would result in directional properties. These last statements help explain some of the interest in composites in recent years by materials scientists who can, with the new technologies of materials, design their own material to fit the specific requirements of a particular application that cannot be met by existing technology or conventional homogeneous materials.

The characteristic properties of individual constituents of a composite interact in various ways to produce the collective properties or behavior of the composite. Some properties obey the *rule of mixtures* in that the composite properties are the sum of the values of the individual constituents. Density, specific heat, thermal and electrical conductivities, and some mechanical properties such as modulus of elasticity follow this rule (see Section 8.6.2). In some composites the properties of the components are somewhat independent and supplement each other to produce a collective performance by the composite. Other composites have resultant properties that are the net results of the interaction or interdependence of the components upon each other. By far, this type of composite behavior is the most important in that the end result of combining materials into a materials system is a performance or set of properties that far exceeds the individual properties possessed by the components acting alone. An example of this type of composite can be taken from the fiber-reinforced plastics (FRP) composites, such as glass fiber embedded in plastic resin. The glass fibers possess extremely high tensile strength, but being very brittle they cannot be used alone. A plastic resin, on the other hand, is relatively weak, but very ductile. Once sufficient glass fibers are embedded in a plastic resin matrix in a unidirectional manner parallel to the direction of the load, the two components act together as a unit to withstand the load by deforming equal amounts and sharing proportionately the load such that the composite or materials system achieves higher tensile strengths than otherwise is possible by the individual components.

8.6.1 Nondestructive Evaluation (NDE)

Using techniques that can detect and characterize hidden flaws without adversely affecting the material part involved, NDE, concerns all aspects of the quality, maintenance, and operation of materials and their concomitant structures. Conventional

NDE techniques such as holography, thermography, ultrasonics, or acoustic emission are applied to composites to detect flaws such as delaminations, inclusions, or fiber bunching. In addition, these techniques reveal material flaws in the microstructure of composites such as intermittent fiber–resin bonding or resin inhomogeneities that have a profound effect on the strength and useful life of composites. The importance of NDE in the continual development of composites cannot be overemphasized when it is recalled that composites are unique, in that the composite material is usually produced while making the finished part. This is in contrast to the usual manufacturing procedure with conventional wrought alloys, in which the product is assembled from previously validated materials that no doubt underwent extensive testing, both destructive as well as nondestructive.

8.6.2 Mixture Laws

Using fiber glass fiber filler and a polyester resin matrix as our composite, the stress–strain curves for each material are sketched in Figure 8-16. The modulus of elasticity (E) for glass is about 69 GPa and for polyester, around 69 MPa. Tensile strength for glass (single fiber) is about 3.45 GPa.

The curves show that there are great differences in the stiffness (slope of the curves) between the two materials. Glass, on the other hand, is extremely brittle and shows little evidence of any elastic strain up to the point of fracture. Polyester, as indicated by its values of modulus and the slope of its stress–strain curve, is a very ductile material. If each material were deformed an equal amount, the stress carried by the glass would be over 1000 times greater than the stress on the polyester.

If these two materials were combined in one ideal material, the glass would act as continuous fibers completely surrounded by a matrix of polyester, which forms a bond between the fibers and the matrix. The bond would allow both to deform (elongate) the same amount under a uniaxial load.

Research shows that the modulus of a composite material in which (1) the load is applied parallel to the continuous fiber, and (2) the bond between the fiber and the matrix material is strong enough to permit both materials to deform equally is the

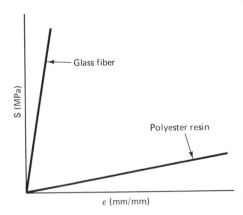

Figure 8-16 Stress-strain curves for glass fiber and polyester resin.

weighted sum of the modulus of the fiber and the modulus of the matrix. This last statement is an example of a *simple mixture law*. Mixture laws for physical properties such as specific heat, density, and conductivity for a composite material can also be developed. For such composite materials, most of the flow of thermal energy or electrical current will be through the component that is the best conductor.

The preceding discussion assumed that the chemical bond between the fibers and the matrix was greater than the strength of the matrix materials. This is not always the case. Much research goes into determining the conditions necessary to improve the bonding so that the stress transfer between the matrix and the fibers can be made without fracture. Such studies of metal matrix composite materials have led to the development of composite materials that can withstand temperatures in excess of 670 K and retain great stiffness. Such materials are represented by tungsten, boron, graphite, or silicon carbide fibers in aluminum, cobalt, or nickel matrices (see Figure 8-17).

8.6.3 Fatigue

The design of a new or improved product for industrial and/or technical application entails the detailed analyses of many factors. The design process links the scientific knowledge of materials with modern industrial methods to reach an overall objective of producing a better product at a lower cost. The interrelationship among materials, process, and design is nowhere more pronounced than in the use of composite materials to replace traditional materials such as steel in both structural and nonstructural components. Figure 8-17a is a flow chart showing the interaction of design, materials, and manufacturing processes, which must exist to ensure the success of any design project.

Composite materials owe their success in large part to the fact that they provide the engineer with considerable design latitude compared to metals. This is in addition to their many other desirable properties. Fatigue in metals, discussed in Module 5, differs from fatigue in composites (see Figure 8-17b). In metals the mechanisms of crack initiation and crack propagation result in the failure of metals under cyclic, repeated loads at stress levels well below their ultimate strengths (static loading). Crack initiation is caused by stress concentrations, inclusions, or voids, which are then acted on by the mechanism of crack propagation, reducing the net load bearing area to the point where the fatigue stress exceeds the metal's ultimate strength. Final failure in metals usually results in a relatively clean fracture surface. In many cases the failure is of catastrophic proportions. Finally, metals exhibit an endurance limit, a stress level below which fatigue failure does not occur regardless of the fatigue cycle applied.

In composites, crack initiation and propagation produce a simultaneous growth of cracks that may (1) extend through the matrix, (2) be stopped at a fiber, or (3) propagate along a fiber–matrix interface. Cracks initiate by such things as filler or fiber debonding, voids, or fiber discontinuities. The crack propagation results in cracks joining each other to the extent that the matrix is unable to perform its basic function of transferring the load from one fiber to the next in FRP composites. The fracture

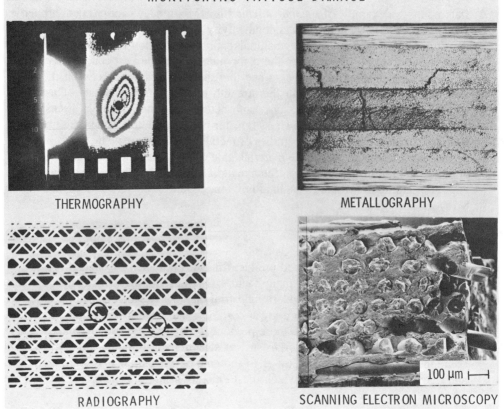

Figure 8-17 Metal reinforced composites. Photomicrographs of metal reinforcing fibers in metal matrices. (NASA)

surface usually shows evidence of a complex assortment of matrix failure, fiber failure, and fiber pullout.

Other fatigue failure characteristics of FRP composites are as follows:

1. A gradual decrease in the slope of the stress–strain curve (modulus of elasticity, E) occurs during cyclic loading.
2. Elevated temperatures normally reduce the performance of matrix-dominated composites (stress carried mainly by the matrix material). Fiber fatigue properties are better than those of the polymer matrix.
3. Moisture and chemical exposure greatly affect the polymer matrix.
4. Notch sensitivity in high-cycle tensile fatigue conditions is less than for metals.
5. Matrix-dominated properties are considerably less than fiber-dominated properties. Hence tensile, compressive, and shear stresses in the interlaminar (through-the-thickness) direction, being matrix dominated, should be kept to a minimum.

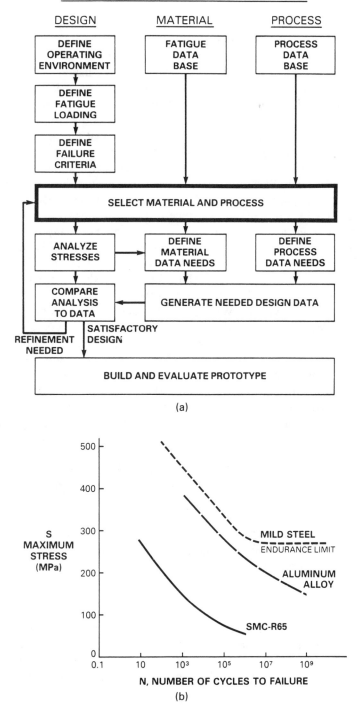

Figure 8-17 (continued). (a) Flowchart for fatigue design with FRP composites. (General Motors.) (b) Typical flexural fatigue curves for metals and a FRP composite. (General Motors.)

8.6.4 Toughness and Impact Strength

As measured by the ASTM D 256 test on unnotched Izod specimens, the impact strength at room temperature for "tough" plastic molding materials is at least 8 J/cm (15 ft-lb/in.). A value of at least 10 J/cm^2 (50 ft-lb/in.^{2}) using the high-speed tensile impact test (ASTM D 1822) also qualifies such materials as being tough materials. Considerable impact testing of such materials using glass and/or carbon fiber reinforcement in varying amounts with different resin materials has been done. Only recently, however, have such tests been done at low temperatures (down to 200 K). From our study of other materials we know that the impact strength of a notched specimen is less than that of an unnotched specimen because notches produce stress concentrations. We also know that it takes energy to not only start a crack but to propagate one. With notched specimens the crack is already provided, and depending on the sharpness of the notch, less energy is usually needed to propagate it. The tremendous effect of surface imperfections such as cuts, tool marks, and scratches on the toughness of materials is well known. For many reinforced thermoplastics, toughness drops off and the notch sensitivity increases at low temperatures. However, there are exceptions. Polypropylene, as well as acetal copolymer, becomes tougher as evidenced by higher test values for unnotched specimens. Glass-reinforced polycarbonates show little change in toughness although their notch sensitivity is greater at lower temperatures. With Du Pont's "super-tough" nylons (Zytel ST resin), notch sensitivity decreases with increased glass/carbon fiber content, while at low temperatures notched Izod impact strength readings increase with fiber content. Using unnotched specimens, the toughness of these nylon resins increases with increasing fiber content both at room temperature and low temperatures (Figure 8-18).

Impact testing measures the material's resistance to fracture under certain prescribed test conditions when a standard specimen is struck at high velocity. In addition to the Izod and Charpy notched and unnotched specimens using a swinging pendulum, other impact tests such as the falling ball, falling dart, high-velocity tensile stress, and fracture toughness tests are also used to measure toughness. *Fracture toughness,* in units of stress times the square root of a crack length (MPa $\cdot$ m$^{1/2}$), is an indicator of a material's resistance to the extension of a preexisting crack. The high-speed tensile test uses unnotched specimens and defines impact strength as the area under the stress–strain curve or the energy required to break a material (toughness). Impact strength using this test setup is expressed in units such as kilojoules per square meter (kJ/m^2). Those tests using notched Izod–Charpy specimens express impact strength in terms of the energy per length of notch or kJ/m as one example.

8.6.5 Test Values and Data

Test values and data presented in this module and elsewhere are typical values for the material. They are offered as an aid to understanding. The properties of parts fabricated from a particular material are too contingent on many factors, such as part configuration, molding techniques, or curing times, to guarantee reproducibility of the data.

Figure 8-18 "Tuff Wheel II" made of glass reinforced nylon (Zytel™) weighs less than aluminum or steel but is stiffer and tougher than either material. (Dupont.)

8.7 TYPES OF COMPOSITES

The products resulting from the different manufacturing techniques, different reinforcing components, and the specialized nature of the parts themselves result in many types of composites. We have divided this ever-increasing variety of composites into fiber, laminated, particulate, flake, and filled composites.

8.7.1 Fiber Composites

8.7.1.1 Glass-reinforced plastics (GRP). GRP represents the earliest and the most widely used (over three-fourths of total fiber composite production) fiber resin composites. With glass fibers in various forms coupled with either a thermosetting or thermoplastic resin, these composites can be produced without the need for high curing temperatures or pressures. The product contains a very good balance of properties, has high corrosion resistance, and has low cost for a multitude of uses as structural, industrial, and consumer-related products, ranging in size from minute

Figure 8-19 Fiberglass composite wind turbine blades (layers of glass mat and rovings in a vinyl resin matrix) GRP provides high strength, light weight, low cost, extended service life, and increased weatherability. (Morrison Molded Fiberglass Co.)

circuit boards to boat hulls. Using 20% to 40% fiber loadings, the composites will, in general, double the strength and stiffness of the plastic resin used alone. Continuous fibers will increase these same properties fourfold with accompanying desirable decreases in thermal expansion, creep rate and with increases in impact strength, heat deflection temperatures, and dimensional stability. These fibers may take the form of continuous filaments (monofilaments) or yarn. Figure 8-19 shows an application for GRP.

The disadvantages accompanying these composites arise, in the main, from the fact that they are essentially two-phase structures. This leads to a degree of environmental degradation greater than experienced by either component material alone. Residual stresses and electrochemical effects result from the marriage of two dissimilar materials. Furthermore, the diffusion of fiber materials into the matrix materials, and vice versa, may take place at several stages. Variation in the thermal expansion of these two materials leads to thermally induced stresses that result in warping, plastic deformation, cracking, or combinations thereof.

Naturally, the advantages of such composites outweigh the disadvantages. Some of these major advantages can be briefly summarized as follows:

1. Greater matrix strength and stiffness that permits little or no cross-plying and greater joint and stress concentration load strength.
2. Increased operating temperatures and increased durations of time spent at those temperatures.
3. Increased resistance to high temperatures and humidity environments.
4. Better conformability with existing metal fabrication techniques.

8.7.1.2 Fiber-reinforced metal matrix composites.

As with plastic resins, many fibers are used with metal matrices to form composites. Volume fraction of reinforcing materials, degree of alignment of the fibers, fiber diameter, and aspect ratio, plus interface–interphase reactions, take a leading role in determining the ultimate success of any metal matrix composite to meet the demanding requirements of a particular design. Figure 8-17 shows photomicrographs of a metal reinforcing fiber embedded in a metal matrix.

The major rationale for the development of such composite materials is to satisfy the important need for tough, strong materials capable of maintaining their special properties under high-temperature operating conditions. Figure 8-20 is a plot showing how the tensile yield strength of some materials varies with temperature.

Reinforcing fibers are similar to those used with resin matrices and include such materials as tungsten, silicon carbide, sapphire, boron nitride, glass and steel piano wire. Metal matrix materials can be represented by the aluminum and titanium alloys used as matrices with carbon reinforcing fibers (C–Al or C–Ti). Industrial uses for

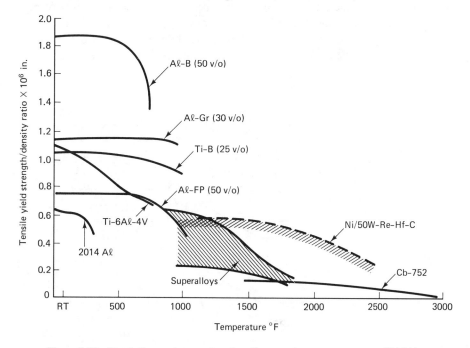

Figure 8-20 Metal alloys and composites. Specific strength versus temperature. (NASA)

metal matrix composites systems are many and varied. Aircraft uses are in structural and engine components. Various components of turbine engines operate at high temperatures with accompanying high stress loads. Blades and stator vanes are of particular interest. Fiber-reinforced superalloy composites (FRS) have undergone 200°C increases in blade temperatures compared to current superalloy (see Module 5) blades. One such FRS uses tungsten fibers of varying strengths, which can provide greater strength, higher thermal conductivity, and lower thermal expansion than conventional directionally solidified superalloy blades (see Figure 8-20). Automotive uses include Wankel engine seals, turbine fan blades, and truck body rails. Electrical bus bars, generator rotors, transformer windings, and pressure tanks are also representative uses. Each use reflects the application of one or more distinct properties of these composite materials, from high specific modulus to high temperature strength to increased wear resistance.

8.7.2 Laminated Composites

Laminated composites (Figures 8-21 and 8-22) consist of layers (lamina) of at least two different solid materials bonded together. Lamination allows the designer of this composite material to utilize the best properties of each layer in order to achieve a more useful material. Such properties as wear resistance, low weight, corrosion resistance, strength, stiffness, and many more can be accented by a wise selection of different constituent layers. A well-known example of a laminated composite is plywood, which has isotropic properties in the plane of a sheet of plywood due to the layers of wood bonded with a thermoset resin such that the longitudinal direction is at right angles in adjacent plies. Only their orientations differ. Clad metals such as bimetallic strip or copper clad stainless steel are composed of two different ma-

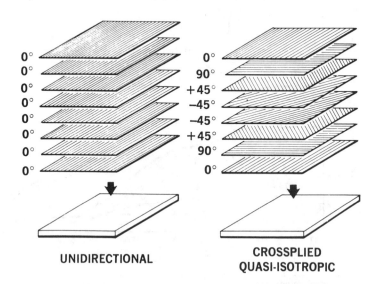

Figure 8-21 Fiber orientation within laminates. Laminates can be tailored to meet specific requirements with savings in time and material. (Hercules, Inc.)

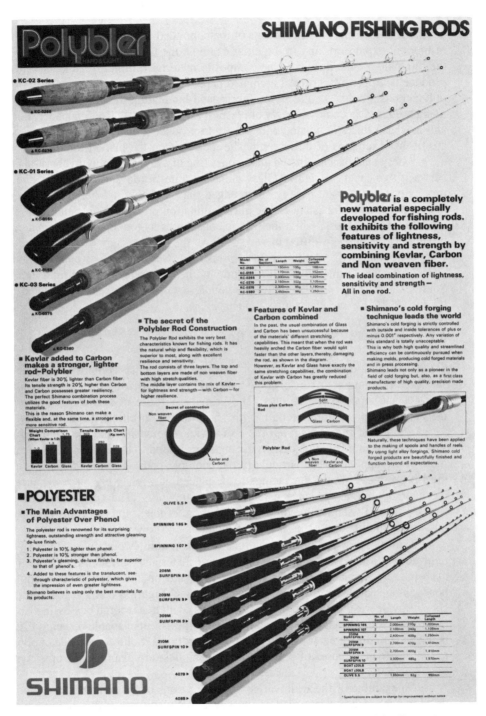

Figure 8-22 Laminated fishing rod using a variety of different fiber materials. (Dupont.)

terials. In the former, the two sheets of metal bonded together have different amounts of thermal expansion, and in the latter example the thermal conductivity property is the rationale for the design of the composite material. Similar treatments with paper produce isotropic properties. Laminated paper with plastic film or metal foil or using polymers as a base, such as nylon fabric, and laminating the polymer with layers of metal produces an endless variety of plastic-based laminates. Roofing paper, Formica, and Kevlar™ fibers laminated with a resin to produce a material for bulletproof vests are further examples of these composite materials on the market today.

Safety glass is a laminated glass consisting of a layer of polyvinyl butyral bonded between two layers of glass. Glass alone is quite brittle and dangerous due to its proclivity to shatter into many sharp-edged pieces. Polyvinyl butyral is very tough but weak in scratch resistance. The glass lamination protects the plastic from scratching and gives stiffness, while the plastic contributes its toughness. Each material, by protecting the other in different ways, allows the other material to contribute to the composite material's vastly improved properties over those of its constituents.

Laminates formed by high pressure and heat are referred to as rigid laminates. Some examples of rigid laminates are counter tops, rods, tubes, and printed circuit boards (PCB) for the electronics industry.

8.7.3 Sandwich (Stressed-Skin) Composites

Sandwich composites may also be classified as laminar composites. Their outer surfaces or facings are made of some material higher in density than the inner material or core that supports the facings. The primary purpose of sandwich composites is the achievement of high strength with less weight, specifically a high *strength-to-weight ratio* or *specific strength*. A sandwich composite may be compared to a structural I beam. The high density facings correspond to the I-beam flanges carrying most of the applied load, particularly the bending loads. The low-density core, like the I-beam web, allows the facings to be placed at a relatively large distance from the neutral place to produce a large section modulus. The core carries the shear stresses. Overall sandwich composites are more efficient than I beams because they possess a combination of high section modulus and low weight.

Typical sandwich facing materials are aluminum, wood, vinyl, paper, glass-reinforced plastics, and stainless steel. Core materials representing all families of materials are primarily cellular in form and take on the configuration known as honeycomb, waffle, corrugated, tubes, and cones (Figure 8-23a and b). These rigid cores provide the greatest strength and stiffness, with honeycomb possessing isotropic properties. Metals that can be made into very thin sheets capable of being diffusion bonded are finding greater use as cellular core materials. Solid cores made of plywood, as well as foamed cores of polystyrene or ceramics, are examples of other types of core materials.

An example of the wide variety of applications of these materials is the recent development of a security wall, a five-layer sandwich barrier that is impervious to cutting tools and oxyacetylene torches. The composite consists of an epoxy-impregnated ceramic fire brick sandwiched between layers of a rubberlike filler and enclosed

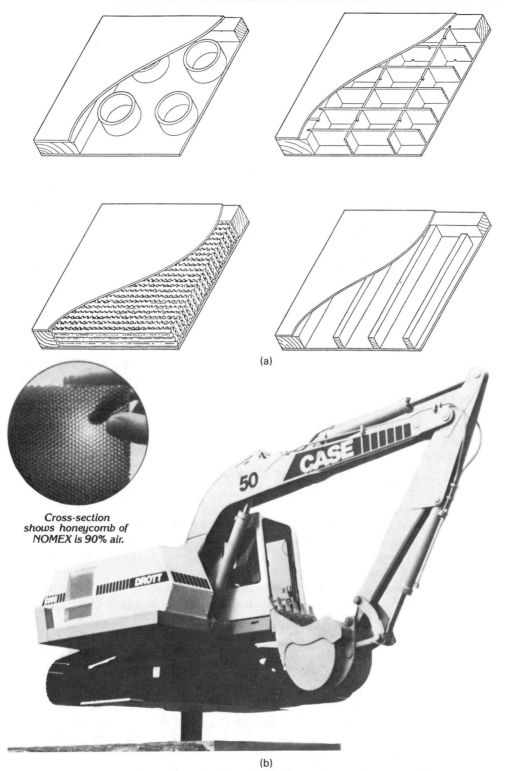

Cross-section shows honeycomb of NOMEX is 90% air.

(a)

(b)

Figure 8-23 (a) Sandwich composites and hollow core doors. (Forest Products Laboratory, Forest Service, USDA.) (b) High strength-to-weight ratio. A 10-inch square by 22-inch high column of honeycomb of Nomex™ aramid weighing 11 pounds supports a load of 27 tons. (Dupont.)

with facings of mild steel. Each of the constituent materials offers its particular resistance to penetration by certain processes, from diamond cutoff wheels to torches.

8.7.4 Particulate Composites

These composite materials contain particles of one or more materials suspended in a matrix of a different material. The particles, either metallic or nonmetallic, by definition, do not chemically combine with the metallic or nonmetallic matrix material. As with nearly all materials, structure determines properties, and so it is with these composite materials. The size, shape, spacing of particles, their volume fraction, and their distribution all contribute to the properties of these materials. Particulate composites are many times divided into subclasses by using some characteristic or combination of particle characteristics. This category of composites does not include particulates flat in shape, which possess sufficiently different properties to warrant a special classification.

8.7.4.1 Cermets. Cermets provide an excellent example of particulate composite material. Nonmetallic cermets are produced by sintering a mixture of ceramic and metal powders to form a structure that consists of a dispersion of ceramic particles in a continuous metallic matrix. Carbides of tungsten, chromium, and titanium are widely used in combination with cobalt, nickel, and stainless steel matrices to provide materials with very high hardness for wire-drawing dies, for very high corrosion resistance for valves, or for very high temperature applications such as turbine parts. Uranium oxide and boron carbide cermets suspended in stainless steel find several uses in nuclear reactor fuel elements and control rods.

8.7.4.2 Dispersion-hardened alloys. Dispersion-hardened alloys are similar to cermets and to precipitation-hardened alloys. Differences among these three categories are spelled out in terms of the constituent particles. The particles in dispersion-hardened alloys are smaller and of a lesser volume fraction (at 3% by volume) than found in cermets. Dispersion-hardened alloys are formed by the mechanical dispersion of particles, as opposed to precipitation-hardened alloys, in which compounds are precipitated from the matrix by heat treatment.

Cold solder is a metal powder suspended in a thermoset to provide a metal that is not only hard and strong, but a good conductor of heat and electricity.

8.7.5 Flake Composites

Flakes of mica or glass in a glass or plastic matrix form a composite material that has a primarily two-dimensional geometry with corresponding strength and stiffness in two dimensions. Flakes tend to pack parallel to and overlap each other and provide properties such as decreasing wear, low coefficient of thermal expansion, and increased thermal and electrical conductivity, which are dependent on higher densities in materials. Aluminum in flake form suspended in paint provides good coverage. Silver flakes similarly used provide good electrical conductivity.

Metal particles suspended in metal matrices such as lead added to copper and steel alloys lead to increased machinability or reduced bearing wear resistance. These usually brittle metal particles are not dissolved in their metal matrix as in the alloying of metals, but instead the metal matrix material is infiltrated around the brittle particles in a liquid sintering process.

8.7.6 Filled Composites

A filler is a material added to another material to significantly alter its physical and mechanical properties or to decrease its costs. Some fillers, when added to polymer materials, improve their strength by reducing the mobility of the polymer chains, much as an appropriate amount of gravel, as a filler, improves the strength of concrete. Fillers are added to polymer materials for a number of other reasons (e.g., to improve frictional characteristics, to control shrinkage, to improve moldability, to reduce dielectric properties, to lower resistivity, to reduce the moisture absorption characteristics, or to enhance wear rate). Celluloid and Bakelite™ are filled polymers, as are electric circuit boards and counter tops made from phenolics, a thermosetting resin. See Module 6 for more details.

Since the accidental discovery early in this century that natural rubber would accept large amounts of carbon black, thus improving its mechanical properties, great advances in the technology of fillers are continuing. Just as with multiphase microstructures such as steel, whose structures (micro) can be changed by the relative amounts of the phases present, the phase distribution, and the size of the individual grains making up the phases, fillers are components that are purposely added to other materials to change their properties. Glass fibers are added as fillers to plastics. Not only the type of filler but their size, shape, and distribution all play an important role in determining the desired properties of the filled composite.

8.7.7 Hybrid Composites

A hybrid composite is one made up of two or more different fiber materials in a matrix. One technique to reduce the cost of composites, particularly the high-performance or advanced composites containing carbon, boron, and graphite, is to use a mix of fibers placing the high-performance fibers in locations where they can most effectively accept the loads and stresses. An example of such hybridization was the mixing of glass fibers with graphite fibers in an epoxy resin matrix in the fabrication of a composite aircraft turbine blade. The result was an increase in the impact resistance of the blade with attendant decrease in cost. Mixing carbon fibers and glass fibers together in a composite and selectively placing the stronger higher modulus carbon fibers at critical locations permits fabrication of a part with the necessary properties with a minimal use of carbon fiber, thereby retaining cost effectiveness. An I beam subjected to bending would have carbon fibers placed along the flanges, with possibly glass fibers used to make up the web. A significant change in stiffness can be realized by a small amount of high-modulus fiber. Hybridization is possible because, generally, all fabrication techniques available for glass fibers and fabrics are compatible with carbon fibers.

Another hybrid composite can be produced by combining continuous and chopped fibers in a common matrix. With continual advances in the design of composites, additional composites can be categorized as hybrids. The first, *interply hybrids,* have two or more plies of unidirectional composites stacked in a particular order or sequence. An example of such a hybrid might be aramid–graphite–aramid. Usually, the matrix material of each ply is the same. If not, the lamination of the composite is fabricated using a curing procedure that is compatible with both systems. *Intraply hybrids* consist of two or more different fibers in the same ply. As with interply hybrids, these hybrids have plies laminated together in a specific sequence. A third category is combination hybrids, known as *interply* and *intraply hybrids.* The last, *superhybrids,* consists of resin matrix composite plies and metal matrix composite plies. This composite is fabricated by adhesively bonding metal foils, the metal matrix composites, the resin matrix composites, and the resin–fiber prepreg tape. A superhybrid composite, consisting of plies of graphite, glass, and Kevlar in an epoxy resin, combined with plies of boron and aluminum metal composite, is used to fabricate, with conventional techniques, aircraft turbine engine fan blades. Such a hybrid provides, among other advantages, more balanced properties.

Hybrid composites are finding greater use in all industries. In the automobile industry, hybrid driveshafts are nearing production stage in their development. One such composite has continuous graphite fibers oriented along the axis of the shaft to provide stiffness and glass fibers to carry the torsional load. The sporting goods industry has turned to hybrids to fabricate tennis racquets and golf clubs. Hybrid tennis racquets of graphite–aramid, graphite–boron prove to be stronger and more rigid than similar products built using conventional laminating techniques and materials (Figure 1-1). The possible number of combinations of various fibers and matrices is extremely large. Add the ability of the materials specialist to adjust the fiber content, as well as the fiber orientation, and the number of composites with different structures and properties is almost limitless.

We have referred to certain industries in this treatment of the various process types of composites. In the next section, several industries have been chosen from literally hundreds to illustrate the spin-off of composite technology, developed mainly through the efforts of government research and development, in the civilian sector. With the added impetus provided by the people's awakening concerns for conserving the world's limited energy sources, composite technology in the form of new equipment, processes, tooling concepts, and innovations in both manufacturing and quality assurance is being developed to not only provide greater productivity but to meet the goal of increased fuel-efficient transport vehicles and energy-efficient new materials.

8.8 FABRICATING OF COMPOSITES

While ordinarily not a part of the study of materials, the numerous manufacturing processes involved in the production of composites are intimately connected with materials, since design affects manufacturing and the particular process affects the properties of the final product. Finally and most importantly, because of the one-piece forming capability of composites in which the material is made at the same time that the end product is manufactured, a brief mention of the major manufacturing processes as a major factor in evaluating the costs involved is justified.

Two stages in the processing of most fiber composites are the lay-up or the combining of the reinforcement and the matrix materials and the molding or curing stage. *Curing* is the drying or polymerization of the resinous matrix to form a permanent bond between fibers and between laminae. It occurs unaided as with contact molding or by the application of heat and/or pressure using vacuum bags, autoclaves, pressure bags, or conventional metal stamping machinery (see Figure 8-24). Basically, there are eight processing methods for producing fiber-reinforced plastic products.

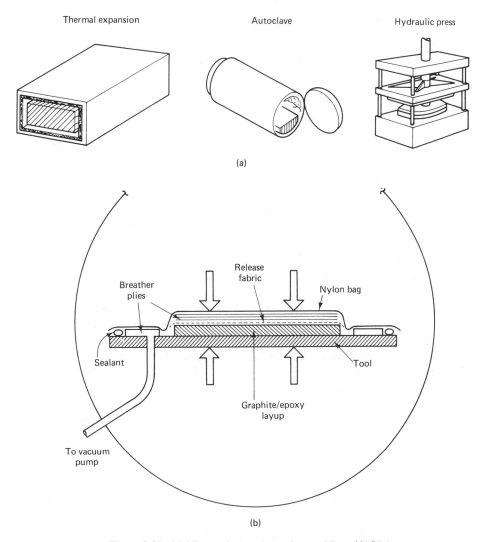

Thermal expansion   Autoclave   Hydraulic press

(a)

Breather plies

Release fabric

Nylon bag

Sealant

Tool

Graphite/epoxy layup

To vacuum pump

(b)

Figure 8-24 Molding methods and autoclave molding. (NASA.)

8.8.1 Contact Molding

Hand lay-up is the simplest of all the methods. Using a single inexpensive mold, reinforcing mat or fabric is placed in the mold and saturated with resin by hand. Layers are built up to the desired thickness to form a laminate that is cured at room temperature. This form of curing without the application of heat is called *contact molding*. *Spray-up* molding using a single mold combines short lengths of reinforcing fibers and resin in a spray gun, which deposits them simultaneously on the mold surface. Contact molding follows.

8.8.2 Matched Metal Die Molding

Matched metal die molding for mass producing high-strength parts limited in size by the press equipment uses pressures about 1.72 MPa and temperatures of about 120°C. The materials are in the form of SMC or BMC premixes. Most automobile panels are now being fabricated from GRP in SMC form by compression molding.

8.8.3 Injection Molding or Resin Transfer Molding

Injection molding or resin transfer molding (using reinforced plastics) is another high-volume molding process for both thermoplastics and thermosetting plastic resin.

8.8.4 Filament Winding or Tape Winding

Filament winding or tape winding, which produces the highest specific strength and glass content by weight of composite parts (fiber loading up to 85% by weight), is generally limited to parts with round, oval, or tapered inner surfaces. External shapes are unlimited. The continuous glass strand or filament is usually passed through a resin bath prior to winding onto a revolving mandrel. The mechanical as well as other properties of a filament wound product can be changed by altering the wind angle (α) shown in Figure 8-25. This angle is measured between the axis of the mandrel and the lay of the filaments. The tangent of this angle equals the ratio of the circumference of the mandrel and the pitch of the filaments. As the angle increases to 90°, the hoop tensile strength increases and the axial tensile strength decreases. Figure 8-25 shows the angle, the pitch, and the axial and hoop directions. The change in modulus of elasticity in the two principal directions is sketched in Figure 8-26. Figure 8-27 shows some internal shapes possible with filament winding, along with a few special design and winding problems.

8.8.5 Pultrusion

Pultrusion uses a bundle of continuous strands, sometimes combined with mat or woven fibers that are impregnated with resin and pulled (drawn) through a long die heated to a temperature around 400 K. This process produces such items as structural

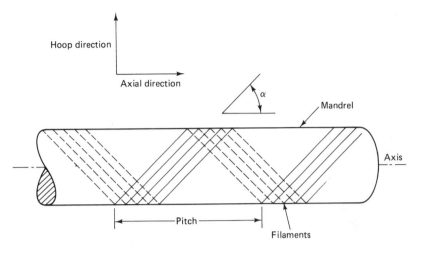

Figure 8-25 Filament winding wind angle.

I beams, solid rods, pipe, insulating panels, and other standard structural forms (Figure 8-28). Other processes not being described here include *continuous laminating* and *centrifugal molding* of cylindrical shaped products.

8.8.6 Superplastic Forming/Diffusion Bonding (SPF/DB)

The property of superplasticity joined with diffusion bonding provide for a spaceage composite. Superplasticity describes a property of a metal alloy that is not considered very ductile at or near room temperatures but at certain higher temperatures possesses extremely high ductility. Titanium is such a metal with normal plastic deformation around 20%, but at certain high temperatures it can be deformed as much as 2000%.

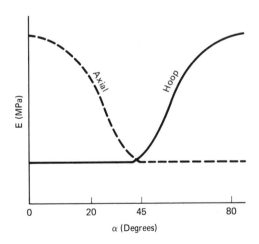

Figure 8-26 Modulus of elasticity versus wind angle.

THESE INTERNAL SHAPES CAN BE FILAMENT WOUND

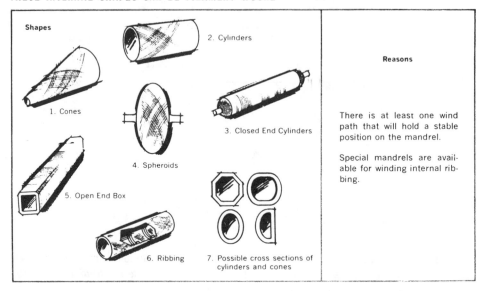

Shapes	Reasons
1. Cones 2. Cylinders 3. Closed End Cylinders 4. Spheroids 5. Open End Box 6. Ribbing 7. Possible cross sections of cylinders and cones	There is at least one wind path that will hold a stable position on the mandrel. Special mandrels are available for winding internal ribbing.

THESE INTERNAL SHAPES CREATE SPECIAL DESIGN AND WINDING PROBLEMS

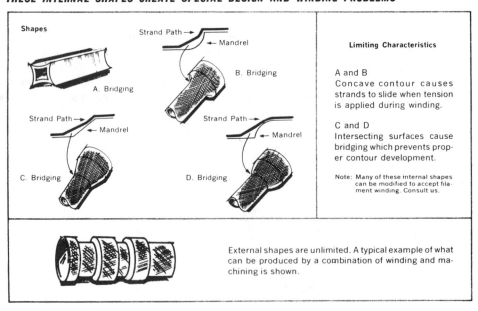

Shapes	Limiting Characteristics
A. Bridging B. Bridging C. Bridging D. Bridging	A and B Concave contour causes strands to slide when tension is applied during winding. C and D Intersecting surfaces cause bridging which prevents proper contour development. Note: Many of these internal shapes can be modified to accept filament winding. Consult us.
	External shapes are unlimited. A typical example of what can be produced by a combination of winding and machining is shown.

Figure 8-27 Design and winding problems with some internal shapes. (Permali, Inc.)

Figure 8-28 Pultrusion shapes. (Morrison Molded Fiberglass.)

Diffusion bonding or hot pressing is a widely used technique for metal fiber composite fabrication. Examples of such composites are aluminum and titanium alloys reinforced with boron or Borsic fibers. The matrix is usually in sheet form, with the reinforcing filaments or wires mechanically spaced and oriented between the sheets to form alternating layers of filaments and matrix. Surface coatings such as carburizing, nitriding, and bimetallic castings are further examples of diffusion bonding, which do not use metal filament reinforcement.

The first step in diffusion bonding is to wind the fiber over a metal-foil-covered drum. The resulting mats are cut and made into tapes by diffusion bonding under

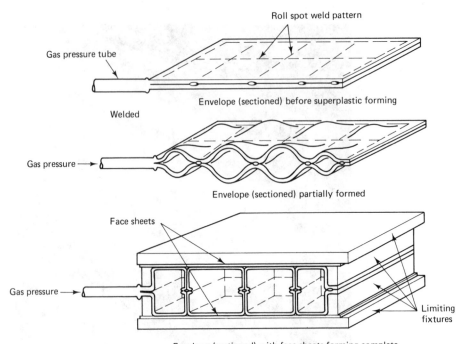

Figure 8-29   Superplastic forming with diffusion bonding. (NASA.)

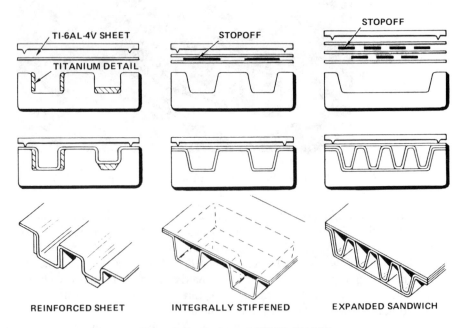

Figure 8-30   Products of SPF/DB (NASA.)

high temperature and pressure. The tapes, in turn, are cut into plies, stacked in a die. and consolidated by further diffusion bonding.

Superplastic forming/diffusion bonding (SPF/DB) uses titanium alloys to fabricate major structural components of aircraft. Titanium alloys have a high strength-to-weight ratio, particularly at elevated temperatures, as compared to steel or aluminum. One main drawback to their use is difficulty in cutting, milling, and forming. This new process, with many of the details still secret, uses one or more titanium sheets laid on top of each other and held together by a jig while being heated in a furnace (Figure 8-29). Argon gas is blown through tiny holes drilled in the sheets at desired locations. When the titanium reaches a soft and supple superplastic state, the gas assists the titanium to expand into cells, with the outer sheets remaining flat. The internal structure takes on whatever shape is intended. Through this process, the outer titanium sheets flow together under pressure to form a complex expanded sandwich unit that has the strength of a single piece of titanium alloy (Figure 8-30).

APPLICATIONS AND ALTERNATIVES

Aerospace and aircraft industries

A government evaluation of the advanced composite technology in late 1972 concluded that two principal barriers, confidence and cost, needed to be overcome if composites were to mature as a new class of engineering materials ready for production. Both of these factors have been addressed with vigorous efforts such that today we see steadily improving manufacturing methods, fewer parts and tools needed as a result of simpler composite engineering designs, steadily decreasing costs of such designs, and decreasing materials' costs as the volume of reinforcing materials increases in the face of increasing metals' costs. Many of these factors are the direct result of the programs mounted to increase the aerospace and aircraft industries as well as the public's confidence in composite materials through extensive service testing, materials evaluation, and a further integration of the many technologies involved.

Various flight service programs (fly and try) to develop confidence in composites and gain manufacturing data represent various degrees of composite utilization from the selected reinforcement of existing metallic components, such as those shown in Figure 8-31, through the substitution of composites for metals to the redesign of the aircraft component for composite utilization (Figure 8-32).

The X-15 series vehicles (1958–1968) pioneered hypersonic flight as they operated near the fringes of space. Capable of flight at Mach 6, where the aircraft surface temperatures often rise above 650°C, the X-15 had to withstand brief surges of heating to only 500°C. The X-15 frame was made of stainless steel and titanium and the body was covered with Inconel X, a nickel–steel alloy having great strength and heat resistance.

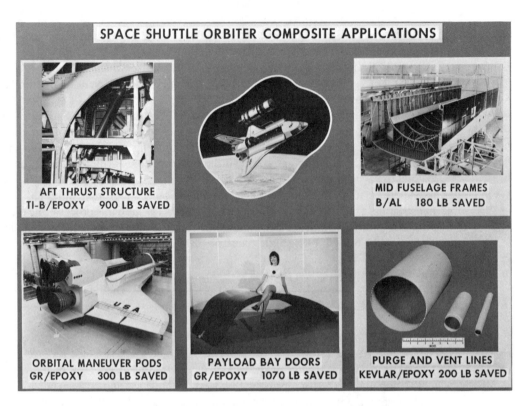

SPACE SHUTTLE ORBITER COMPOSITE APPLICATIONS

AFT THRUST STRUCTURE
TI-B/EPOXY   900 LB SAVED

MID FUSELAGE FRAMES
B/AL   180 LB SAVED

ORBITAL MANEUVER PODS
GR/EPOXY   300 LB SAVED

PAYLOAD BAY DOORS
GR/EPOXY   1070 LB SAVED

PURGE AND VENT LINES
KEVLAR/EPOXY 200 LB SAVED

Figure 8-31 Space shuttle component development. (NASA.)

Much of what was learned in this program was used in the design and development of the Space Shuttle orbiters designed to return to earth, reentering the earth's atmosphere near Mach 25 (see Figure 7-24). The underbody of the orbiter reaches temperatures up to 1260°C during reentry. At some points the temperatures range from −110°C to 1648°C. Therefore, the orbiter's body is covered by thermal protective ceramic tiles that permit the orbiter to be made with lower-strength aluminum, internal structural members.

The thermal protection system (TPS) for some 70% of the exterior of the Space Shuttle is another example of a man-made composite materials system. The tiles are a composite consisting of short staple, amorphous, silica fibers derived from common sand bonded in a slurry with a colloidal silica binder solution. Coated with a 15-mil silica frit the tiles on the upper surface of the shuttle (temperatures from 315° to 1185°C) receive a layer of white silica compound and shiny alumina oxide to reflect the sun's rays while in orbit. The underbody tiles (temperatures up to 1260°C) receive a layer of black reaction-cured borosilicate glass. The upperbody tiles are also treated with a waterproofing polymer to help eliminate water ab-

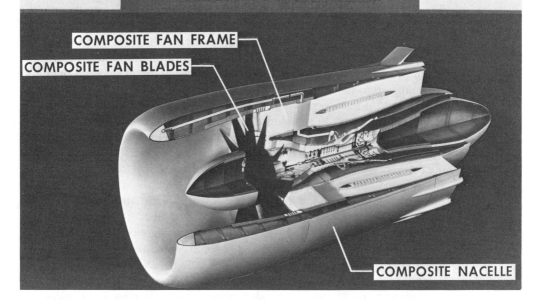

Figure 8-32 Use of advanced composites in new engine design. (NASA.)

sorption. Tile thicknesses range from 5 to 125 mm. The design specifications require these tiles to withstand repeated heating and cooling, undergo a high degree of acoustic noise (165 dB at launch), and survive great thermal shock. With the surface heated to 1260°C in an oven, these tiles dissipate the heat so quickly that they can be held by their edges with an ungloved hand seconds after they are removed from the oven, even while the interior of the tiles glow red hot (see Figure 7-2).

In addition to these properties, low weight (weight density is 9 lb/ft^3), low thermal conduction and expansion, and high temperature stability are called for. The tiles are bonded to another major component of the system, the NOMEX™ felt strain isolator pad (SIP), which is a laminated nylon composite that separates the aluminum from the tiles since each material has a different thermal coefficient of expansion (see Figure 7-25). The SIP is then bonded to the aluminum with the same room temperature vulcanizer (RTV) adhesive used to bond the tile to the SIP. Another man-made composite, reinforced carbon–carbon (RCC), a carbon cloth impregnated with additional carbon, heat treated and coated with silicon carbide, is used in the nose and leading edges of the wing to protect these ultrahot areas. NOMEX felt used alone will protect some areas such as the cargo bay doors up to temperatures of 680°C. These doors are, in turn, made of a graphite epoxy (see Figure 8-31).

Automobile and trucking industry

The automobile industry is now actively engaged in research into the use of composites. The recent emphasis is on *structural* parts since the industry has been using some composites for purely decorative parts. Ford Motor Company experimented with GFRP on a trunk lid in 1941. Glass-fiber-reinforced plastics (GFRP) have seen a sharp rise in use in auto and truck parts since the first all-FRP body Corvette appeared in 1953. With the exception of a few sports cars, this use had not been repeated until the recent development of an electric car—the Copper Development Association's Electric Runabout with its GFRP body. In the late 1970s, GFRP was used extensively in cars and trucks for front end panels, fender liners, hoods, and truck cabs. More and more applications of composites are found in the functional area, such as disc brake pistons, battery trays, oil filler caps, fans, and one-piece steering columns. This latter example on many GMC cars is a production part made of glass fiber in a nylon matrix representing a weight savings of 50% over the conventional metal parts.

The substitution of strong, lightweight composites in vehicles is under consideration in order to shed up to 20% of a vehicle's weight to comply with government's fuel economy regulations. These regulations have contributed to the first production of composite parts in an automobile. Ford Motor Company went into production with a hybrid composite mounting bracket for an air-conditioning compressor in 1980 models. This bracket is the result of a two-year development effort in the compression molding of a hybrid composite material containing 50% chopped glass and 20% continuous graphite fibers in a polyester resin matrix. The second major breakthrough in the production application of composite materials to automobiles is a glass fiber–epoxy resin composite transverse leaf spring for the rear support of the 1981 Corvette. This one-piece filament-wound leaf spring is thin at its ends and narrow at its center and weighs only 8 lb, compared to the multileaf steel spring it replaces weighing 45 lb (Figure 8-34).

Another advantage of composites is in the area of parts consolidation, that is, the replacement of an assembly containing several parts with one made of a single piece. A typical aircraft assembly contains numerous individual parts that must be assembled into one piece with numerous holes drilled for rivets. The time consumed and the labor expended are excessive and costly. In the auto industry a typical car door consists of outer and inner panels and a multitude of fittings and attachments that need to be fastened into one piece by bolting, welding, or riveting. A

Figure 8-33 Composite spring mounted on a Corvette. Spring provides a low stiffness with a relatively high design strength and weighs about 60% less than steel springs. (General Motors.)

lock, hinge seats, and trim, could save almost half the weight. Secondary finishing and heat treatments would also be reduced if not eliminated.

Finally, composite materials are energy efficient. It takes more energy to produce metal parts than plastic parts. The energy required to propel a conventional car made from metal parts for its normal lifetime leads to the conclusion that lightweight composite parts certainly have the advantage in helping to conserve the world's conventional energy supplies.

The proper design of composite parts, particularly engine parts, is one of the first steps in gaining the confidence of people in industry and the public sector. Manufacturing expertise and speed for full production will see the greatest improvement in such parts as connecting rods, pistons, piston pins, push rods, and valves, many of which have already been produced in limited quantities and operationally tested in Formula I racing engines. Valve springs and driveshafts are in a similar state of development as are crankshafts. The GFRP technology that has been discussed to this point applies equally well to other fiber reinforcements, including graphite. Hybrid composites have the potential to continue to drive down costs of graphite and carbon fibers. Figure 8-34 shows examples of a Ford Motor Company six-passenger prototype car using GFRP to the greatest extent possible. This experimental car weighs 2500 lb, or 1250 lb less than the company's 1979 intermediate six-passenger cars built with conventional materials.

Applications and Alternatives

Figure 8-34 Experimental composite car. A weight comparison; a graphite truck spring (left) versus steel spring (right). (Ford Motor Co.)

These developments in the application of composite materials typify recent efforts to conserve the world's readily available energy resources. The appropriateness of such efforts is underscored by the knowledge that the energy requirements per pound of material are only half as much for plastics as for aluminum or steel. Furthermore, added emphasis can be gleaned from the fact that the iron and steel industry is the largest single energy-consuming industry in the world, accounting for about 11% of total world energy consumption. Design flexibility of FRP makes it very attractive for those interested in design for automation.

Recreational industry

Composite materials have made inroads in the sporting or recreational industry, as well as in the aerospace and aircraft industries. Composites are used not only in large structures such as boat hulls, but for such items as archery bows, golf clubs, fishing rods, skis, bicycles, and tennis racquets. A composite pole-vaulting pole made of glass fibers and epoxy helped Bob Mathias win the Olympic Decathlon.

How many additional applications can you list where composites would make good material alternatives? Why do you think that substitution has not been made?

8.9 SELF-ASSESSMENT

8-1 Identify and list the composite materials present in a typical American kitchen; include furniture, equipment, and tools usually found in the room.

8-2 Reinforced concrete is a composite material. List its constituents that must act in concert with each other to produce this widely used engineering material.

8-3 Name at least three composite materials found in the human body.

8-4 A natural polymeric composite material, wood, has directional properties. Explain what is meant by this and cite an application of wood that takes this particular property into account.

8-5 The terms specific strength and specific modulus are often used in describing composites. What does specific modulus refer to and what does it tell about a particular material?

8-6 Oilite bearings are an example of what type of composite material?

8-7 In speaking of interface and interphase, which term represents the larger area or volume in a composite material?

8-8 Which fiber material has the greater specific stiffness, graphite or carbon?

8-9 Name two polymorphs of carbon.

8-10 State a major disadvantage of graphite as a fiber material.

8-11 What is the trade name for aramid fibers?

8-12 Compare the tensile strength of Kevlar with nylon. Which material would make better ropes?

8-13 The abbreviation NDE stands for what?

8-14 Using Figure 8-13, with a bending load that produces a stress of 750 MPa, which material listed would produce a bending strain of 1%?

8-15 Verify the units of specific strength (inches) using Figure 8-17.

8-16 Describe three tests used to measure toughness.

8-17 The properties of a reinforcing fiber are dependent on their size and length. How are these two qualities related and used in describing the strength of such a fiber?

8-18 Regardless of the name given to the processing method for producing fiber-reinforced plastic products, two stages are included in most of them. Name these two stages.

8-19 Which molding process provides the highest specific strength and glass content by weight?

8-20 Describe the effect of a decreasing wind angle on the hoop stress.

8-21 When a part that is being filament wound contains a transition (concave contour), bridging is likely to occur. Explain what transition-bridging means.

8-22 Laminations in plywood are known by what name?

8-23 List two examples of a rigid laminate.

8-24 Sandwich composites have cores of various configurations. List three such configurations.

8-25 A cermet is basically a composite made from a ceramic and a metal. Which of the two materials is the softest?

8-26 What thermal protection system protected space capsules from the intense heat of reentry into the earth's atmosphere prior to the flights of the Space Shuttle?

8-27 Identify a possible use for a laminated composite consisting of two bonded sheets of different metals each having a different coefficient of thermal expansion.

8-28 A popular kitchen cooking utensil contains a composite base made up of copper clad to stainless steel. What is the purpose of this composite application?

8-29 Discuss the reasons why the emphasis in materials science over the past three decades has been on the research and development of fiber composite materials as opposed to efforts to develop or refine traditional load bearing materials such as an alloy steel.

8-30 If present materials technology could produce crystals of practical size free from defects, such high-strength whiskers could be used as reinforcing material in a composite that would outstrip the performance of existing composites. Discuss what problems could arise, if any, in combining this filler material with a suitable matrix material.

8.10 REFERENCES AND RELATED READING

AGARWAL, BHAGIVAN D., and LAURENCE J. BROUTMAN. *Analysis and Performance of Fiber Composites*. New York: Wiley, 1980.

Annual Book of ASTM Standards Part 36—Plastics—Materials, Film, Reinforced and Cellular Plastics; High Modulus Fibers and Their Composites. Philadelphia: American Society of Testing and Materials, 1983.

ASTM Phila, 1977, ASTM STP 617. *Composite Materials, Testing and Design*. Englewood Cliffs, N.J.: Prentice-Hall, 1977.

ASTM Phila, 1975, ASTM STP 593. *Fracture Mechanics of Composites*. Englewood Cliffs, N.J.: Prentice-Hall, 1975.

BROSTOW, WITOLD. *Science of Materials*. New York: Wiley, 1979.

BROUTMAN, LAURENCE J., and RICHARD H. KROCK, eds. *Composite Materials*. Vol. 1, Interfaces in Metal Matrix Composites; Vol. 2, Mechanics of Composite Materials; Vol. 3, Engineering Applications of Composites; Vol. 4, Metallic Matrix Composites; Vol. 5, Fracture and Fatigue; Vol. 6, Interfaces in Polymer Matrix Composites; Vols. 7 and 8, Structural Design and Analysis Parts 1 and 11. New York: Academic Press, 1974.

BROUTMAN, L. J., and R. H. KROCK, eds. *Modern Composite Materials*. Reading, Mass.: Addison-Wesley, 1977.

DELMONT, J. *Technology of Carbon and Graphite Fiber Composites*. New York: Van Nostrand Reinhold, 1981.

FOLKES, M. J. *Short Fiber Reinforced Thermoplastics*. New York: Wiley, 1982.

GORDON, J. E. *The New Science of Strong Materials*. New York: Penguin, 1976.

HANNANT, D. J. *Fibre Cements and Fibre Concretes*. New York: Wiley, 1979.

KREIDER, K. G., ed. *Composite Materials*. New York: Academic Press, 1974.

LENOE, E. W., D. W. OPLINGER, and J. L. BURKE, eds. *Fibrous Composites in Structural Design*. New York: Plenum Press, 1980.

LUBIN, G., ed. *Handbook of Fiber Glass and Advanced Plastics Composites*. New York: Van Nostrand Reinhold, 1969.

Lynch, C. T., and J. P. Kershaw. *Metal Matrix Composites*. Cleveland: Chemical Rubber Co., 1972.

Modern Plastics Encyclopedia. New York: McGraw-Hill, 1971.

NASA SP 5974 (03), "Composite Materials—A Compilation," Jan. 1976.

Nielson, Laurence E. *Mechanical Properties of Polymers and Composites*. New York: Marcel Dekker, 1974.

Piggott, Michael R. *Load Bearing Fiber Composites*. Elmsford, N.Y.: Pergamon Press, 1980.

Richardson, M. O., ed. *Polymer Engineering Composites* (Applied Science Publishers, Elsevier). Englewood Cliffs, N.J.: Prentice-Hall, 1976.

Periodicals

Machine Design   *Plastics Engineering*
Manufacturing Engineering   *Plastics World*
Materials Engineering

9

ELECTRONIC-RELATED

MATERIALS

9.1 PAUSE AND PONDER

The electronics industry employs many types of materials in the circuitry and devices that operate on electronic impulses. Earlier modules cover some of these materials, including discussions of electrical properties, metal conductors, ceramic substrates, optical glass fibers, polymer dielectrics, and ceramic magnets. Research and development efforts of both materials scientists and electronics engineers aim for reduction of the size of electronic components, elimination of moving parts, reduction of energy to operate electronic components, and an increase in speed of their operation. This module covers several materials important to the revolutionary changes in electronics or solid-state electronics, including a section on piezoelectric materials. Solid-state electronics, which got its start in the early 1950s, deals with devices in which electron flow goes through solid materials rather than the vacuum used in vacuum tubes. Vacuum tubes dominated electronics prior to solid-state devices and created problems of size and heat but are nearly eliminated in modern electronic devices. Vacuum tube computers filled whole buildings, while today's computers replace the former with a memory chip less than a square centimeter in size. Printed circuit boards use transistors, resistors, diodes, and capacitors mounted on reinforced plastic boards with etched copper conductors to connect the components.

Integrated circuits (IC) developed from solid-state electronics bring together many electronic devices, such as transistors, conductors, diodes, and resistors, into a single chip. Semiconducting materials form the backbone for IC devices. Integrated circuits allow the reduction of complex circuit boards into small microprocessor chips, which perform the functions of a circuit board that can be hundreds of times larger.

The IC chip may contain several different types of electronic circuits. During their initial stages of development, IC memory chips stored about 1000 bits of information. The information is stored in the form of simple electronic charges. Through progress in IC technology, LSI (large scale integration) evolved so that memory chips only one centimeter square have a memory capacity up to 10,000 bits of information. In May 1984, IBM announced the development of an experimental one megabit chip. This dynamic random-access memory (DRAM) chip is another advance in very large scale integration (VLSI) techniques. Silicon and aluminum metal oxide semiconductor (SAMOS) technology can produce these one megabit chips with existing manufacturing facilities.

Microprocessors hold promise for many technological developments in all phases of society. Improved computer capabilities such as robotics, low-cost home microcomputers, automotive controls, vending machine controls, traffic signals, space vehicle controls, environmental monitoring devices, numerous communications devices, and manufacturing equipment controls represent but a few areas that profit from microprocessor technology (Figure 9-1).

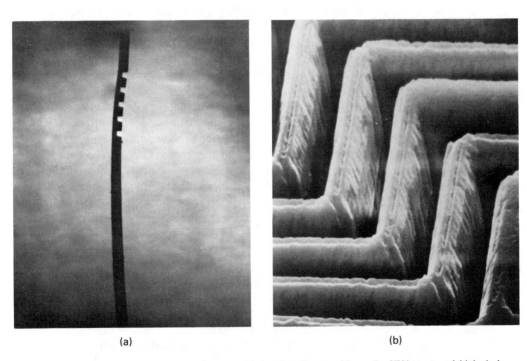

(a)   (b)

Figure 9-1 (a) A new phenomenon, ablative photodecomposition, using UV lasers to etch biological and polymeric materials without heating effects. Photograph of a single human hair showing clean etching cuts. (IBM.) (b) A SEM micrograph showing etched plastic film with line widths of 5 μm. Technique has potential for fabricating ICs. (IBM.)

As you study this module relate the materials to new stories that you hear about advances in electronic controls. Look for these materials in new products. When you read about robotized manufacturing, new communications systems, and entertainment devices, note the application of the materials discussed here. How do these materials affect your life today and in the future?

9.2 SEMICONDUCTOR MATERIALS

The basic differences among conductors, insulators, and semiconductors were outlined in the discussion of the energy band theory of the electronic structure of materials (see Module 2). The properties of semiconductor materials make possible the transistor, solar cells, tiny lasers, and the multitude of devices using integrated miniature circuits (Figure 9-1b). This new technology has as its basis the development of man-made materials. Such materials permit the generation and precise control of electron motion within the confines of a tiny crystal of semiconductor material to which has been purposely added (in most cases) a minute amount of impurity atoms forming a solid solution.

In our previous discussion of energy band diagrams, we defined the energy gap for semiconductors and listed the width of this gap in the case of silicon and germanium. These materials are known as "pure," ideal, or *intrinsic* semiconducting materials. They possess a minimum of impurity atoms, certainly none added purposely. We recall that it is a very rare occurrence to find anywhere in a natural state a pure substance. Second, we have learned that impurity atoms can effect profound changes in the properties of materials. So it is with semiconducting materials. If selected impurity atoms in accurately measured amounts are added to certain pure semiconducting materials, highly significant changes in behavior will occur. These materials are now known as *extrinsic* semiconducting materials.

Intrinsic semiconducting materials use the elements in Group IV in the periodic table, germanium (Ge) and silicon (Si). These elements form crystal structures like that of carbon in diamond (Figure 9-2a). Each atom is bonded to its four neighboring atoms with covalent double bonds. There are no free electrons in the conduction band. How can these bound electrons be excited to act as charge carriers?

9.2.1 Electrically Charged Particles

Electricity is defined in terms of the flow or movement of electrically charged particles. This movement to other locations results in increased kinetic energy, which can be converted to some form of work for our benefit. Several forms of charged particles can satisfy these conditions, although some are not as efficient as others. The word *charged* refers to the electrical charge carried by these particles. All particles have this same charge, which is equal to 1.6×10^{-19} coulombs. This is the charge that all electrons carry. Its symbol is e^- and it is the standard unit of charge for measuring electrical charge. The magnitude of the charge in one proton is the same. However,

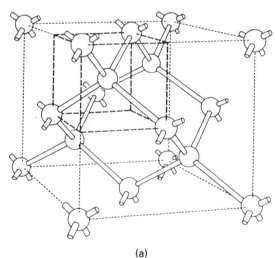

(a)

Figure 9-2 A new high operating speed IC chip capable of storing 72,000 bits of data compares in size to a 3¢ postage stamp. (IBM.) (a) Representation of the silicon crystal lattice arrangement. (U.S. Dept. of Energy.)

the proton is a positive charge. Its symbol is e^+. The measure of electric current I is the amount of charge q that passes a given point per unit of time t.

$$\text{Current } (I) = \frac{\text{charge } (q)}{\text{time } (t)} \quad \text{or} \quad I = \frac{q}{t}$$

An ampere (A) is defined as an electric current of 1 coulomb (C) of charge passing a point in 1 second (s). Knowing that each electron contains an electric charge of

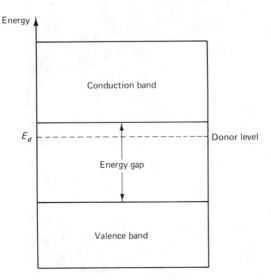

Energy

Conduction band

E_d – – – – – – – – – – – – – – – – – – – Donor level

Energy gap

Valence band

E_d — Energy needed by electrons in upper portion
of valence band to permit them to move into
the conduction band

(a)

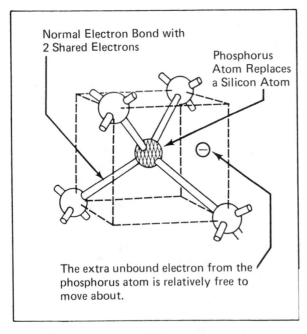

Normal Electron Bond with
2 Shared Electrons

Phosphorus
Atom Replaces
a Silicon Atom

The extra unbound electron from the
phosphorus atom is relatively free to
move about.

(b)

Figure 9-3 (a) Energy band diagram of an
n-type semiconductor. (b) Substitution of a
phosphorus atom with five valence electrons
for a silicon atom leaves an unbonded electron. (U.S. Dept. of Energy.)

1.6×10^{-19} C and using a proportion to determine the number of electrons in 1 C of electric current, the results are 6.25×10^{18} electrons in 1 C. Therefore, 1 A of electric current is the passage of 6.25×10^{18} electrons in 1 s. One microampere (1 μA) equals a current of 6,250,000 million electrons per second (6.25×10^{12}).

A hole, the absence of an electron, is considered as having the equivalent of an electron's charge only with a positive sign. Holes in the valence band, as you recall, are produced by supplying electrons with energy sufficient to cause them to be excited to the conduction band. Free electrons have greater mobility than holes. For comparison purposes, the mobility of electrons in pure germanium at 300 K is about 3950 cm^2 per volt second, and for holes it is about 1950.

9.2.2 Donor Doping

If a Group V element such as phosphorus (P) with its five valence electrons is added to germanium (Ge), only four outer electrons are needed to perfect a covalent bond with its neighboring germanium atoms to form a substitutional solid solution. The extra electron not needed for bonding will be attracted to the impurity atom (phosphorus) by a weak coulomb force of attraction of the nucleus. Since this electron is held much less tightly than the four bonding electrons, it and similar electrons contributed by other impurity atoms can be raised into the conduction band by the absorption of much smaller energies than those required for raising electrons from the valence band into the conduction band. This situation is depicted in Figure 9-3a and b. The extra electron is shown lying just below the conduction band at an extra energy level (E_d) called the *donor level*. This type of impurity atom makes the material *n*-type (possesses free electrons). The process of adding impurities is called *doping*. Figure 9-4 shows a sketch of one phosphorus atom substituting for a germanium atom in the crystal lattice of an *n*-type germanium semiconductor material. Other donor impurities that can be used for donor doping from Group V are arsenic (As) and antimony (Sb). Note that these atoms, having five valence electrons, are referred to as *pentavalent* atoms.

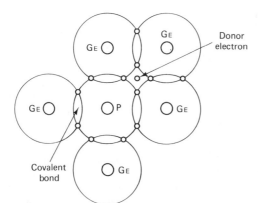

Figure 9-4 Effect of a dopant phosphorus atom substituting for a germanium atom in the crystal structure of an *n*-type semiconductor material.

9.2.3 Holes

From our study of the electronic structure of atoms we have learned that any orbit of most atoms, other than the K shell or first shell (which is completely filled with two electrons), is stable if it contains eight electrons. Silicon (see Figure 9-5) has only four electrons in its outermost shell (M shell). Being its outermost shell, it contains space for four additional electrons. These vacancies or absences of an electron in the valence shell are called *holes*. Holes are created in the valence band when

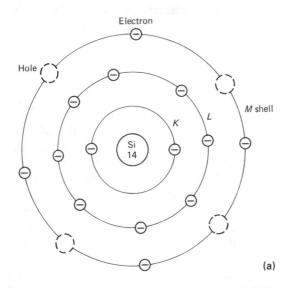

(a)

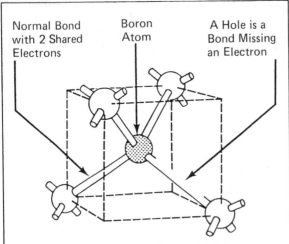

Normal Bond with 2 Shared Electrons

Boron Atom

A Hole is a Bond Missing an Electron

The hole can move relatively freely as a nearby electron leaves its bond and pops into it, moving the hole to the bond from which the electron came.

(b)

Figure 9-5 (a) Two-dimensional representation of the silicon (Si) atom. (b) Boron, a three valence electron atom in a silicon crystal, is normally bonded except one of the bonds is missing an electron creating a hole. (U.S. Dept. of Energy.)

valence electrons receive sufficient energy to move up into the conduction band. In the case of intrinsic semiconductors, the valence electrons with the highest energies (outer shell electrons have different energies) will be freed first when energy is added to the material as a result of an increase in temperature. The transfer of these electrons to the upper band creates holes in the lower (valence) band, which can then be occupied by electrons lower down or deeper in the valence band. When an electron receives an input of energy from some outside source, the electron will be loosened from its bonded position in an atom and be free to move about in the crystal structure (in the conduction band). If the energy received is insufficient to loosen the electron, it may just vibrate, which gives off heat. Once the electron breaks loose, it leaves behind a bond missing an electron. The bond missing an electron may be also called a hole (see Figure 9-5a). Electrons and holes freed from their positions in the crystal

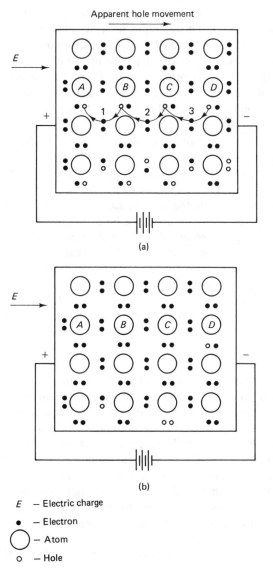

(a)

(b)

E — Electric charge

• — Electron

◯ — Atom

o — Hole

Figure 9-6 Conduction by holes in valence band.

structure are said to be electron–hole pairs. A hole, like a free electron, is free to move about in the crystal (Figure 9-6).

9.2.4 Conduction in Solids

Electricity flow (conduction) occurs in a solid when an applied voltage (electric field) causes charge carriers within the solid to move in a desired direction. In the absence of an electric field, the movements of charge carriers are random and result in zero charge transport. If a metal is placed in an electric field (in an electrical circuit), the free electrons in the conduction band will move toward the positive terminal. As they do so, they receive additional momentum (mass × velocity) and hence more energy. Those electrons moving toward the negative terminal lose momentum and reduce their energy.

Hole transfer involving electrons takes place when an electron from one atom jumps to fill the hole in another atom. This electron jump leaves a hole behind it. Or we can describe this movement not in terms of what the electron does but in terms of the hole movement. The hole moves in the opposite direction to the electron. Therefore, the flow of electric current is brought about by the movement of free electrons in the conduction band and/or holes in the valence band. Free electrons are easier to move through a semiconductor material than holes, hence they have greater mobility (drift velocity of the carrier).

With the help of Figure 9-6a, the movement of one hole can be clarified. The hole initially starts with atom A. As an electric field (E) is applied, a valence electron breaks free and moves to its left to fill the vacancy (hole) shown by arrow 1. The hole is now at atom B. A similar effect is felt by one of the valence electrons at atom C, which also moves to its left and fills the hole at atom B. Arrow 2 shows this movement. Each time an electron moves, it creates a hole. Figure 9-6b shows the final location of the hole at atom D. Note that the hole movement, in this case from left to right, is toward the negative-charged terminal of the battery, opposite to the flow of electrons as they move toward their left to fill the holes. In other words, this hole movement is a measurement of positive charge in the direction of the applied field (E). The action of the electrons in their moves, labeled 1, 2, and 3, in filling the just-created holes is known as *recombination,* which will be discussed later.

9.2.5 Acceptor Doping

When a Group III or trivalent element such as boron (B), aluminum (Al), gallium (Ga), or indium (In) is introduced as an impurity into an intrinsic semiconductor such as silicon (or other Group IV elements), a mismatch in the electronic bonding structure occurs. Using aluminum as an example of the dopant, this element lacks one electron to satisfy the covalent bonds of the Group IV matrix element such as silicon. This mismatch is illustrated in Figures 9-7 and 9-5a. One covalent bond near each dopant atom (Al) is incomplete; that is, it is missing an electron, or it contains a hole. If an external electronic field is applied to this solid, one of the neighboring electrons from another covalent bond can acquire sufficient energy to move into the hole. This, in turn, causes the hole to move to the position formerly occupied by the electron. In

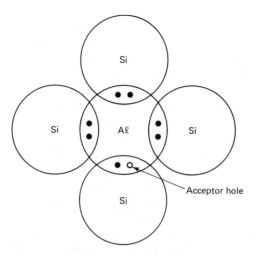

Aℓ = Acceptor atom

Figure 9-7 Structure of silicon with aluminum atom added as an impurity (dopant).

this manner the hole moves through the solid as a positive charge carrier. The electric current produced is mainly the result of hole movement (positive charge carrier). This structure is called *p*-type, because of the presence of free positive charges (the moving holes).

In terms of the energy band diagram for this *p*-type structure (Figure 9-8), the aluminum atom has provided an energy level that is only slightly higher than the upper limit of the valence band. This puts it in the forbidden gap. Thus, a nearby valence electron could be easily excited into this intermediate level. This level is called an *acceptor level*. As with *n*-type extrinsic semiconductor materials, the extremely small concentration of dopant is measured in parts per million (ppm). In

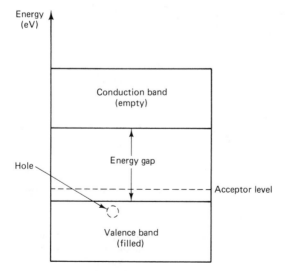

Figure 9-8 Energy band diagram for a *p*-type semi-extrinsic semiconductor material.

summary, the addition of acceptor atoms to a semiconductor material increases the number of holes in the valence band without an increase in the number of electrons. These positive-charge carriers are termed *majority carriers,* and the electrons are called *minority carriers* in *p*-type semiconductors, just the reverse of their designations in *n*-type semiconductors.

Figure 9-9 represents an *n*-type semiconductor material in a circuit with a switch in the open position. No electrons will flow in the external circuit. However, within the doped *n*-type semiconductor material, the free electrons contributed by the dopant atoms and the electrons that break away from their parent atoms in the valence band are diffusing throughout the material in a random fashion. In addition, each time a valence electron breaks free of its atom an electron deficiency in the valence shell of that atom is created, which makes the atom a positive ion with an electric charge of equal magnitude to that of the free electron. The vacancy in the valence shell (energy band) of the electronic structure of the atom is called a hole. The phrase *electron–hole pair* refers to this energy transfer process. When a free electron collides with a hole, it is captured by it, which produces a balanced atom in its lowest equilibrium state. This process is known as *recombination.* Taking into consideration the remaining atoms that make up the material, we recognize that overall equilibrium must be maintained, and thus a new electron–pair is generated with every recombination. (Figure 9-10).

Next we close the switch in the external circuit in Figure 9-9. This causes electron flow to occur traveling from the negative terminal of the source through the semiconductor material and back to the positive terminal of the source. Figure 9-9 attempts to show that only electrons are flowing in the external circuit within the

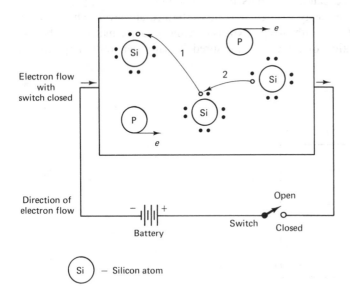

Electron flow with switch closed

Direction of electron flow

Open
Switch   Closed

Battery

Si — Silicon atom

P — Phosphorus donor atom

• — Electron

o — Hole

Figure 9-9 Random movement of electrons and holes in an *n*-type semiconductor material.

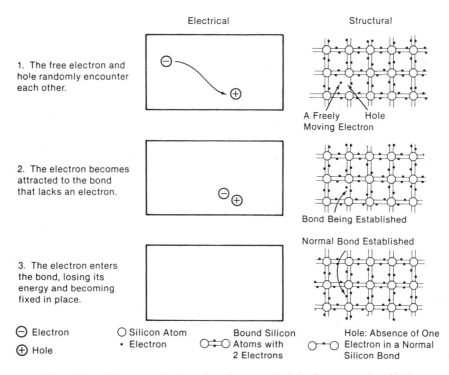

Electrical   Structural

1. The free electron and
hole randomly encounter
each other.

A Freely   Hole
Moving Electron

2. The electron becomes
attracted to the bond
that lacks an electron.

Bond Being Established

Normal Bond Established

3. The electron enters
the bond, losing its
energy and becoming
fixed in place.

⊖ Electron   ◯ Silicon Atom   Bound Silicon   Hole: Absence of One
                    • Electron   ◯═◯ Atoms with   ◯•◯ Electron in a Normal
⊕ Hole   2 Electrons   Silicon Bond

Figure 9-10   Direct recombination of an electron and a hole. Excess energies of both
the electron and hole are lost to heat. (U.S. Dept. of Energy.)

material. The action of the electric field gives direction and movement (drift) to the
flow of both free electrons donated by the phosphorus atoms in this case and the
holes (minority carriers) in their movement from right to left. Both the free electrons
donated by the donor atoms and the electrons supplied through the external circuit
combine with the holes to accomplish the conduction process through the processes
of generation and recombination.

When a piece of *n*-type material is brought into contact with a piece of *p*-type
material, a *pn* junction is formed (Figure 9-11). A junction of this type produces a
diode, which is the simplest of the semiconductor junction devices. Before observing
the diode in a simple circuit, it would be helpful to study the interaction of the
electrons at the *pn* junction.

The *n*-type material contains an excess of electrons, whereas the *p*-type material
has an excess of holes. When the *pn* junction is formed, there is a diffusion of charges
between the two materials. The electrons from the *n*-type will move across the junction
to the *p*-type. This will create holes or positively charged particles in the *n*-type. The
reason for this diffusion of charges is to counteract the electrical imbalance in the
two materials caused by the formation of the junction. The diffusion process will
stop once an equilibrium of charge is established between the two materials. This
equilibrium is reached when no additional charge carriers have the energy to overcome
the electric field that has been built up at the junction. It can be seen in Figure 9-

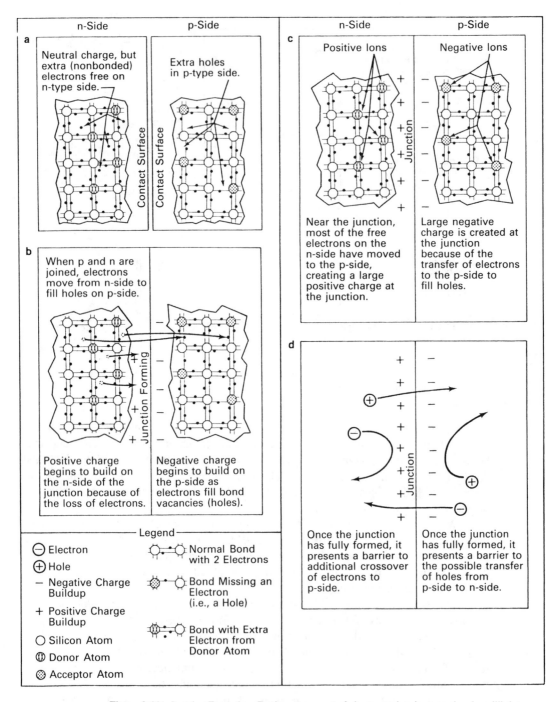

n-Side   p-Side

a

Neutral charge, but extra (nonbonded) electrons free on n-type side.

Extra holes in p-type side.

Contact Surface

Contact Surface

b

When p and n are joined, electrons move from n-side to fill holes on p-side.

Junction Forming

Positive charge begins to build on the n-side of the junction because of the loss of electrons.

Negative charge begins to build on the p-side as electrons fill bond vacancies (holes).

─── Legend ───

⊖ Electron

⊕ Hole

— Negative Charge Buildup

+ Positive Charge Buildup

○ Silicon Atom

⦿ Donor Atom

⊛ Acceptor Atom

Normal Bond with 2 Electrons

Bond Missing an Electron (i.e., a Hole)

Bond with Extra Electron from Donor Atom

n-Side   p-Side

c

Positive Ions

Negative Ions

Junction

Near the junction, most of the free electrons on the n-side have moved to the p-side, creating a large positive charge at the junction.

Large negative charge is created at the junction because of the transfer of electrons to the p-side to fill holes.

d

Junction

Once the junction has fully formed, it presents a barrier to additional crossover of electrons to p-side.

Once the junction has fully formed, it presents a barrier to the possible transfer of holes from p-side to n-side.

Figure 9-11 Junction Formation. Further movement of charge carriers is stopped and equilibrium is established. (U.S. Dept. of Energy.)

12, that a boundary is formed across the *pn* junction. This boundary is usually referred to as the *depletion region* because all mobile charge carriers have been depleted from this area.

The electric field set up by the stationary charges in the depletion region creates a difference of potential between the two materials. This electric field represents the equilibrium potential difference from *n*-type to *p*-type in direction. This voltage is called the contact potential of the junction. For silicon it is about 0.6 V.

All the previous discussion has been aimed at the *pn* junction (a diode) as a closed system. Next the junction will be observed with an externally applied voltage (Figure 9-12b).

The *p*-type side of the diode is called the anode and the *n*-type side is called the cathode. Applying a voltage to the diode is called *biasing*.

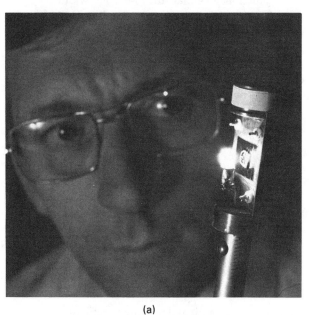

(a)

Figure 9-12 (a) Speedy electrons. Scientist prepares to test experimental semiconductor crystal shown at right of light. By increasing the speed at which electrons move through the crystal a fundamental advance in solid-state technology was achieved. (b) A *p-n* junction (diode) showing no bias, forward bias, and reverse bias conditions. (Bell Labs.)

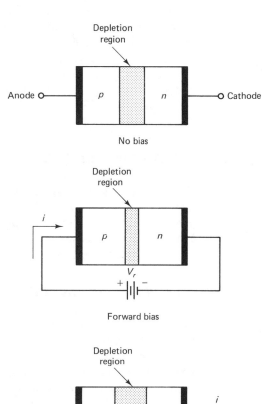

(b)

If, as in Figure 9-12b, a battery is connected across the diode so that the positive lead is attached to the anode and the negative lead is attached to the cathode, a large current will flow. This is the forward-biased condition of the diode. The reason for the large current is that the junction potential (contact potential and applied potential) has been lowered due to the forward bias. Electrons from the battery that enter the anode of the device will see little resistance and can flow easily across the *narrowed* depletion region to the cathode.

In Figure 9-12, the leads of the battery have been reversed so that the positive side of the battery is across the cathode and the negative side is across the anode. This is the *reverse-biased* condition of the diode, and very little (only leakage current) will flow. The reason for the extremely small current in this condition is that the battery is acting in the same direction as the electric field across the junction; thus the junction potential is larger. When the junction potential increases, the width of the depletion region and the amount of stationary charges will increase. This means that the charge carriers must gain an extremely large amount of energy before they can cross from cathode to anode.

The *pn* junction is the basic building block for all semiconductor junction devices. The preceding discussion can be applied to the transistor, SCR (silicon control rectifier), FET (field effect transistor), or any other junction device.

9.3 OPTOELECTRONIC DEVICES

Optoelectronic devices are (1) operated by light (photoelectric), (2) produce or emit light, or (3) modify light. Photoelectric devices, in turn, can be categorized as photoconductive, photovoltaic, or photoemissive. A review of optical properties in Module 4 as well as semiconductor materials in this module is appropriate prior to studying the following treatment of these devices.

9.3.1 Photoconductive Devices

In photoconductive devices, the conductivity of the semiconductor material will vary provided the energy supplied by the light (visible, infrared, or ultraviolet) is sufficient to raise the electrons into the conduction band. A photomultiplier is such a device which can produce an image that is visible when illuminated by a weak light source. A light meter is another example which operates in accordance with the same basic principle.

9.3.2 Photovoltaic Devices

Photovoltaic cells are semiconductor junction devices that convert electromagnetic energy in the form of light directly into electrical energy. The amount of electrical current (flow of electrons) is directly proportional to the amount of light that is incident on the semiconductor material. A *solar* or *photovoltaic cell* is a photodiode that is used to extract electrical energy directly from sunlight. This direct conversion of sunlight to electricity differs from the solar thermal conversion process used in solar heating of homes and offices, which uses panels to absorb sunlight and uses the energy to heat water or other medium, including air for heating a building.

A typical solar cell (Figure 9-13a) contains two extremely thin layers of silicon connected externally by a wire to the load where light energy is converted to work. When light is absorbed in silicon, it creates electron–hole pairs (Figure 9-13b). Electron–hole pairs that reach the junction are separated, holes going to the p side and electrons to the n side. This buildup of charge on either side of the junction creates a voltage that drives current through the external circuit. Reflection of sunlight from an untreated silicon cell can be as great as 30%. By chemical treatments and texturing of the surface, this figure can be reduced to around 5%. Light with a certain wavelength (energy) is required to interact with specific materials before the optimum electron–hole generation is achieved (Figure 9-13c). This characteristic energy is called the material's band gap energy. (Silicon requires 1.1 eV, gallium arsenide, 1.4 eV.) Overall, these mismatches of light with a solar cell's material waste some 55% of the energy from sunlight. Research is proceeding to find better ways to process sunlight to make it monochromatic (one wavelength) and possess an exact energy match with that required by the cell material to make the conversion of incoming sunlight more efficient.

Theoretically, the silicon solar cell should be able to convert about 25% of the sun's energy into electricity. To achieve greater conversion of sunlight to electricity, two different solar cells can be used, one of aluminum gallium-arsenide and the other of silicon. These two cells absorb a wider range of wavelengths from the solar spectrum, but various optical losses reduce the overall system efficiency to about 25%. The term *efficiency* is the percentage of energy in sunlight striking the cell that is converted into electricity. Individual solar cells have limited power. To produce electricity for most applications, they must be joined together electrically to form modules to become building blocks for larger arrays (Figure 9-13d).

Solar cells are very expensive primarily because of their fabrication costs. Many breakthroughs in materials technology are needed before solar cells can be developed to compete on the mass market to produce electricity costing around 50 cents per peak watt or about six to eight cents per kilowatt hour. Ultrapure (99.999% purity) semiconductor-grade silicon costs around $15 a pound compared to $0.15 a pound for metallurgical silicon with a purity of 99.5%. Approximately 30% of the cost per peak watt of solar-produced electricity represents this processing cost. Peak watt is a term used to express the amount of power produced by a solar cell in full sunlight at 25°C. A 4-in. cell, for example, can produce 1 peak watt at noon on a sunny day in the U.S. Southwest. This is a description of ideal conditions. A more relevant term is *average wattage*, which is about one fifth the peak wattage. See Table 9-1 for some current applications.

Most common solar cells, first invented at Bell Laboratories in 1954 with a 4% efficiency, are made from a single crystal of pure silicon artificially grown in the form of an ingot that is sliced into wafers 0.012 in. thick, which are then polished, trimmed, and doped in an oven. The finished cells are mounted in arrays or panels

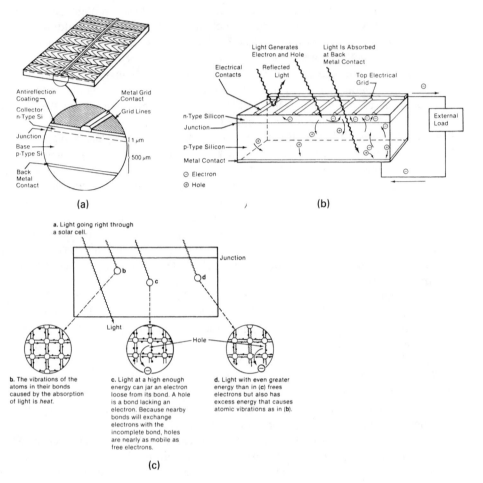

(a)

(b)

a. Light going right through a solar cell.

Junction

Light

Hole

b. The vibrations of the atoms in their bonds caused by the absorption of light is *heat*.

c. Light at a high enough energy can jar an electron loose from its bond. A hole is a bond lacking an electron. Because nearby bonds will exchange electrons with the incomplete bond, holes are nearly as mobile as free electrons.

d. Light with even greater energy than in (c) frees electrons but also has excess energy that causes atomic vibrations as in (b).

(c)

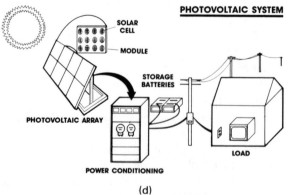

PHOTOVOLTAIC SYSTEM

SOLAR CELL

MODULE

STORAGE BATTERIES

PHOTOVOLTAIC ARRAY

LOAD

POWER CONDITIONING

(d)

Figure 9-13   (a) A typical p-n junction single crystal silicon solar cell outlining its components. (b) Light striking a cell creates electron-hole pairs which are separated by the potential barrier creating a voltage that drives a current through an external circuit. (c) What happens to light entering a cell? (d) A schematic of a photovoltaic system. (U.S. Dept. of Energy.)

TABLE 9-1 EXAMPLES OF PRESENT TERRESTRIAL APPLICATIONS OF PHOTOVOLTAIC POWER UNITS

Application	Peak rating (W)
Warning lights:	
Airport light beacon	39
Marine light beacon	90
Railroad signals	
Highway barrier flashers	1.2
Tall structure beacon	
Lighthouse	
Communications systems:	
Remote repeater stations for	
microwaves	50
radio	109
TV	78
Remote communications stations	3,500
Mobile telephone communication station	2,400
Portable radio	50
Emergency locator transmitter	
Water systems:	
Pumps in desert regions	400
Water purification	10,800
Scientific instrumentation:	
Telemetry—collection and transmission platforms for environmental, geological, hydrological, and seismic data	
Anemometer	100
Remote pollution detectors (H_2S, noise)	3
Industrial:	
Remote machinery and processes, e.g., copper electrolysis installation	1,500
Cathodic protection of underground pipeline	30
Electric fence charger	
Domestic water meter	20
Off-shore drilling platforms	
Forest fire lookout posts	
Battery charging:	
Boats, mobile homes and campers, golf carts	6–12
Construction site equipment	
Ni-Cd-powered military equipment	74
Recreational and educational:	
Educational TV	35
Vacation home	
lighting, TV	
refrigerators	200
Sailboats	
lighting, ship-shore communication	
automatic pilot	66
Portable TV camera	
Camping lighting	
Electronic watches, calculators	
Recreational center sanitary facility	168
Security systems:	
Closed-circuit TV surveillance	150
Intrusion alarms	6

Figure 9-14 Semiconducting crystals grown by controlled composition of individual atomic layers. Background screen displays an electron microscope image of an atomic layered structure magnified 2 million times similarly grown. (Bell Labs.)

containing dozens of individual cells whose diameters may range up to 4 in. One manufacturing innovation produces cells continuously in the form of thin ribbons. This thin-film technology, similar to that used in making electronic integrated circuits, could cut manufacturing costs substantially and permit the fabrication of large-area cells that would make solar cells more cost effective.

One solar cell device using thin-film techniques developed by Bell Laboratories (Figure 9-14) consists of a layer of polycrystalline cadmium sulfide deposited on a single crystal substrate of copper indium phosphide. Reportedly, its efficiency is about 12.5%. This new device can also be used quite effectively as a *photodetector* for converting light impulses into electrical signals. Many applications of this device can be found in optical communications systems that detect the presence of infrared and visible light transmissions through optical glass fibers.

9.3.3 Photoemissive Devices

In a photoemissive device, the light (or photons) generated by the recombination of electron-hole pairs is emitted from the surface of the device. Electrons striking the TV screen, which is coated with a semiconductor material doped with copper, raise the electrons in the coating into the conduction band where they recombine and emit energy in the form of light (photons). This is known as electroluminescence.

Fluorescent lamps depend on ultraviolet light striking electrons in the inner coating of the lamps which emit light in the visible region of the electromagnetic spectrum. To round out our coverage of these devices, a light-emitting diode will be treated in more detail in the following paragraphs. It is important to repeat at this point that regardless of the device, the underlying mechanisms involve the mutual interaction of electrons with electromagnetic radiation.

9.3.3.1 Light emitting diode (LED).

Figure 9-15 shows a cross-section of a LED. When forward bias is placed across the junction, electrons cross from the *n*-side of the junction to recombine with the holes in the *p*-side giving off energy in the form of heat and light. The light will be emitted assuming the semiconductor material is translucent and the gold film cathode effectively reflects the light to the surface. Gallium phosphide (Ga P) and gallium arsenide phosphide (GaAsP) are typical semiconductor materials used in LEDs to produce red, yellow, or green light. The relatively large amounts of current consumed by LEDs is their primary disadvantage, but their ruggedness and long life tend to make up for this shortcoming.

Silicon, the second most abundant element in the earth's crust, exists as silica or silicon dioxide (SiO_2). The processing of quartzite, a mineral (almost 99% silica) and the source of pure silicon, raises the cost of silicon to about \$70/kg. Pure polycrystalline silicon is further processed into single-crystal or large-grain (5 mm) polycrystalline silicon (Figure 9-16a). The single-crystal silicon technology, the oldest and most well established, is slowly giving way to newer technologies using polycrystalline and amorphous silicon.

The latest fabrication techniques form silicon directly into usable wafers (0.5 mm thick) without intermediate steps, such as sawing, thus reducing processing costs considerably. More advanced methods produce ribbons (long ultrathin, rectangular sheets). One such method produces ribbons 33 m long, 5 cm wide, and 0.5 mm thick at a linear rate of 55 cm/min, with cell efficiencies of over 10% (Figure 9-16a). *Polycrystalline silicon* is cheaper to produce and fabricate than single-crystal silicon, but at present single-crystal cells have maintained their lead in efficiency. In poly-

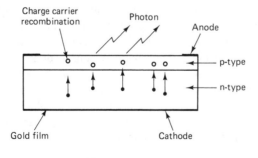

Figure 9-15 Cross section of an LED.

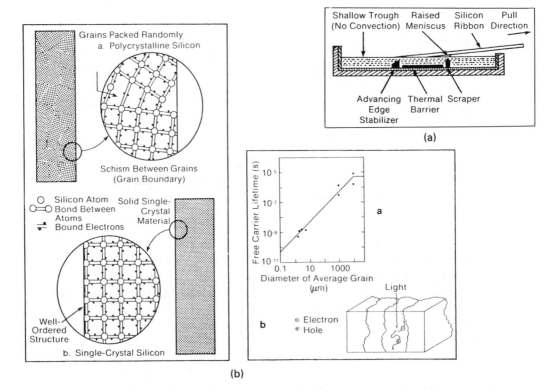

Figure 9-16 (a) Low-angle silicon sheet (LASS) growth is a very fast method of drawing a silicon ribbon from a shallow trough of molten silicon. (b) Lifetime of free charge carriers dependent on grain size and orientation. (c) A p-i-n device. Electric field sweeps the charge carriers to opposite ends of cell. (d) Silicon atoms build a 3-D tetrahedral structure which in amorphous silicon is rotated randomly producing a disordered atomic structure. (U.S. Dept. of Energy.)

crystalline silicon the grain boundaries impose numerous restraints on the movement of charge carriers (Figure 9-16b). However, with the constant growth of new technologies the effect of the grain boundaries is being reduced.

A third form of silicon, *amorphous silicon* (abbreviated aSi) is being used since 1974 to produce cells with high output voltage (0.8 V), currents greater than 10 mA/cm², and efficiencies of 6% using *p-i-n* cells (*p*-type, intrinsic, and *n*-type layers). Figure 9-16c is a schematic of this cell pointing out the ultrathinness of the layers of amorphous silicon. To overcome amorphous silicon's disordered atomic structure (Figure 9-16d) with its incomplete bonds, hydrogen is added (doped) to complete this bonding and increase the cell's efficiency. A distinct advantage of amorphous silicon is its ability to absorb light about 40 times more strongly than crystalline silicon. Amorphous silicon cells are used to power hand-held calculators because they are more efficient and cost effective under fluorescent light than either single-crystal or polycrystalline silicon. This characteristic of amorphous silicon allows such cells to be extremely thin, one-fortieth of the thickness required to absorb the same light. The low mobility of the charge carriers means relatively rapid recombination,

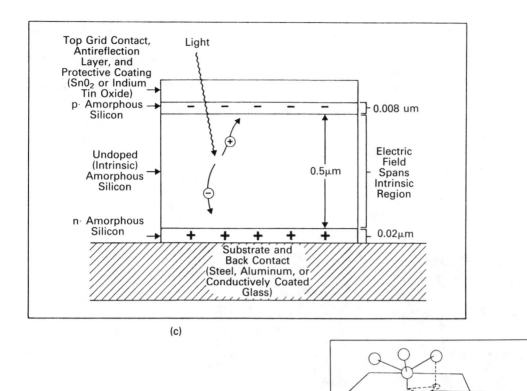

Top Grid Contact, Antireflection Layer, and Protective Coating (SnO₂ or Indium Tin Oxide)

Light

p· Amorphous Silicon

0.008 um

Undoped (Intrinsic) Amorphous Silicon

0.5μm

Electric Field Spans Intrinsic Region

n· Amorphous Silicon

0.02μm

Substrate and Back Contact (Steel, Aluminum, or Conductively Coated Glass)

(c)

The Random Variation from 60° is Less Than 10°

○ Silicon Atom

(d)

which in turn means that the charge carriers must be separated by the p-i-n junction in the short time during illumination. In addition thinness translates into less material, and depositions of the material tend to be easier.

9.4 LIQUID-PHASE EPITAXY AND MOLECULAR-BEAM EPITAXY (MBE)

Liquid-phase and molecular-beam epitaxies are methods developed by Bell Laboratories to produce new semiconducting, crystalline materials not found in nature with tailor-made atomic structures that provide for an array of new, built-in electronic, mechanical, and optical properties. The word *epitaxy* describes the overgrowth in layers of a crystalline material deposited in a definite orientation on a base material of different crystal structure and chemical composition. In this world of man-made materials this scientific research and development effort by Bell Laboratories demonstrates once more our ability to exert precise control over the creation of a new

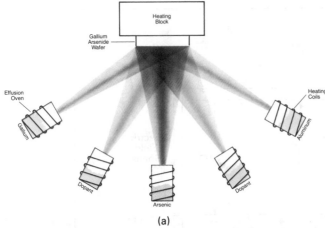

Figure 9-17 (a) Atomic or molecular beams formed by heating elements in effusion ovens aimed at a base metal. (Bell Labs.) (b) "Rough" surface of wafer (left) smoothed after MBE growth of one layer (right) permitting growth of layers of uniform thickness. (Bell Labs.) (c) Patterned layer. Island of gallium arsenide doped with elements with desired properties can be grown in a "sea" of semi-insulating gallium arsenide. (Bell Labs.) (d) Rectangular glass tubing. Such tubing has made tritium lighting practical for LCDs in digital watches. (Corning Glassworks.)

(a)

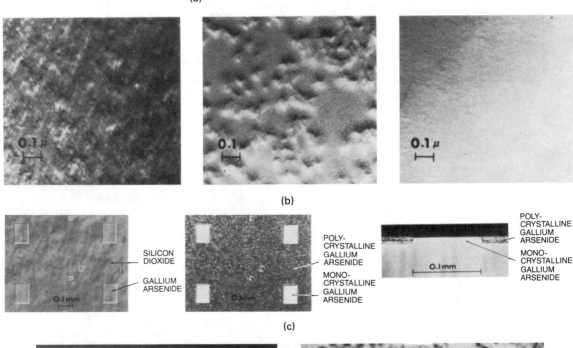

(b)

(c)

(d)

material by methods that ensure the exact composition and structure of the alternating atomic layers that make up a material.

Starting with a substrate of gallium arsenide, a semiconductor material, atomic beams of controlled intensity are aimed at the base material in an ultrahigh vacuum. (Figure 9-17a and b). Shutters are used to turn the beams on and off as they are directed at the base from a heated oven. A layer of gallium atoms followed by layers of arsenic, aluminum, arsenic, then gallium atoms, repeated many times, produces a crystal resembling a thin, highly polished mirror. Liquid-phase epitaxy, an older technique, produces satisfactory wafers, but MBE appears to be the most promising, Using MBE, ultrathin multilayer crystals with different structures have been grown. Quantum well structures with up to 100 alternating layers represent one class of such crystals, with each layer as thin as several hundredths of a micrometer. Monolayer structures with alternating layers of atoms in a stack, each layer two atoms planes thick, are being produced. Multilayer structures consisting of several thousands of individual monolayers have been grown with a total thickness of 1 μm. Research has revealed that due to the thinness of these layers the electrons and holes normally confined to each layer will interact with their counterparts in adjacent layers and build *superlattice* structures with even more desirable properties (Figure 9-17c). Refer to the Gardiner and Tills book listed in the references at the end of this module for more depth on these advanced manufacturing processes.

9.5 LIQUID CRYSTALS

Liquid crystal displays (LCD) currently used in watches, signs, and other similar applications offer a display system that does not require as much energy as LED (light emitting diodes) (Figure 9-17d). The image familiarly seen on LCD produces a silver display, while the LED is the familiar red display. Images for LCD that are generated by lasers have potential for multicolored video displays for information systems. The LCD offers many possibilities for easier information retrieval.

Most known substances can exist, given certain conditions, in a gaseous, liquid, or solid state. A very few compounds can exist in a fourth state in between the liquid and solid phases known as the *liquid-crystal phase*. This mesomorphic (intermediate) phase or anisotropic liquid was discovered nearly 100 years ago (1888) but has remained a physical oddity until recent advances in the technologies of electrooptics and thin-film components have permitted several important applications of these materials, particularly in the visual display of information.

Liquid crystals (LC) are organic compounds that flow like a liquid while maintaining a long-range orderliness of a solid. Normally, crystals of a pure compound when heated show a well-defined and characteristic melting point at which the ordered crystalline lattice structure breaks down and the material becomes a liquid. In the liquid phase the individual molecules show no preferred spatial orientation. The feature of liquid crystals is that during the melting process the well-ordered three-dimensional crystalline structure transforms into a one- or two-dimensional state of order. This results in a material that has some optical properties of a solid combined with the fluidity of liquids, resulting in a mix of unique properties.

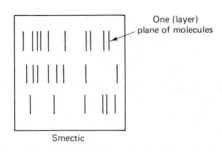

One (layer) plane of molecules

Smectic

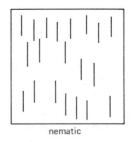

nematic

Figure 9-18 Model of the smectic mesophase. **Figure 9-19** Model of nematic mesophase.

The molecular aggregates (collection of molecules) of an LC compound are in the form of long cigar-shaped rods. The orientation of these rodlike polar molecules forms the basis for classifying three basic types of liquid crystals: smectic, nematic, and cholesteric. The *smectic* phase consists of flat layers of cigar-shaped molecules with their long axes oriented perpendicular to the plane of the layer. This is the most ordered phase. The molecules within each layer remain oriented within each layer and do not move between layers. Figure 9-18 shows a structural model of this smectic mesophase. The molecules lie with their long axes parallel in layers. The molecules may move relative to each other, and consequently several types of smectic structures may be formed, depending on the inclination of the long axes of the molecules to the plane of the layers.

The *nematic* phase also has molecules with their long axes parallel, but they are not separated into layers. Rather, their structural arrangement is similar to the ordinary packing of toothpicks in a box. Figure 9-19 depicts such a nematic mesophase structure. While maintaining their orientation, the individual molecules can move freely up and down.

Cholesteric mesophase (Figure 9-20) may be defined as a special case of the nematic in which the thin layers (one molecule thick) of mostly parallel molecules have their longitudinal axes twisted (rotated) in adjacent layers at a defined angle. Each layer is basically a nematic structure. The axes of alignment of contiguous layers differ by a small angle and produce a helix or progressive rotation of many layers in an LC material.

The nematic liquid crystals, because their molecules can be aligned by electric and magnetic fields to produce changes in their optical properties, find increasing use in electrooptical devices. Additionally, several mechanical influences such as pressure, impact, or temperature induce a change in the structure of nematic liquid crystals similar to those brought about by electromagnetic fields. This cholesteric structure exhibits anisotropic optical properties. When such a material is illuminated with white light at certain temperatures, iridescent (having shifting or changing colors) colors are observed in the material brought about by a fraction of the incident light. The much larger portion of light is transmitted by the LC material. This reversible color phenomenon functions over a temperature range of about $-20°$ to $250°C$. These properties of LC materials are finding many applications in heat-transfer studies (thermal mapping), nondestructive testing (NDT), toys and games, holography, medical diagnosing of vascular diseases, and in flat panel, full-color display panels technology.

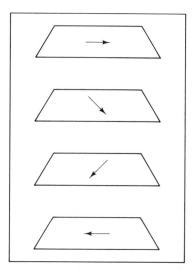

Figure 9-20 Model of cholesteric meso-phase.

9.5.1 Liquid Crystal Cell

A LC cell consists of a layer of nematic mesophase LC material between two glass plates that are glued or fused together. The thickness of the LC material is about 10 to 25 μm. The two glass plates have transparent electrodes deposited on their inside faces made of a transparent and conducting material such as tin or indium oxide. Figure 9-21 shows a sketch of the construction of a LC cell.

9.5.2 Homeotropic and Homogeneous Orientations

Two preferred orientations of the LC molecules, *homeotropic* and *homogeneous*, are used within LC display devices (LCDs). Figure 9-22 shows the homeotropic orientation with the long axes of the molecules perpendicular to the surface of the glass plates and electrodes. This orientation can be achieved by chemical doping of the nematic phase. The long, cigar-shaped rod orientation of the molecules allows the groups of molecules to act like dipoles in the presence of an electric field.

Figure 9-23 shows the orientation of the molecules to be parallel to the glass plates. This parallel orientation can be produced in the LC material by mechanically

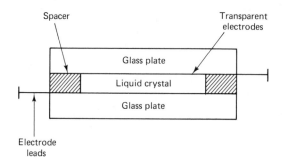

Figure 9-21 Cross section of a LC cell.

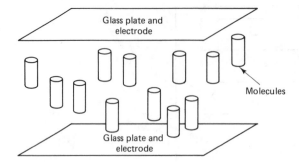

Figure 9-22 Homeotropic molecular orientation.

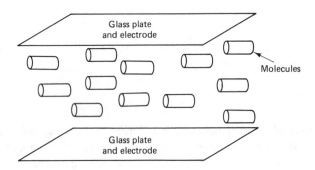

Figure 9-23 Homogeneous molecular orientation.

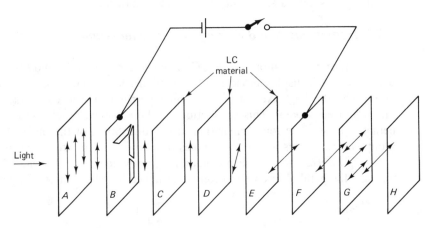

A — Vertical polarizer
*B — Front glass plate electrode
C — Liquid crystal field effect cell
D — Liquid crystal field effect cell
E — Same as *C* and *D*
*F — Back glass plate electrode
G — Horizontal polarizer
H — Reflector

* Numerical character segments
contained in *B* and *F*.

Figure 9-24 Reflective type LCD (electric field not applied).

rubbing, unidirectionally, the glass plates with a leather cloth prior to assembly of the LCD. Another means of accomplishing this orientation is by the deposition of a layer of dielectric over the transparent electrodes. Each plate is then rotated 90° in relation to each other. In effect, the LC material acts now as a set of polarizers that cause the light passing through the LC cell to be rotated 90°. Another way of altering the alignment of these layers in the LC material is to apply an electric field across the material. The molecules then align themselves in the direction of the electric field, which prevents the 90° twist of the molecular layers. The incident light cannot be transmitted through the cell. By judicious selection of the LC material, polarizers, glass plate treatment, and the electric field, the incident light can be controlled through the cell; that is, it may be reflected, rotated, transmitted, or extinguished.

9.5.3 Reflective-Type LC Cell Operation

Although there is a transmittive-type LC cell, the reflective-type cell is chosen to explain the operation of a numerical display LC device (LCD). Figure 9-24 is a sketch of the components of a reflective-type LCD. The main components of the LC cell consist of the two glass plates, electrodes, and LC material. A vertical polarizer and a horizontal polarizer, plus a reflector, are added to the cell. With the plates treated and arranged to produce the 90° twist described previously, unpolarized light enters the vertical polarizer, rotates 90° through the LC material, passes through the horizontal polarizer, and on to the reflector. The reflector reflects the light back through the same path. The face of the reflector appears light all over its surface.

If we wish to show one numeral in the display, then several LC cells are placed together in a particular configuration; each controls one segment of the digital display, as shown in Figure 9-25. An electric field activates the selected segments, which acting together form the particular numerical character desired. The digital display can be impressed on the electrode on the front glass plate in Figure 9-25, with the

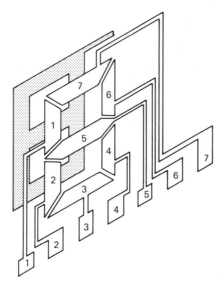

Figure 9-25 Seven segment numeric display.

electrode on the back plate acting as the base of the common electrode. The molecules in the area of the energized segments will align with the electric field and prevent rotation of the polarized light. Consequently, the vertically polarized light passes through the cell everywhere but in the region of the energized pattern elements. The end result is that the energized display elements appear as black images against a light background. By properly choosing the correct combination of segments to be activated, any digit can be displayed from 0 to 9.

9.6 PIEZOELECTRIC MATERIALS

Piezoelectric crystals are physically uniform solids that are bonded together by ionic bonds. We have learned that ionic bonding is a result of the electrostatic forces of attraction between the ions of opposite charge. Normally, the number of positive-charged ions equals the number of negative-charged ions. Figure 9-26a is an attempt to show the symmetric, crystalline structure of a crystal using three representative

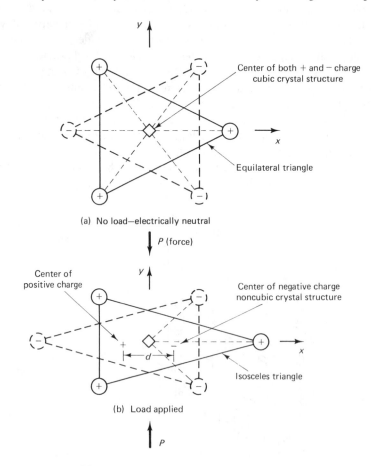

Figure 9-26 Model of piezoelectric crystal structure.

positive ions located equidistant from each other (at the vertices of an equilateral triangle). Similarly located are three negative-charged ions. The centers of charge both occupy a common center of symmetry. Being of equal magnitude, these centers of charge not only coincide in location but they cancel each other leaving the electric charge of these six ions neutralized.

Figure 9-26b is a similar representation of the same structure of a crystal with the addition of a mechanical load or force (P) that produces a stress assumed to be uniformly distributed over the face of the crystal upon which it acts. This force, in this case, acting along the Y or mechanical axis is tending to compress the crystal, resulting in a compressive, elastic strain or unit deformation. The elastic deformation, in turn, results in the offsetting of the centers of electric charge from each other. In effect, this offsetting of centers of charge creates electric dipoles throughout the material, which combine to produce a measurable electric potential along the X or electric axis of the crystal. Conversely, if an electric potential were to be applied to the X axis, a detectable deformation would be observed along the Y axis. The prefix *piezo-* comes from the Greek word meaning "to sit on, or press." The word *piezoelectricity* refers to electricity generated by exerting pressure on an ionic crystal.

Figure 9-27 is a model of such a crystal with an electric field impressed along the X axis. This particular arrangement of electric charge acting as the input of energy would produce a mechanical deformation (contraction) similar to that of the force (P) in Figure 9-26. Note that the ions of negative charge are attracted to the right toward the positive terminal, with a similar movement of the positive ions and their center of charge to the left, thus increasing the separation of the centers of charge (increasing the dipole length). If the polarity of the terminal were reversed, the deformation produced would be distorted in the opposite direction (expansion). An ac voltage, when impressed on the crystal, would cause the crystal to expand and contract (oscillate) at its driven frequency (maximum amplitude at resonance), which

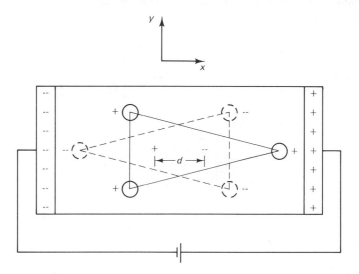

Figure 9-27 An electric field across a piezoelectric crystal.

could be transmitted as a wave of sound energy into the surrounding medium. In other words, if this potential were produced by an alternating current, it would cause the crystal to vibrate. Each crystal has its own natural, mechanical frequency and the ac potential could be adjusted to match this frequency and produce resonance (maximum amplitude of vibration).

In summary, if a mechanical stress is applied to two opposite ends of a crystal (mechanical energy supplied), the remaining ends of the crystal are charged with electricity (output in electrical energy); if an electric voltage is supplied to one set of faces of a crystal, the other two faces of the crystal are deformed (either they contract or they expand); and if an ac potential is supplied, the output is crystal oscillation at its driven frequency, which can be transmitted into a surrounding medium as sound energy at a constant wavelength. A piezoelectronic crystal is a *transducer* in that it not only transfers energy from one system to another, but it converts energy from one form into another.

Quartz crystals (SiO_2), possessing piezoelectric properties, are both naturally occurring and grown commercially. In its uncut state the crystal is in the form of a hexagonal prism as sketched in Figure 9-27. The X axis passing through the corners of the hexagonal cross section is the *electrical axis;* the Y axis perpendicular to the faces of the hexagonal cross section is the *mechanical axis;* and the Z axis is the *optical axis.* The crystal is cut in a variety of ways, depending on the desired characteristics. The sections cut from the crystal are known as *blanks.* One such blank, sketched in Figure 9-27, is called an X cut and has a thickness that is parallel to the X axis and a length parallel to the Y axis. A mechanical stress applied to its faces along the Y axis produces a voltage along the X axis. The dipoles created in quartz crystals do not rotate; thus there is a permanent orientation. This characteristic permits a quartz crystal, once cut and polished to specified dimensions, to maintain a resonance frequency with an extremely high degree of accuracy. This feature finds great use in controlling the frequency of radio broadcast signals and in chronometers (time pieces) of great accuracy.

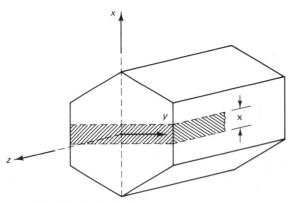

z — Optical (longitudinal) axis
x — Electrical axis
y — Mechanical axis

"x" cut blank shown in section

Figure 9-28 Uncut hexagonal quartz crystal.

A piezoelectric crystal, an electromechanical transducer, finds many uses. The ultrasonic inspection of solid materials for the presence of defects is an example. The piezoelectric crystal sends out pulses of ultrasonic waves into the material. Upon striking a discontinuity, these waves are reflected back to the crystal, which converts the mechanical impulses into electrical signals for viewing on a screen. In addition to its use in nondestructive inspection (NDI), the combination of piezoelectric crystals and ultrasonics can be seen in such applications as cleaning of parts, mixing of paints, and the homogenization of milk. Sonar (sound ranging and navigation) and depth-finding equipment, as well as certain bats, use ultrasonic vibrations beyond the auditory limit (greater than 20,000 Hz) to determine the location of near and distant objects by the reflection and conversion of sound energy into electrical energy.

Ceramics, by nature, have been the traditional source of piezoelectric materials. However, the National Bureau of Standards (NBS) has recently (1978) developed a man-made piezoelectric polymer film from polyvinylidene. Because piezoelectricity is not a natural property of polymers, a thin film of polyvinylidene is first coated on both sides with aluminum to act as electrodes. The composite is then heated to about 400 K, and an electric film is applied and maintained while the material is cooled to room temperature. Stabilized piezoelectric properties are induced into the film, and are, in many cases, superior to those of ceramic crystals. Applied like a Band-Aid, these thin-film polymers, 0.5 cm thick, are ideal for monitoring the pulse rate and heart beat of humans. Future applications of these man-made films appear unlimited.

9.7 MATERIAL DEVELOPMENTS

Numerous developments in electronic materials are forthcoming and many stagger the imagination. Developments related to *lasers* have a wide range of applications, from weaponry to three-dimensional television. *Fiber optics* made tremendous strides during the 1970s and offers solutions as efficient conductors to replace the limited copper materials and less efficient aluminum conductors. Lasers couple with fiber optics to provide a new dimension in communications while using glass, which has a vast raw material resource in sand. *Superconductors,* which have no electrical resistance at temperatures around absolute zero (0 K), have the potential for high-speed switching of minute currents for computers, as well as for conducting large currents for power transmission or rail transportation. The problem with superconductivity is the very low temperatures necessary for the property to exist. Superconductors hold promise for small-scale power electromagnets and electric motors. Practical application of superconductivity seems to require new materials that will act as superconductors at temperatures near or at room temperature.

Magnetic bubble technology and light bubbles are radically different technologies that have applications in communication systems. *Light bubbles* are microscopically small sources of light in thin films of manganese-doped zinc sulfide that become mobile under the proper conditions. By application of oscillating high-frequency voltage across the thin film, tiny light-emitting filaments appear to pour out from isolated points in the material and to move randomly about it. Each filament is about 1 μm in diameter.

Magnetic bubbles were discovered in the early 1960s by scientists of Bell Laboratories. It was discovered that tiny magnetized bubblelike areas in their films of magnetic material could be controlled and ordered in a "digital" format much like that used by computers to process information. The key problem was in obtaining proper materials: special crystalline materials only a few thousandths of an inch thick to achieve a consistent magnetic bubble effect and fabricated to precise standards. Magnetic bubbles are actually tiny magnetized areas of domains within a crystalline material. The domains are highly mobile and can be moved about by using the same forces that cause magnets to attract or repel one another (Figure 9-29). Bubbles follow precisely defined tracks, deriving power and being guided and switched from track to track much as are trains in a subway system (Figure 9-30). Materials scientists devised a growing technique for special rare-earth magnetic film on a nonmagnetic substrate or base. The product, thin films of magnetic garnet, makes an ideal material for magnetic bubbles due to their magnetic properties, uniformity, and thickness. Growing the magnetic garnet on a nonmagnetic substrate resembles integrated-circuit manufacture. But whereas integrated circuits require a multistep exposure and etching process, nearly all the detail work for the bubbles is performed in one operation, using masks that are generated on an electron beam exposure system (EBES). In one

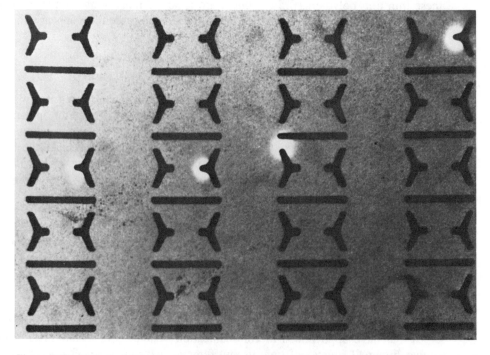

Figure 9-29 Magnetic bubble domains (magnified—the light circles) are shown moving through a circuit pattern formed on a thin epitaxial film of uniaxial garnet. The Y and bar circuit elements are part of an experimental shift register. One bubble, somewhat elongated, can be seen in transition from one pole to the next. The bubbles are 0.0003 inches in diameter. (Bell Labs.)

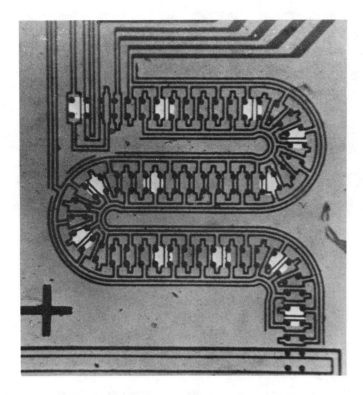

Figure 9-30 Tiny computers and switching systems of the future may accomplish counting, switching, memory, and logic functions all within one solid magnetic material. This actual circuit, a photolithographic pattern on the surface of a sheet of thulium orthoferrite, can move magnetic "bubbles" (large white dots) through a shift register. (Bell Labs.)

pass over a light-sensitive mask, the EBES uses an electron beam to define the pattern through which the bubbles will travel. Magnetic bubbles have potential as reliable, low-cost alternatives to other mass memory devices. A magnetic bubble memory can store 70,000 bits, compared to a large-scale integrated circuit of the same size, which stores 16,000 bits. A bit, or binary digit, is a unit of data such as yes, no, on or off. Magnetic bubbles will not lose stored information should power fail.

The charge-coupled device (CCD) has an advantage over integrated circuits. Developed in the late 1960s, these devices store information as packets of charges within a semiconductor chip. The application of voltage to a routing pattern inscribed on the chip's surface controls the charge (Figure 9-31). The circuitry necessary for storing, moving, and accessing the packets is contained on the same chip. CCDs perform many of the functions of more highly complex integrated circuits, yet they are simple, three-layered structures. They also find application as imaging devices in solid-state cameras. With the removal of power supplies, CCDs lose stored information. For accessing information, they are about ten times faster than magnetic bubbles and store about 65,000 bits of information.

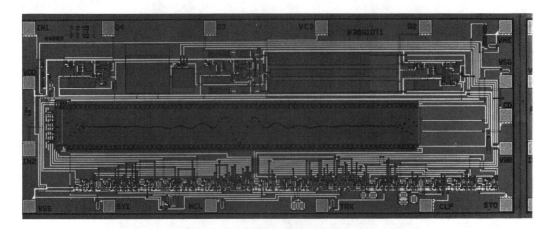

Figure 9-31 Semiconductors-CCD's. Enlarged view of CCD filter showing chip features in detail. (Bell Labs.)

9.8 SELF-ASSESSMENT

9-1 Integrated circuits use many types of solid-state devices. Name three such devices.

9-2 What is the purpose of doping a semiconductor material?

9-3 Compare the energy required to raise a donor electron to the conduction band with the energy required to raise a valence electron.

9-4 Name one method by which an ideal semiconductor having no impurity atoms can be made to conduct electrons.

9-5 Once a donor electron has been raised into the conduction band by the receipt of energy from some external source, conduction occurs as a result of the combined actions of what charged bodies?

9-6 A Group VI element, selenium, is a semiconductor element. From what group in the periodic table would you choose a dopant to convert selenium to (a) a *p*-type material, and (b) an *n*-type material?

9-7 Identify what are termed the majority carriers in a *n*-type semiconductor.

9-8 Explain what is meant by the term "biasing."

9-9 The depletion zone in a *p-n* junction differs in size when a forward bias is applied from a condition of zero bias. Specify what this difference is in terms of area.

9-10 What device operates in a backward fashion to a solar cell and produces light from electrical energy?

9-11 What creates the voltage in a typical solar cell that drives electrical current through the external circuit?

9-12 Reflection of sunlight from the surface layers of a photovoltaic cell is maximized by treating the surface with some form of antireflective coating. True or false? Explain.

9-13 Man has dreamed of converting sunlight directly into electricity down through the ages. How efficient is this process today? Define efficiency as used in this process.

9-14 Name four possible things that can happen to sunlight entering a solar cell.

9-15 Explain the terms solar cell, module, and array.

9-16 What is the major cause of poor efficiency of polycrystalline silicon cells in relation to single crystal silicon cells?

9-17 Amorphous silicon used in hand-held calculators is more efficient under fluorescent lighting than either single or polycrystalline silicon. Name one characteristic that amorphous silicon has that explains this advantge.

9-18 Define the word "epitaxy" and describe why it is being used in the semiconductor industry.

9-19 Both LED and LCD devices produce light. Which one requires more energy and consequently needs a battery source of power to function?

9-20 Optoelectronic devices all operate in accordance with a basic underlying scientific principle. Explain this principle.

9-21 The three types of liquid crystals are based on what characteristic of liquid crystals?

9-22 Name three influences that can change the structure of liquid crystals.

9-23 A homogeneous orientation of LC molecules can be accomplished in several ways. Name two such methods other than use of an electric field.

9-24 In a reflective type LC cell, the light source and the scattered light are on the same side of the cell. One glass plate is given a reflective coating. A transmittive-type cell's light source is on one side of the cell and the scattered light appears on the other. For this to happen, how are the glass plates modified, if at all? Draw a sketch of a transmittive-type cell similar to Figure 9-23.

9-25 Referring to Figure 9-24 to display the numeral "7," which segments need to be activated?

9-26 Referring to Figure 9-27, which axis (x or y) is known as the electric axis?

9-27 Name several applications of piezoelectric crystals.

9-28 The highest frequency of sounds that can be detected by the human ear is about how many cycles per second? Express in units of Hz.

9-29 A certain pickup for a record player converts the movements of a stylus into corresponding electrical signals. Other devices such as a loudspeaker use a similar element, a crystal. What material property peculiar to certain materials do these devices depend on to perform their functions?

9-30 Superconductors have great technological potential for many applications in this energy-dependent world. What is one difficulty that limits their development?

9.9 REFERENCES AND RELATED READING

Annual Book of ASTM Standards Part 43: Electronics, Philadelphia: American Society for Testing and Materials, 1983.

ANSELM, A. *Introduction to Semiconductor Theory.* Englewood Cliffs, N.J.: Prentice-Hall, 1982.

Basic Photovoltaic Principles and Methods, SERI/SP-290-1448. Washington, D.C.: Solar Energy Research Institute, Department of Energy, 1982.

BDH Liquid Crystals. Carle Place, N.Y.: Gallard Schlesinger Chemical Manufacturing Company.

BRODY, HERB. "Superconductors," *High Technology,* vol. 3, no. 2, Feb. 1983, pp. 42–48.

BROWN, GLENN H., and others, *A Review of the Structure and Physical Properties of Liquid Crystals.* Cleveland: Chemical Rubber Co., 1971.

FREE, JOHN R. *"Flat Screen TV," Popular Science,* 29 March 1975, pp. 94–98.

GARDINER, KEITH M. (ed). *Systems and Technology for Advanced Manufacturing,* Dearborn, MI: Society of Manufacturing Engineers, 1983.

JAMES, HERBERT I. *How a LC Display Works.* Webster, N.Y.: Xerox Corp.

MORT, J., and G. PFISTER. *Electronics Properties of Polymers.* New York: Wiley, 1982.

NASHELSKY, LOUIS, and ROBERT L. BOYLESTAD. *Devices: Discrete and Integrated.* Englewood Cliffs, N.J.: Prentice-Hall, 1981.

PASCOE, K. J. *Properties of Materials for Electrical Engineers.* New York: Wiley, 1973.

Photovoltaics: Solar Electric Power Systems, SERI/SP-433-487. Washington, D.C.: Solar Energy Research Institute, Department of Energy, Feb. 1980.

Polycrystalline and Amorphous Silicon Photovoltaic Cells, SERI/SP-281-1704. Washington, D.C.: Solar Energy Research Institute, Department of Energy, 1982.

RIORDAN, KEITH. *"What Is an LCD?" Progress.* Washington, D.C.: Fairchild Camera and Instrument Corp., 1978.

ROOP, RAY. "Trends in High Voltage Integrated Circuit Technology," *Solid State Technology,* May, 1984, pp. 147–151.

TELL, WILLIAM C., and JAMES T. LUXON. *Integrated Circuits: Materials, Devices and Fabrications.* Englewood Cliffs, N.J.: Prentice-Hall, 1982.

UMAN, MYRON F. *Introduction to the Physics of Electronics.* Englewood Cliffs, N.J.: Prentice-Hall, 1974.

Periodicals

Design News

High Technology

IEEE Design and Test of Computers

IEEE Micro

IEEE Potentials

Microelectronics Manufacturing and Testing

Popular Electronics

Semiconductor International

Solid State Technology

(Annual Desk Manual offers numerous terms, definitions, materials, testing, and processing techniques)

10

APPENDIX
OF TABLES

TABLE 10-1A PERIODIC TABLE OF THE ELEMENTS

TABLE 10-1B   TABLE OF PERIODIC PROPERTIES OF ELEMENTS

Percent Ionic Character of a Single Chemical Bond

Difference in electronegativity	0.1	0.2	0.3	0.4	0.5	0.6	0.7	0.8	0.9	1.0	1.1	1.2	1.3	1.4	1.5	1.6	1.7	1.8	1.9	2.0	2.1	2.2	2.3	2.4	2.5	2.6	2.7	2.8	2.9	3.0	3.1	3.2
Percent ionic character %	0.5	1	2	4	6	9	12	15	19	22	26	30	34	39	43	47	51	55	59	63	67	70	74	76	79	82	84	86	89	91	92

SARGENT-WELCH
SARGENT-WELCH SCIENTIFIC COMPANY
7300 LINDER AVENUE, SKOKIE ILLINOIS 60077

Catalog Number S-18806

NOTES: (1) For representative oxides (higher valence) of group. Oxide is acidic if color is red, basic if color is blue and amphoteric if both colors are shown. Intensity of color indicates relative strength.

(2) ◻ Cubic, face centered; ◻ cubic, body centered; ◻ cubic;
◇ hexagonal; ◇ rhombohedral; ◻ tetragonal; ◻ orthorhombic; ◻ monoclinic.

(3) At 300 K (27°C)
(4) At boiling point
(5) At melting point
(6) Generally at 293 K (20°C)
(7) Quantum mechanical value for free atom
(8) From density at 300 K (27°C) for liquid and solid elements; values for gaseous elements refer to liquid state at boiling point

© Copyright 1962
© Copyright 1964
© Copyright 1965
© Copyright 1966
© Copyright 1968
© Copyright 1979
© Copyright 1980

SARGENT-WELCH SCIENTIFIC COMPANY

TABLE 10-2 SYMBOLS OF THE ELEMENTS AND THEIR ATOMIC WEIGHTS

Name	Symbol	Atomic number	Atomic weight[1]	Name	Symbol	Atomic number	Atomic weight[1]
Actinium	Ac	89	(227)	Lawrencium	Lr	103	(257)
Aluminum	Al	13	27.0	Lead	Pb	82	207.2
Americium	Am	95	(243)	Lithium	Li	3	6.94
Antimony	Sb	51	121.8	Lutetium	Lu	71	175.0
Argon	Ar	18	39.9	Magnesium	Mg	12	24.3
Arsenic	As	33	74.9	Manganese	Mn	25	54.9
Astatine	At	85	(210)	Mendelevium	Md	101	(256)
Barium	Ba	56	137.3	Mercury	Hg	80	200.6
Berkelium	Bk	97	(247)	Molybdenum	Mo	42	95.9
Beryllium	Be	4	9.01	Neodymium	Nd	60	144.2
Bismuth	Bi	83	209.0	Neon	Ne	10	20.2
Boron	B	5	10.8	Neptunium	Np	93	(237)
Bromine	Br	35	79.9	Nickel	Ni	28	58.7
Cadmium	Cd	48	112.4	Niobium	Nb	41	92.9
Calcium	Ca	20	40.1	Nitrogen	N	7	14.01
Californium	Cf	98	(251)	Nobelium	No	102	(254)
Carbon	C	6	12.01	Osmium	Os	76	190.2
Cerium	Ce	58	140.1	Oxygen	O	8	16.00
Cesium	Cs	55	132.9	Palladium	Pd	46	106.4
Chlorine	Cl	17	35.5	Phosphorus	P	15	31.0
Chromium	Cr	24	52.0	Platinum	Pt	78	195.1
Cobalt	Co	27	58.9	Plutonium	Pu	94	(242)
Copper	Cu	29	63.5	Polonium	Po	84	(210)
Curium	Cm	96	(247)	Potassium	K	19	39.1
Dysprosium	Dy	66	162.5	Praseodymium	Pr	59	140.9
Einsteinium	Es	99	(254)	Promethium	Pm	61	(147)
Erbium	Er	68	167.3	Protactinium	Pa	91	(231)
Europium	Eu	63	152.0	Radium	Ra	88	(226)
Fermium	Fm	100	(253)	Radon	Rn	86	(222)
Fluorine	F	9	19.0	Rhenium	Re	75	186.2
Francium	Fr	87	(223)	Rhodium	Rh	45	102.9
Gadolinium	Gd	64	157.3	Rubidium	Rb	37	85.5
Gallium	Ga	31	69.7	Ruthenium	Ru	44	101.1
Germanium	Ge	32	72.6	Samarium	Sm	62	150.4
Gold	Au	79	197.0	Scandium	Sc	21	45.0
Hafnium	Hf	72	178.5	Selenium	Se	34	79.0
Helium	He	2	4.00	Silicon	Si	14	28.1
Holmium	Ho	67	164.9	Silver	Ag	47	107.9
Hydrogen	H	1	1.008	Sodium	Na	11	23.0
Indium	In	49	114.8	Strontium	Sr	38	87.6
Iodine	I	53	126.9	Sulfur	S	16	32.1
Iridium	Ir	77	192.2	Tantalum	Ta	73	180.9
Iron	Fe	26	55.8	Technetium	Tc	43	(99)
Krypton	Kr	36	83.8	Tellurium	Te	52	127.6
Lanthanum	La	57	138.9	Terbium	Tb	65	158.9

[1]The values given in parentheses are mass numbers of the principal isotopes of unstable elements.

TABLE 10-2 (continued)

Name	Symbol	Atomic number	Atomic weight[1]	Name	Symbol	Atomic number	Atomic weight[1]
Thallium	Tl	81	204.4	Vanadium	V	23	50.9
Thorium	Th	90	232.0	Xenon	Xe	54	131.3
Thulium	Tm	69	168.9	Ytterbium	Yb	70	173.0
Tin	Sn	50	118.7	Yttrium	Y	39	88.9
Titanium	Ti	22	47.9	Zinc	Zn	30	65.4
Tungsten	W	74	183.9	Zirconium	Zr	40	91.2
Uranium	U	92	238.0				

[1]The values given in parentheses are mass numbers of the principal isotopes of unstable elements.

TABLE 10-3 GREEK SYMBOLS AND THEIR PRONUNCIATIONS

Γ	gamma (cap.)	η	eta (lc)
Δ	delta (cap.)	θ	theta (lc)
Θ	theta (cap.)	ϑ	theta (lc)
Λ	lambda (cap.)	κ	kappa (lc)
Ξ	xi (cap.)	λ	lambda (lc)
Π	pi (cap.)	μ	mu (lc)
Σ	sigma (cap.)	ν	nu (lc)
Υ	upsilon (cap.)	ξ	xi (lc)
Φ	phi (cap.)	π	pi (lc)
Ψ	psi (cap.)	ρ	rho (lc)
Ω	omega (cap.)	σ	sigma (lc)
α	alpha (lc)	τ	tau (lc)
β	beta (lc)	φ	phi (lc)
γ	gamma (lc)	φ	phi (lc)
δ	delta (lc)	χ	chi (lc)
ε	epsilon (lc)	ψ	psi (lc)
ζ	zeta (lc)	ω	omega (lc)

10.1 INTERNATIONAL SYSTEM OF UNITS (SI)

10.1.1 Usage and Computations

SI was created in 1960 by international agreement and represents a worldwide measurement system far superior to earlier measurement systems (gravitational) in expressing scientific and technical data. All gravitational systems using force as a fundamental dimension, including the American engineering system with its pound of mass a fundamental unit, are considered obsolete, and the changeover, though not mandatory, is proceeding at a steady pace with some industries completing the changeover in record time. For most technology users, this means they will continue to be confronted with problems that arise when two systems of measurement exist. Con-

TABLE 10-4 NAMES AND SYMBOLS OF SI UNITS

Quantity	Name of unit	Symbol	Expressed in base units where applicable
BASE UNITS			
Length	meter	m	
Mass	kilogram	kg	
Time	second	s	
Electric current	ampere	A	
Thermodynamic temperature	kelvin	K	
Luminous intensity	candela	cd	
Amount of substance	mole	mol	
DERIVED UNITS			
Area	square meter	m^2	
Volume	cubic meter	m^3	
Frequency	hertz, cycles per second	Hz	s^{-1}
Density (mass)	kilogram per cubic meter	kg/m^3	
Velocity (linear)	meter per second	m/s	
Velocity (angular)	radian per second	rad/s	
Acceleration (linear)	meter per second squared	m/s^2	
Acceleration (angular)	radian per second squared	rad/s^2	
Force	newton, kilogram-meter per second squared	N	$kg \cdot m \cdot s^{-2}$
Permeability	henry per meter	H/m	$m \cdot kg \cdot s^{-2} \cdot A^{-2}$
Permittivity	farad per meter	F/m	$m^{-3} \cdot kg^{-1} \cdot s^4 \cdot A^2$
Pressure (mechanical stress)	pascal, newton per square meter	Pa	$N \cdot m^{-2}$
Kinematic viscosity	square meter per second	m^2/s	
Dynamic viscosity	newton-second per square meter	$N \cdot s/m^2$	$m^{-1} \cdot kg \cdot s^{-1}$
Work, energy, quantity of heat	joule, newton-meter	$J, N \cdot m$	
Power	watt, joule per second	W, J/s	
Quantity of electricity, electric charge	coulomb	$C, A \cdot s$	
Potential difference, electromotive force	volt	V, W/A	
Electric field strength	volt per meter	V/m	$m \cdot kg \cdot s^{-3} \cdot A^{-1}$
Electric resistance	ohm	Ω, V/A	
Capacitance	farad	F, A · s/V	
Magnetic flux	weber	$Wb, V \cdot s$	
Inductance	henry	H, V · s/A	
Magnetic flux density	tesla	$T, Wb/m^2$	
Magnetic field strength	ampere per meter	A/m	
Magnetomotive force	ampere	A	
Luminous flux	lumen	lm	
Luminance	candela per square meter	cd/m^2	
Illuminance	lux	lx	
Wave number	1 per meter	m^{-1}	
Entropy	joule per kelvin	J/K	
Specific heat capacity	joule per kilogram kelvin	$J/(kg \cdot K)$	$m^2 \cdot s^{-2} \cdot K^{-1}$
Thermal conductivity	watt per meter kelvin	$W/(m \cdot K)$	$m \cdot kg \cdot s^{-3} \cdot K^{-1}$
Conductance	siemens	S, A/V	
Torque	newton-meter	N/m	$m^2 \cdot kg \cdot s^{-3} \cdot K^{-1}$
SUPPLEMENTARY UNITS			
Plane angle	radian	rad	

sequently, they must be very familiar with both systems and demonstrate ability to convert from one to another upon completion of their calculations.

Units. SI units are grouped into three general classes: base or fundamental units, derived units, and supplementary units. Table 10-4 lists the base units as well as some of the more common derived units used in this text. Special names are given to some of the derived units (see Table 10-5). For example, hertz (Hz) is the special name given to the SI unit for frequency. Units of force, stress, power, and energy also have special names and therefore need not be expressed in their base units. The amount of force required to accelerate one kilogram of mass one meter per second squared is given the special name of newton (N). Thus $1 \text{ N} = 1 \text{ kg} \cdot \text{m/s}^2$.

TABLE 10-5   SPECIAL NAMES FOR UNITS

Quantity	Unit	Symbol	Formula
Frequency (of a periodic phenomenon)	hertz	Hz	l/s
Force	newton	N	$\text{kg} \cdot \text{m/s}^2$
Pressure, stress	pascal	Pa	N/m^2
Energy, work, quantity of heat	joule	J	$\text{N} \cdot \text{m}$
Power, radiant flux	watt	W	J/s
Quantity of electricity, electric charge	coulomb	C	$\text{A} \cdot \text{s}$
Electric potential, potential difference, electromotive force	volt	V	W/A
Capacitance	farad	F	C/V
Electric resistance	ohm	Ω	V/A
Conductance	siemens	S	A/V
Magnetic flux	weber	Wb	$\text{V} \cdot \text{s}$
Magnetic flux density	tesla	T	Wb/m^2
Inductance	henry	H	Wb/A
Luminous flux	lumen	lm	$\text{cd} \cdot \text{sr}$
Illuminance	lux	lx	lm/m^2

Prefixes. Prefixes corresponding to powers of 10 are attached to the units discussed above in order to form larger or smaller units. In technical work the powers of 10 divisible by 3 are preferred. Table 10-6 contains the authorized prefixes.

Rules for usage. For standardized usage, the following rules should be observed:

1. Uppercase (capitals) and lowercase letters are never interchanged: kg, not KG.
2. The same symbol is used for plurals: N, not Ns; 14 meters or 14 m.
3. No space is left between the prefix and its unit symbol: GHz, not G Hz.
4. To form products, a raised dot is preferred (or a dot on a line): kN · m, or kN.m. The dot may be dispensed with if confusion is not created by its absence.

TABLE 10-6 PREFIXES

Multiplication factor		Prefix	Symbol
1 000 000 000 000 000 000 =	10^{18}	exa	E
1 000 000 000 000 000 =	10^{15}	peta	P
1 000 000 000 000 =	10^{12}	tera	T
1 000 000 000 =	10^{9}	giga	G
1 000 000 =	10^{6}	mega	M
1 000 =	10^{3}	kilo	k
100 =	10^{2}	hecto	h
10 =	10^{1}	deka	da
0.1 =	10^{-1}	deci	d
0.01 =	10^{-2}	centi	c
0.001 =	10^{-3}	milli	m
0.000 001 =	10^{-6}	micro	μ
0.000 000 001 =	10^{-9}	nano	n
0.000 000 000 001 =	10^{-12}	pico	p
0.000 000 000 000 001 =	10^{-15}	femto	f
0.000 000 000 000 000 001 =	10^{-18}	atto	a

5. To form quotients, one solidus (an oblique line), a fraction line (horizontal), or a negative power is used to express derived units: m/s, $\dfrac{m}{s}$, or $m \cdot s^{-1}$.

   Note: The solidus must not be repeated on the same line: m/s^2, not $m/s/s$. Also, $kg/(m \cdot s)$, $\dfrac{kg}{m \cdot s}$, or $kg \cdot m^{-1} \cdot s^{-1}$, but not $kg/m/s$. Note also the use of the parenthesis to avoid ambiguity.

6. An exponent affixed to a symbol containing a prefix indicates that the multiple or submultiple of the unit is raised to the power expressed by the exponent.

$$1\ mm^3 = 10^{-9}\ m^3, \quad \text{not } 10^{-3}\ m^3$$
$$1\ cm^3 = 10^{-6}\ m^3, \quad \text{not } 10^{-2}\ m^3$$
$$1\ cm^{-1} = 10^{2}\ m^{-1}, \quad \text{not } 10^{-2}\ m^{-1}$$

7. A period is used as a decimal marker. It is not used to separate groups of digits. A space is left for this purpose: 5 279 585 J, and 0.000 34 s.

8. Numbers are preferably expressed between the limits 0.1 and 1000, using the appropriate prefix to change the size of the unit: 5.23 GN.

9. For decimal numbers less than 1, the leading zero is never omitted: 0.625, not .625.

10. When units are written in words, they always start with lowercase letters except at the beginning of a sentence. If the unit is derived from the name of an individual, the symbol is capitalized. Plurals of special names are written in the usual manner.

125 watts or 125 W

0.25 newtons or 0.25 N

58.6 hertz or 58.6 Hz

11. A space or hyphen may be used to form the product expressed in words: newton-meters or newton meters.

12. For quotients, the word *per* may be used: newton per meter squared, kilogram per cubic meter.

13. The kelvin (symbol K) is the standard unit of temperature. In writing this absolute temperature, the word *degree* or its symbol (°) is not used: 472 K. In addition, K may be used to express an interval or a difference in temperature. Celsius temperature is expressed in degrees Celsius with symbol °C. The unit degree Celsius is equal to the unit kelvin and may also be used to represent an interval or a difference of Celsius temperature: 25°C. Temperature in K = temperature in °C + 273.15.

Computations.   The SI system of units makes computations relatively simple because 1) a single unit is used to represent a particular physical quantity, 2) the system is coherent in that the factor of 1.0 replaces many conversion factors, and 3) SI is based on the decimal system.

Prior to computation, all prefixes should be replaced by their respective powers of 10. The final step is to select a suitable prefix to express the answer once the resulting answer is rounded to the appropriate number of significant digits.

Two examples illustrate the solution of typical problems using SI units and the preceding information.

Problem 1

Given: A metal rod under tensile load of 356 kN is allowed to withstand a unit stress of 110 MN/m^2.
Required: Find the diameter of the rod in millimeters.
Solution:

1. Convert SI prefixes to powers of 10:

$$356 \text{ kN} = 356 \times 10^3 \text{ N}$$

$$110 \text{ MN/m}^2 = 110 \times 10^6 \text{ N/m}^2$$

2. Using the direct stress formula ($s = P/A$) and solving for the area (A),

$$A = \frac{P}{s} = \frac{356 \times 10^3 \text{ N}}{110 \times 10^6 \text{ N/m}^2} = 3.24 \times 10^{-3} \text{ m}^2$$

3. Using the formula for circular area $(A = \dfrac{\pi}{4} D^2)$ and solving for diameter (D),

$$D = \sqrt{\dfrac{4}{\pi}\,(3.24 \times 10^{-3}\ m^2)}$$

or

$$D = \left[\dfrac{4}{\pi}\,(3.24 \times 10^{-3}\ m^2)\right]^{1/2}$$

$$= \left[\dfrac{4}{\pi}\,(32.4 \times 10^{-4}\ m^2)\right]^{1/2}$$

$$= 6.429 \times 10^{-2}\ m$$

$$= 64.29 \times 10^{-3}\ m = 64.3\ mm \quad \text{(rounding up}$$

and using three significant digits)

Problem 2

Given: Modulus of elasticity in tension (E) for steel is 29,120,000 psi.
Required: Express E in GPa.
Solution:

1. Express E in terms of powers of 10:
$$E = 29.12 \times 10^6\ \text{psi}$$

2. Locate conversion ratio (see Table 10-7):
$$1\ \text{psi} = 6.895 \times 10^{-3}\ \text{MPa}$$

3. Express prefixes in terms of powers of 10 (see Table 10-6):
$$1\ \text{psi} = 6.895 \times 10^3\ \text{Pa}$$

4. Multiply E by conversion ratio:
$$29.12 \times 10^6\ \text{psi}\left[\dfrac{6.895 \times 10^3\ \text{Pa}}{1\ \text{psi}}\right] = 200.8 \times 10^9\ \text{Pa}$$

5. Express answer using required SI prefix (see Table 10-6):
$$200.8\ \text{GPa} \quad \text{(using four significant digits)}$$

TABLE 10-7 CONVERSIONS

Quantity	U.S. customary to SI
Acceleration	$1 \text{ ft/s}^2 = 3.048 \times 10^{-1} \text{ m/s}^2$
Area	$1 \text{ in.}^2 = 6.452 \times 10^2 \text{ mm}^2$
	$1 \text{ ft}^2 = 9.290 \times 10^{-2} \text{ m}^2$
Density (mass)	$1 \text{ lb/in.}^3 = 2.768 \times 10^4 \text{ kg/m}^3$
	$1 \text{ lb/ft}^3 = 1.602 \times 10^1 \text{ kg/m}^3$
Energy, work	$1 \text{ BTU} = 1.055 \text{ kJ}$
	$1 \text{ in.-lb} = 1.129 \times 10^{-1} \text{ J}$
	$1 \text{ ft-lb} = 1.356 \text{ J}$
Force	$1 \text{ lbf} = 4.448 \text{ N}$
	$1 \text{ kgf} = 9.807 \text{ N}$
Impulse	$1 \text{ lb-s} = 4.448 \text{ N} \cdot \text{s}$
Length	$1 \text{ Å} = 1 \times 10^{-1} \text{ nm}$
	$1 \text{ microinch} = 2.540 \times 10^{-2} \text{ μm}$
	$1 \text{ mil} = 2.540 \times 10^1 \text{ μm}$
	$1 \text{ in.} = 2.540 \times 10^1 \text{ mm}$
	$1 \text{ ft} = 3.048 \times 10^{-1} \text{ m}$
Modulus of elasticity, E	$1 \text{ lb/in.}^2 = 6.895 \times 10^{-6} \text{ GPa}$
Moment of force, torque	$1 \text{ lb-in.} = 1.130 \times 10^{-1} \text{ N} \cdot \text{m}$
	$1 \text{ lb-ft} = 1.356 \text{ N} \cdot \text{m}$
Moment of inertia, I (of area)	$1 \text{ in.}^4 = 4.162 \times 10^5 \text{ mm}^4$
Momentum, linear	$1 \text{ lb-ft/s} = 1.383 \times 10^{-1} \text{ kg} \cdot \text{m/s}$
Power	$1 \text{ BTU/min} = 1.758 \times 10^{-2} \text{ kW}$
	$1 \text{ ft-lb/min} = 2.259 \times 10^{-2} \text{ W}$
	$1 \text{ hp} = 7.457 \times 10^{-1} \text{ kW}$
Stress (pressure)	$1 \text{ lb/in.}^2 = 6.895 \times 10^{-3} \text{ MPa}$
	$1 \text{ ksi} = 6.895 \text{ MPa}$
Temperature	$1°\text{F (difference)} = 0.555°\text{C}$
	$1.8°\text{F} = 1°\text{C (difference)}$
Thermal expansion, linear coefficient, α	$\text{in./in./}°\text{F} = 1.8 \text{ K}^{-1}, \text{K} = °\text{C} + 273.15$
Thermal conductivity	$1 \text{ BTU/ft} \cdot \text{hr} \cdot °\text{F} = 1.729 \text{ W/m} \cdot \text{K}$
Velocity, linear	$1 \text{ in./s} = 2.540 \times 10^1 \text{ mm/s}$
	$1 \text{ ft/s} = 3.048 \times 10^{-1} \text{ m/s}$
	$1 \text{ in./min} = 4.233 \times 10^{-1} \text{ mm/s}$
	$1 \text{ ft/min} = 5.080 \times 10^{-3} \text{ m/s}$
Velocity, angular	$1 \text{ rev/min} = 1.047 \times 10^{-1} \text{ rad/s}$
Volume	$1 \text{ in.}^3 = 1.639 \times 10^4 \text{ mm}^3$
	$1 \text{ ft}^3 = 2.832 \times 10^{-2} \text{ m}^3$
	$1 \text{ yd}^3 = 7.646 \times 10^{-1} \text{ m}^3$
Electric current	$1 \text{ ampere} = 1 \text{ C/s}$
Magnetic flux	$1 \text{ maxwell} = 10^{-8} \text{ Wb}$
Magnetic flux density	$1 \text{ Wb/M}^2 = 1 \text{ T (TESLA)}$
Magnetic field strength	$1 \text{ oersted} = 79.58 \text{ ampere turns per meter}$

$1 \text{ joule} = 10^7 \text{ ergs} = 0.625 \times 10^{19} \text{ eV}$

$1 \text{ gauss} = 10^{-4} \text{ Wb/m}^2 = 1 \text{ T}$

$1 \text{ weber (Wb)} = 1 \text{ T/m}^2$

TABLE 10-8 CONSTANTS

Quantity	Symbol	Value
Acceleration of gravity	g	$9.80 \text{ m} \cdot \text{s}^{-2}$
Atomic mass unit	amu	$1.66 \times 10^{-27} \text{ kg}$
Avogadro's number	N_A, N_0	6.022×10^{23} molecules/mole
Bohr radius	a_0	$5.292 \times 10^{-11} \text{ m}$
Electron charge	$q, -e$	$1.60 \times 10^{-19} \text{ C}$
Electron mass	m	$9.11 \times 10^{-31} \text{ kg}$
Electron volt	eV	$0.160 \times 10^{-18} \text{ J}$
Speed of light (vacuum)	c	$3.00 \times 10^{8} \text{ m} \cdot \text{s}^{-1}$
Permittivity of free space	ϵ_0	$\dfrac{10^7}{4\pi c^2} = 8.854 \times 10^{-12} \text{ c/v} \cdot \text{m}$
Magnetic permeability of free space	μ_0	$4\pi \times 10^{-7} \text{ H/m} = 1.257 \times 10^{-6} \text{ H/m}$
Planck's constant	$\hbar$	$6.63 \times 10^{-34} \text{ Js}$

10.1.2 REFERENCES

ASTM E 380: *Standard for Metric Practice.*

ANSI Z 210.1: *American National Standard for Metric Practice.*

ISO 1000: *SI Units and Recommendations for the Use of Their Multiples and of Certain Other Units.*

NBS Special Publication 330: *The International System of Units (SI).*

10.2 PROPERTIES AND USES OF SELECTED MATERIALS

TABLE 10-9 REPRESENTATIVE PLASTICS

Common name (chemical name)	ASTM Abbreviations	Trade names	Common structure	Grouping*	Typical uses
ABS (acrylonitrile butadiene styrene)	ABS	Absinol Abson Cycolac Royalite	Amorphous terpolymer	TP, EP	Pipe, toys, luggage, boat hulls, football helmets, chrome-plated plumbing, gears, and auto parts
Acetal (polyoxymethylene) (polymerized formaldehyde)	POM	Delrin Celcon Formaldafil	Highly crystalline homopolymer and copolymers	TP, EP	Gears, bearings, fan blades, shower heads, auto parts, and aerosol bottles
Acrylics (polymethyl methacrylate)	PMMA	Plexiglas Lucite Acrylite	Amorphous	TP, GP	Lenses, windows, signs, sculpture, light pipes, and skylights
(polyacrylonitrile)	PAN				

TABLE 10-9 (continued)

Common name (chemical name)	ASTM Abbreviations	Trade names	Common structure	Grouping*	Typical uses
Alkyd plaskon (modified polyester resins)		Premix Dyal Glaskyd	Crosslink network	TS, GP	Coatings: enamel, lacquer and paint, molded electrical parts
Allylics (diallyl phthalate) (diallyl isophthalate) (diethylene glycol bisallyl/carbonate)	DAP DAIP CR39™- allyl diglycol carbonate	Diall Poly-Dap	Crosslink network	TS, GP TS	Electronic parts, pump impellers, glass fiber impregnate, dinnerware, watch crystals
Aminos (urea-formaldehyde) (melamine-formaldehyde)	UF MF	Plaskon Cymel	Crosslink network	TS, GP	Electrical parts, particle board binders, coatings, dinnerware, paper impregnate
Cellulosics (cellulose acetate) (cellulose butyrate) (cellulose nitrate) (cellulose propionate) (ethyl cellulose)	CA CAB CN CAP EC	Tenite Uvex Nixonite Forticel Ethocel Methocel	Highly crystalline	TP, GP	Packaging film, pipe, optical frames, flashbulb shields, helmets, rollers

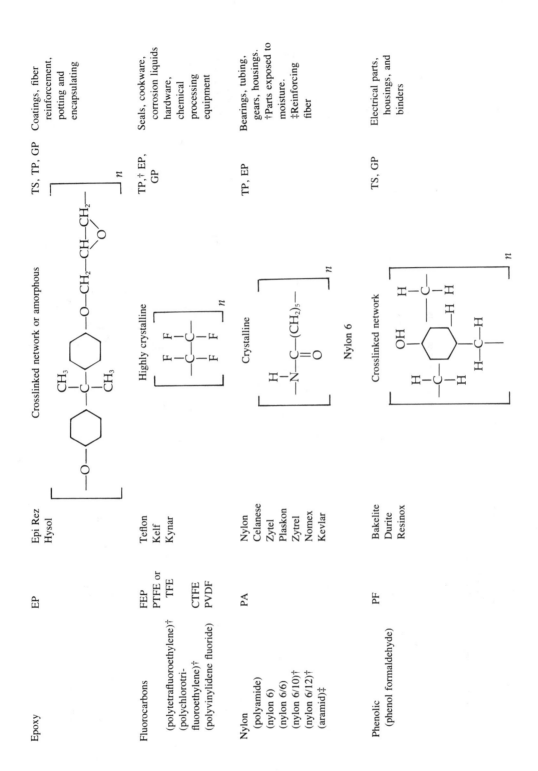

Epoxy	EP	Epi Rez Hysol	Crosslinked network or amorphous	TS, TP, GP	Coatings, fiber reinforcement, potting and encapsulating
Fluorocarbons (polytetrafluoroethylene)† (polychlorotrifluoroethylene)† (polyvinylidene fluoride)	FEP PTFE or TFE CTFE PVDF	Teflon Kelf Kynar	Highly crystalline	TP,† EP, GP	Seals, cookware, corrosion liquids hardware, chemical processing equipment
Nylon (polyamide) (nylon 6) (nylon 6/6) (nylon 6/10)† (nylon 6/12)† (aramid)‡	PA	Nylon Celanese Zytel Plaskon Zytrel Nomex Kevlar	Crystalline Nylon 6	TP, EP	Bearings, tubing, gears, housings. †Parts exposed to moisture. ‡Reinforcing fiber
Phenolic (phenol formaldehyde)	PF	Bakelite Durite Resinox	Crosslinked network	TS, GP	Electrical parts, housings, and binders

TABLE 10-9 (continued)

Common name (chemical name)	ASTM Abbreviations	Trade names	Common structure	Grouping*	Typical uses
Phenoxy (polyhydroxyethers)			Amorphous or crosslinked network	TP, TS, EP	Gas pipe, sports equipment, electrical housing, adhesives and coatings
Phenylene oxide (polyphenylene oxide)	PPO	Noryl	Crystalline	TP, EP	Auto trim, panels, electrical housing, TV cabinets, pump parts
Polycarbonate	PC	Lexan	Amorphous	TP, EP	Optical lenses, bullet resistant windows, housings and cookers
Polyester (polyethylene terephthalate) (polybutylene terephthalate) (aromatic polyesters)	PET (TS) PBT (TP)	Mylar Dacron Kodel / Fortrel Laminac Selectron Gafite tp Ekonol	Amorphous or crosslinked network	TS, TP, GP, EP	Glass-fiber reinforcer, films, fibers

Polymer	Trade names	Structure	Form	Applications
Polyimide	Kapton	Crosslinked network or amorphous	TS, TP, EP	Jet engine vane bushings, seals, ball bearing separators, high-temperature film, fiber matrix
Poly(amide/imide)	Torlon	Amorphous	TP, EP	Engineering plastic gears, structural components, bearings, seals, and valves
Polyolefins (polyethylene, low density)	LDPE Alathon	polypropylene / polyethylene — Low to high crystallinity or crosslinked network	TP, TS, GP, EP	PE: packaging, squeeze bottles, electrical insulation and tubing
(polyethylene, high density)†	HDPE Dylan			
(polyethylene, ultrahigh molecular weight)†	UHMWPE Marlex			PP: packaging, auto battery cases, housings, electrical components and fan blades
(polypropylene)	PP Escon			
(polyallomer)	Tenite			EVA: shoe soles and hypodermic syringes
(ethylene-vinyl acetate)	EVA			
(polybutylene)	TPX			
(polymethyl pentene)	SURLYN A			
(ionomer)				

TABLE 10-9 (continued)

Common name (chemical name)	ASTM Abbre-viations	Trade names	Common structure	Grouping*	Typical uses
Polyphenylene sulfide	PPS	Ryton	Amorphous	TP, EP	Electrical terminal block and connectors, seals, gears
Polystyrene (styrene acrylonitrile)	PS SAN	Styron Stryrofoam Lustrex Dylite Tyril	Amorphous	TP, GP	Packaging, control knobs, TV cabinets, wood substitute, foam insulation
Polysulfone (polyethersulfone) (polyphenylsulfone)		Udel Radel	Amorphous	TP, EP	Electrical insulators, auto distributor caps, tubing and aircraft cabin interiors
Polyurethane (isocyanate-polyester or polyether)	PUR	Estane Flexane Texin Calspan	Amorphous or crosslinked network	TP, TS, EP	Reaction injection molded (RIM) foamed auto parts, solid tires, auto bumpers, synthetic leather

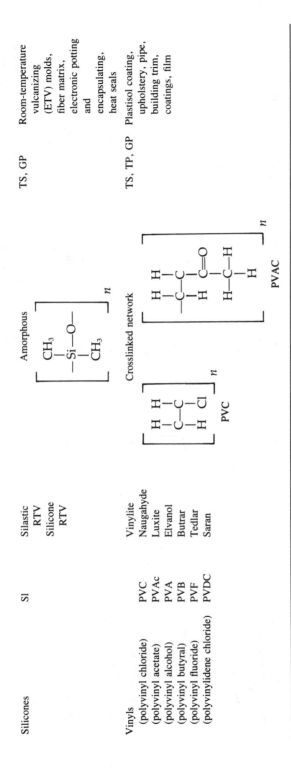

| Silicones | SI | Silastic RTV
Silicone RTV | Amorphous | TS, GP | Room-temperature vulcanizing (ETV) molds, fiber matrix, electronic potting and encapsulating, heat seals |
| Vinyls
(polyvinyl chloride)
(polyvinyl acetate)
(polyvinyl alcohol)
(polyvinyl butyral)
(polyvinyl fluoride)
(polyvinylidene chloride) |
PVC
PVAc
PVA
PVB
PVF
PVDC |
Vinylite
Naugahyde
Luxite
Elvanol
Butrar
Tedlar
Saran | Crosslinked network | TS, TP, GP | Plastisol coating, upholstery, pipe, building trim, coatings, film |

*TP, thermoplastic; TS, thermosetting; EP, engineering plastic; GP, general-purpose or special. ASTM, American Society for Testing and Materials.

†Engineering plastics.

TABLE 10-10 REPRESENTATIVE ELASTOMERS

ABBREVIATION (ASTM):	NR	IR	SBR	IIR	BR	NBR
COMMON NAME:	Natural rubber	Isoprene	GRS or buna S	Butyl	Polybutadiene	Nitrile or buna N
CHEMICAL NAME:	Natural polyisoprene	Polyisoprene	Styrene Butadiene	Isobutene Isoprene	Butadiene	Nitrile butadiene
(1) Tensile strength (kPa)	31	27.5	24	20.6	20.6	24
(2) Hardness Shore A	30–100	40–80	40–90	40–75	45–80	40–95
(3) Specific gravity	0.93	0.93	0.94	0.92	0.94	1.00
(4) Abrasion resistance	A	A	A	B	A	A
(5) Tear resistance	A	B	C	B	B	B
(6) Flexibility at low temperature	B	B	C	B	B	C
(7) Impact resistance	A	A	A	B	B	C
(8) Resiliency	A	A	B	C	A	B
(9) Creep	A	B	B	C	A	B
(10) ASTM/SAE type class	AA	AA	AA–BA	AA–BA	AA	BF, BG, PK, CH
(11) Maximum service temp. (°C)	70	70	100	100	70	100–125
(12) Heat-aging resistance	B–C	B–C	B	B–A	C	B
(13) Flame resistance	D	D	D	D	D	D
(14) Oil and gasoline resistance	X	X	X	X	C	A
(15) Oxidation resistance	C	B–C	C	C	C	C
(16) Ozone resistance	C	C–D	C–D	A	C–D	C–D
(17) Ultraviolet resistance	C–D	C–D	B–C	B	B–C	B–C
(18) Acid and base resistance	B–C	B–C	B–C	B	B–C	B
(19) Water absorption resistance	A	A	B	A	A	C
(20) Permeability to gases	C	C	C	A	C	B
(21) Electrical resistivity	A	A	A	A	A	D
(22) Adhesion to metals	A	A	A	B	A	A
(23) Trade names	—	Natsyn, Isoprene, Ameripol SN	K-Resin	Enjay Butyl, Petro-Tex Butyl	Diene, Ameripol CB	Paracril
(24) Typical uses	Tires, seals, bearings, couplings, shoe soles and heels	Same as natural rubber	Shock absorbers, belts, heels, sponges, gaskets, belts	Truck and auto tires, shock absorbers, inner tubes	Pneumatic tires, gaskets, seals, abrasion resistance belts	Gasoline, chemical and oil seals, gaskets and O-rings, and belting

A, excellent; B, good; C, fair; D, poor; X, not recommended.

TABLE 10-10 (continued)

	ACM Acrylate Polyacrylate	CR Neoprene Chloropene	CSM Hypalom® Chlorosulfanyl Polyethylene	FPM Fluorocarbon Fluorinated hydrocarbon	EPDM EPDM Ethylene propylene	SI Silicone Rubber Polysilicone	UE Urethane Rubber Polyether Polyether urethane	PTR Thikol® Polysulfide
(1)	17	27.5	27.5	17	20.6	4–10	34–55	4–10
(2)	40–90	40–95	50–95	60–90	30–90	25–80	35–100	20–80
(3)	1.10	1.23	1.12–1.28	1.45	0.86	1.14–2.05	1.06	1.34
(4)	B	B	B	B	B	D	B	D
(5)	C–B	C–B	B	B	C	B	B	D
(6)	D	C	B	D	B	A	C	C
(7)	D	B	C–B	B	B	D–C	B–A	D
(8)	C–B	A	C	C	B	D–A	C–A	C
(9)	C	B	C	B	C–B	C–A	C–A	D
(10)	DF, DH	BC, BE	CE	HK	AA, DA, CA	FC, FE, FK	AD, EC	AK
(11)	125	100	125	250	125	200–225	100	70
(12)	A	A	C	A	A	A	B–A	C–B
(13)	D	B	C	A	D	A	D	D
(14)	A	C	C	A	X	D–C	B	A
(15)	A	A	A	A	A	A	A	B
(16)	B	B	A	A	A	A	A	A
(17)	B	B	B	A	A	A	B	B
(18)	C	B	B	A	A	D	D	B
(19)	A	B	A	A	A	A	A	X
(20)	B	B	B	A	C	D	B	B
(21)	B	C	B	B	B	A	B	D
(22)	B	A	A	C–B	C	A	C	C–B
(23)	Hycar, Acrylon	Neoprene, Perbunanc	Hypalon	Viton, Proflo, Fluorel	Nordel, Epcar, Royalene	Adiprene, RTV, Silastic	Kalrez, Estane, Roylar	Thikol
(24)	Oil hose, colored and white parts, pressure and oil O-rings	Belts, hose, extruded goods, molded sheet, adhesives, chemical tank liners	Laminated roofing, tarpaulins, reservoir and pond liners, diaphragms, shoe soles and heels, whitewall tires	Brake seals, ducting connectors, carburetor, needle tips, roll coverings	Garden and industrial hoses, belts, bike tires, electrical wire insulation, paintable auto bumpers	Industrial tires and rolls, mining belts, die pads, gaskets and seals	Chemical O-ring seals, valve seats, gaskets, nuclear, oil, gas, hydraulic and acid seals	Gasoline hose, printing rolls, caulking, adhesives, and binders

TABLE 10-11 REPRESENTATIVE METALS

Material	Nominal composition (essential elements), %		Form and condition	Typical mechanical properties			
				Yield strength (0.2% offset), 1000 psi	Tensile strength 1000 psi	Elongation in 2 in., %	Hardness, Brinell or Rockwell
Copper CA 110 Sheet–ASTM B 152 Rod–ASTM B 124, B 133 Wire–ASTM B1, B2, B3	Cu	–99.90 min	Strip Annealed	10[a]	32	45	R_F40
			Spring Temper	50[a]	55	4	R_B60
Commercial Bronze CA 220 Plate, Sheet, Strip, Bar–ASTM B 36 Wire–ASTM B 134	Cu	–90	Strip Annealed	10[a]	37	45 ·	R_F53
	Zn	–10	Spring Temper	62[a]	72	3	R_B78
Red Brass CA 230 Strip, Sheet, Plate–ASTM B 36 Wire–ASTM B 134 Tube–ASTM B 135	Cu	–85	Strip Annealed	15[a]	40	50	50
	Zn	–15	Hard Temper	60[a]	75	7	135
Copper–Nickel CA 715 Sheet–ASTM B 122 Plate–ASTM B 171 Tube–ASTM B 111	Cu Ni Fe Mn	–bal –30 –0.55 –0.5	Tube Annealed	25[a]	60	45	R_B45
Aluminum Alloy Alclad 2024 Sheet & Plate–ASTM B 209	Core:2024 Al Cu Mn Mg Cladding: Al	–bal –4.5 –0.6 –1.5 –99.3 min	Sheet Annealed Heat Annealed	11 42	26 64	20 19	— —
Aluminum Alloy 3003 Sheet & Plate–ASTM B 209	Al Mn	–bal –1.2	Sheet Annealed Cold Rolled	6 27	16 29	30 4	28 55
Aluminum Alloy 5052 Sheet & Plate–ASTM B 209	Al Mg Cr	–bal –2.5 –0.25	Sheet Annealed Cold Rolled	13 37	28 42	25 7	47 77
Aluminum Alloy 6061 Sheet & Plate–ASTM B 209	Al Si Cu Mg Cr	–bal –0.6 –0.25 –1.0 –0.25	Sheet Annealed Heat Treated	8 40	18 45	25 12	30 95
Aluminum Alloy 707S Bar, Rod, Wire, & Shapes–ASTM B 221	Al Zn Cu Mg Cr	–bal –5.6 –1.6 –2.5 –0.3	Bar Annealed Heat Treated	15 73	33 83	16 11	60 150
Cast Alluminum Alloy 13 Castings–ASTM B 85 Grade S 12A	Al Si	–bal –12.0	Die Casting As Cast	21	43	2.5	—

				Typical physical properties				
Density, lb/cu in	Specific gravity	Melting point, °F	Specific heat (32 to 212°F), Btu/lb/°F	Thermal expansion coefficient (32 to 212°F), in 10^{-6} in./in./°F	Thermal conductivity (32 to 212°F), Btu/sq ft /hr/°F/in.	Electrical resistivity (68°F) ohms /cir mil ft	Tensile modulus of elasticity, in 10^6 psi	Torsional modulus of elasticity, $\times 10^6$ psi
0.322	8.91	1980	0.092	9.4[1]	2512[3]	10.3	17	6.4
—	—	—	—	9.8[2]	—	—	—	—
0.318	8.80	1910	0.09	10.2[2]	1308[3]	23.6	17	6.4
—	—	—	—	—	—	—	—	—
0.316	8.75	1880	0.09	10.4[2]	1104[3]	28	17	6.4
—	—	—	—	—	—	—	—	—
0.323	8.94	2260	0.09	9.0[2]	204[3]	225	22	8.3
0.100	2.77	1180	0.23	12.6[1]	1340[5]	21	10.6	3.75
—	—	—	—	—	840[5]	35	10.6	4.0
0.099	2.73	1210	0.23	12.9[1]	1340[5]	21	10.0	3.75
—	—	—	—	—	1070[5]	26	10.0	3.75
0.097	2.68	1200	0.23	13.2[1]	960[5]	30	10.2	3.75
—	—	—	—	—	960[5]	30	10.2	3.75
0.098	2.70	1205	0.23	13.0[1]	1190[5]	23	10.0	3.75
—	—	—	—	—	1070[5]	26	10.0	3.75
0.101	2.80	1175	0.23	12.9[1]	—	—	10.4	3.9
—	—	—	—	—	840[5]	35	10.4	3.9
0.096	2.65	1080	0.23	11.5[1]	870[5]	34	10.3	3.85

TABLE 10-11 (continued)

| | | | Typical mechanical properties | | | |
Material	Nominal composition (essential elements), %		Form and condition	Yield strength (0.2% offset), 1000 psi	Tensile strength 1000 psi	Elongation in 2 in., %	Hardness, Brinell or Rockwell
Magnesium Alloy AZ 31B Plate & Sheet–ASTM B 90	Mg	–bal	Sheet Annealed	22	37	21	56
	Al	–3.0					
	Zn	–1.0	Hard Sheet	32	42	15	73
	Mn	–0.2 min					
Magnesium Alloy AZ 80A Forgings–ASTM B 91	Mg	–bal	As Forged	33	48	11	69
	Al	–8.5					
	Zn	–0.5	Forged & Aged	36	50	6	72
	Mn	–0.15 min					
Magnesium Alloys AZ 91A & AZ 91B Castings–ASTM B 94	Mg	–bal					
	Al	–9.0	Die Cast	22	33	3	63
	Zn	0.7					
	Mn	–0.2 min					
Titanium Ti–35A Forgings–ASTM B 381 Sheet, Strip, & Plate–ASTM B 265 Pipe–ASTM B 337 Tubes–ASTM B 338 Bars–ASTM B 348	Ti	–bal					
	C	–0.08 max					
	Fe	–0.12 max	Sheet Annealed	30	40	30	135
	N_2	–0.05 max					
	H_2	–0.015 max					
Ti-6 Al-4 V Alloy Sheet, Strip, Plate–ASTM B 265 Bar–ASTM B 348 Forgings–ASTM B 381	Ti	–bal					
	Al	–6.5	Sheet Annealed	130	140	13	R_c-39
	V	–4					
	C	–0.08 max	Heat Treated	165	175	12	—
	Fe	–0.25 max					
	N_2	–0.05 max					
	H_2	–0.015 max					
Nickel 211 ASTM F 290	Ni	–95.0					
	Mn	–4.75	Annealed	35	75	40	140
	C	–0.10					
Nickel (Cast)	Ni	–95.6					
	Cu	–0.5					
	Fe	–0.5	As Cast	25[a]	57	22	110
	Mn	–0.8					
	Si	–1.5					
	C	–0.8					
Duranickel Alloy 301	Ni	–bal					
	Al	–4.5					
	Si	–0.55	Hot Rolled &	132	185	28	330
	Ti	–0.5	Aged				
	Mn	–0.25					
	Fe	–0.15					
	C	–0.15					

Density, lb/cu in	Specific gravity	Melting point, °F	Specific heat (32 to 212 F), Btu/lb/°F	Thermal expansion coefficient (32 to 212°F), in 10^{-6} in./in./°F	Thermal conductivity (32 to 212°F), Btu/sq ft /hr/°F/in.	Electrical resistivity, (68° F) ohms /cir mil ft	Tensile modulus of elasticity, in 10^6 psi	Torsional modulus of elasticity, $\times 10^6$ psi
0.064	1.77	1170	0.245	14.5[1]	657[6]	55	6.5	2.4
—	—	—	—	—	—	—	—	—
0.065	1.80	1130	0.25	14.5[1]	522[2]	87	6.5	2.4
—	—	—	—	—	—	—	—	—
0.065	1.80	1105	0.25	14.5[1]	493[2]	102	6.5	2.4
0.163	4.50	3063	0.124	4.8	108[3]	336	14.9	6.5
0.160	4.42	3000	0.135	4.9	50[3]	1026	16.5	6.1
—	—	—	—	—	—	—	—	—
0.315	8.73	2600	0.11	7.4[17]	306[3]	102	30	11
0.301	8.34	2550	0.13	8.85	410	125	21.5	—
0.298	8.75	2620	0.104	7.2	165[3]	255	30	11

TABLE 10-11   (continued)

Material	Nominal composition (essential elements), %		Typical mechanical properties				
			Form and condition	Yield strength (0.2% offset), 1000 psi	Tensile strength 1000 psi	Elongation in 2 in., %	Hardness, Brinell or Rockwell

Monel Alloy 400
  Rod & Bar–ASTM B 164
  Plate, Sheet, Strip–ASTM B 127
  Tube–ASTM B 165
	Ni
	Cu
	Fe
	Mn
	Si
	C
		–bal
		–31.5
		–1.35
		–0.90
		–0.15
		–0.12
			Rod Hot Rolled Annealed	30	79	48	125

Inconel Alloy 600 Plate,
  Sheet, Strip–ASTM B 168
  Rod & Bar–ASTM B 166
  Pipe & Tube–ASTM B 163 & B 167
	Ni
	Cr
	Fe
	Mn
	Si
	C
	Cu
		–bal
		–15.8
		–7.2
		–0.2
		–0.2
		–0.04
		–0.10
			Rod Hot Rolled Annealed	36	90	47	150

HASTELLOY* Alloy W
  Wire–AMS 5786
  Bar & Forgings–AMS 5755A
	Cr – 5
	Mo–24.5
	Fe – 5.5
	Ni –bal
			Sheet Annealed	53	123	55	—

Ingot Iron	Fe –99.9 plus		Hot rolled	29	45	26	90
			Annealed	19	38	45	67
Wrought Iron Forgings–ASTM A 73	Fe –bal Slag – 2.5		Hot Rolled	30	48	30	100
Carbon Steel–SAE 1020 ASTM A 285	Fe –bal C – 0.20 Mn– 0.45 Si – 0.25		Annealed	38	65	30	130
			Quenched and Tempered at 1000°F	62	90	25	179
300 M Alloy Steel Bar & Forgings–AMS 6416	Fe –bal Mn– 0.80 Si – 1.6 Ni – 1.85 Cr – 0.85 Mo– 0.38 V – 0.08 C – 0.43		Hardened	240	290	10	535
Cast Gray Iron ASTM A 48 Class 30	C – 3.4 Si – 1.8 Mn– 0.8 Fe –bal		As Cast	—	32	—	190
Malleable Iron Castings–ASTM A 47	C – 2.5 Si – 1 Mn– 0.55 max Fe –bal		Annealed	33	52	12	130

Density, lb/cu in	Specific gravity	Melting point, °F	Specific heat (32 to 212 F), Btu/lb/°F	Thermal expansion coefficient (32 to 212°F), in 10^{-6} in./in./°F	Thermal conductivity (32 to 212°F), Btu/sq ft /hr/°F/in.	Electrical resistivity, (68° F) ohms /cir mil ft	Tensile modulus of elasticity, in 10^6 psi	Torsional modulus of elasticity $\times 10^6$ psi
0.319	8.84	2460	0.102	7.7[17]	151[3]	307	26	9.5
0.304	8.43	2600	0.106	7.4[17]	103[3]	620	31	11
0.325	9.03	2400	—	6.3[1]	—	—	—	—
0.284	7.86	2795	0.108	6.8	490	57	30.1	11.8
—	—	—	—	—	—	—	—	—
0.278	7.70	2750	0.11	6.35	418	70	29	—
0.284	7.86	2760	0.107	6.7	360	60	30	—
—	—	—	—	—	—	—	—	—
0.283	7.84	2740	0.107	6.7	360	60	30	11.6
0.260	7.20	2150	—	6.7	310	400	14	—
0.264	7.32	2250	0.122	6.6	—	180	25	—

TABLE 10-11 (continued)

Material	Nominal composition (essential elements), %	Form and condition	Yield strength (0.2% offset), 1000 psi	Tensile strength 1000 psi	Elongation in 2 in., %	Hardness, Brinell or Rockwell
Ductile Iron (Nickel Containing) Grade 60-40-18 Castings–ASTM A 536	Fe –bal C – 3.6 Si – 2.3 Mn– 0.5 Ni – 0.75	Annealed	47	65	24	160
Ductile Iron (Nickel Containing) Grade 120-90-02 Castings–ASTM A 536	Fe –bal C – 3.6 Si – 2.3 Mn– 0.5 Ni – 0.75	Oil Quenched and Tempered	120	140	4	325
Type 201 Stainless Steel (UNS 20100) Plate, Sheet & Strip–ASTM A 412 Bar–ASTM A 429	Fe –bal Cr –17 Ni – 4.5 Mn– 6.5 N_2 – 0.25 max C – 0.15 max	Strip Annealed	55	115	60	R_B90
Type 302 Stainless Steel (UNS 30200) Plate, Sheet & Strip–ASTM A 167 & A 240 Bar–ASTM A 276 & A 314 Wire–ASTM A 313 Forgings–ASTM A 473	Fe –bal Cr –18 Ni – 9 C – 0.15 max	Sheet Annealed Cold Rolled	40 up to 165	90 up to 190	50 5	R_B85 up to R_C40
Type 303 & 303 Se Stainless Steel (UNS 30323) Bar–ASTM A 276 & A 314 Forgings–ASTM A 473	Fe –bal Cr –18 Ni – 9 S – 0.15 min or Se – 0.15 min C – 0.15 max	Bar Annealed	35	90	50	160
Type 314 Stainless Steel (UNS 31400) Bar–ASTM A 276 & A 314	Fe –bal Cr –25 Ni –20 Si – 2.50 C – 0.25 max	Bar Annealed	50	100	45	180
Type 405 Stainless Steel (UNS 40500) Plate, Sheet & Strip–ASTM A 176 & A 240 Tube–ASTM A 268 Bar–ASTM A 276 & A 314	Fe –bal Cr –12.5 C – 0.08 max Al – 0.20	Sheet Annealed	40	65	25	R_B75

Density, lb/cu in	Specific gravity	Melting point, °F	Specific heat (32 to 212°F), Btu/lb/°F	Thermal expansion coefficient (32 to 212°F), in 10^{-6} in./in./°F	Thermal conductivity (32 to 212°F), Btu/sq ft /hr/°F/in.	Electrical resistivity (68°F) ohms /cir mil ft	Tensile modulus of elasticity, in 10^6 psi	Torsional modulus of elasticity, × 10^6 psi
0.250	7.1	2150	—	6.2[1]	276[10]	399[11]	24.5	9.3
0.252	7.2	2150	—	5.9[1]	218[10]	408[11]	24.5	9.3
0.283	7.86	—	—	8.7[27]	—	423	28.6	—
0.29 —	7.9 —	2590 —	0.12 —	9.6 —	113[10] —	432 —	28 —	12.5 —
0.29	—	2590	0.12	9.6	113[10]	432	28	—
0.279	—	—	0.12	8.4[28]	121[10]	462	29	—
0.28	7.7	2790	0.11	6.0	—	360	29	—

(INCO.)

Module 10   Appendix of Tables

TABLE 10-12 PROPERTIES OF SELECTED CERAMICS

	Density (lb/in.³) (kg/m³)	Hardness (M, Moh's) (K, Knoop)	Tensile strength (psi) (MPa)	Thermal conductivity (Btu·in./hr ft²·°F) (W/m·K)		Coefficient of thermal expansion ($10^{-6}·F^{-1}$) ($10^{-6}·K^{-1}$)
Alumina	0.14	M, 9	25,000	25°C	192–255	77–1830°F 4.3
(Al_2O_3)	3.8	K, 2500	172		27.7–36.7	298–1272 K 8.1
Beryllia	0.11	M, 9	23,000	25°C	1741	68–2550°F 5.28
(BeO)	2.92	K, 2000	159		250	293–1672 K 9.5
Boron carbide	0.087	M, 9	22,500	70°F	104–197	0–2550°F 1.73
(B_4C)	2.41	K, 2800	155		—	255–1672 K 3.1
Boron nitride	0.076	—	3,500	70°F	100–200	70–1800°F 4.17
(BN)	2.10		24.1			294–1255 K 7.5
Cordierite	0.065	M, 6.5	3,500	25°C	12–22	68–212°F 2.08
($2MgO·2Al_2O_3·5SiO_2$)	1.8		24.1		1.7–3.2	293–373 K 3.7
Silicon carbide	0.11	M, 9	24,000	70°F	101	0–2552°F 2.4
(SiC)	3.17	K, 2500	165		—	225–1672 K 4.3
Steatite	0.09	M, 7.5	8,700	25°C	20–41	68–212°F 3.99
($MgO·SiO_2$)	2.7	K, 1500	60		2.9–5.9	293–373 K 7.2
Zircon	0.13	M, 8	12,000	25°C	4.9–6.2	68–212°F 1.84
($ZrO_2·SiO_2$)	3.7		82.7			293–212 K 3.3

Data adapted from *Machine Design*, '84 Materials Reference Issue 3M, and Ceramics Bulletin No. 757, *Materials Engineering*, '84 Materials Selection Issue.

Dielectric constant at 10^6 cycles/s (except as noted)	Volume resistivity	Compressive strength (psi) (MPa)	Flexural strength (psi) (MPa)		Impact strength (in.-lb) (N·m)	Modulus of elasticity (psi × 10^6) (GPa)	Safe service temperature (°F) (°C)
			70°F	2,250°F			
8.0–10.0	> 10^{20}	340,000	48,000	31,000	6.5	50	3540
		2,344	331	214	0.73	379	1965
6.4–7.0	> 10^{14}	260,000	33,000	—	—	47	4350
		1,793	228	—	—	324	2414
—		420,000	44,000	—	—	65	1100
		2,896	303	—	—	448	611
4.1–4.8	—	45,000	—	—	—	7	3000
		310	—	—	—	48	1665
4.02–6.23	> 10^{20}	40,000	8,000	—	2.5	7	2282
		276	55	—	0.28	48	1250
—	—	100,000	110,000	80,000	—	62	3200
		690	758	552	—	427	1776
5.9–6.3 6×10^1 cps (6×10^1 Hz)	> 10^{20}	90,000	19,000	—	5.0	15	1832
		620	131	—	0.56	103	1016
8.0–10.0	> 10^{20}	100,000	22,000	—	5.5	23	2012
		690	152	—	0.62	159	1117

TABLE 10-13 REPRESENTATIVE GLASSES

U.S. CUSTOMARY

Type	Color	Principal use	Class	Corrosion resistance Weathering	Water	Acid	Thermal expansion—multiply by 10^{-7} in./in./°F 32 to 572°F	77°F to Setting point	Upper working temperatures (mechanical considerations only) Annealed Normal service °F	Annealed Extreme service °F	Tempered Normal service °F	Tempered Extreme service °F	Thermal shock resistance—plates 6 × 6 in. Annealed 1/8 in. Thick °F	1/4 in. Thick °F	1/2 in. Thick °F
(1) Soda lime	Clear	Lamp bulbs	I	3	2	2	52	58.3	230	860	428	482	149	122	95
(2) Potash soda lead	Clear	Lamp tubing	I	2	2	2	49.8	53.9	230	716	—	—	—	—	158
(3) Aluminosilicate	Clear	Electron tube	I	1	1	3	25.6	30	392	1202	752	842	257	212	158
(4) Borosilicate	Clear	Sealed beam lamps	I	[3]1	[3]2	[3]2	20.4	21.2	446	860	500	500	320	266	294
(5) Borosilicate	Clear	General	I	2	2	2	18.9	20.6	446	842	482	482	320	266	194
(6) 96% Silica	Clear	High temp.	I	1	1	1	4.2	3.1*	1652	2192	—	—	—	—	—
(7) Fused silica	Clear	Optical	I	1	1	1	3.1	1.9*	1652	2012	—	—	—	—	—
(8) Glass-ceramic	White	Missile nose cones	II	—	1	4	31.7	—	1292	—	—	—	572	338	266

TABLE 10-13 (continued)

	Thermal stress resistance °F	Viscosity data				Knoop hardness KHN$_{100}$	Density lb/ft^3	Young's modulus multiply by 10^6 psi	Poison's ratio	Log$_{10}$ of volume resistivity ohm-cm			Dielectric properties at 1MHz, 68°F			Refractive index
		Strain point °F	Annealing point °F	Softening point °F	Working point °F					77°F	482°F	662°F	Power factor %	Dielectric constant	Loss factor %	
(1)	29	883	957	1285	1841	465	154	10.2	.22	12.4	6.4	5.1	.9	7.2	6.5	1.512
(2)	36	743	815	1166	1805	382	190.3	8.6	.22	17.+	10.1	8.0	.12	6.7	.8	1.560
(3)	47	1229	1310	1666	2134	514	164.7	12.5	.24	17.+	13.5	11.3	.16	6.3	1.0	1.547
(4)	86	932	1011	1436	2133	—	140.4	9.3	.19	18.	8.1	6.6	.45	4.85	2.18	1.476
(5)	94	892	973	1436	2188	442	139.8	9.0	.20	17.	9.4	7.7	.18	4.5	.79	1.473
(6)	396	1634	1868	2786	—	487	136	9.8	.19	17.+	9.7	8.1	.04	3.8	.15	1.458
(7)	515	1753	1983	2876	—	489	137.2	10.5	.16	17.+	11.8	10.2	.001	3.8	.0038	1.459
(8)	29	—	—	—	—	657	162.2	17.2	.24	16.7	10.0	8.7	.30	5.6	1.7	—

(Corning Glass Works)

TABLE 10-13 (continued)

Type	Color	Principal use	Class	Corrosion resistance			Thermal expansion—multiply by 10⁻⁷ cm/cm/°C		Upper working temperatures (mechanical considerations only)				Thermal shock resistance plates 15 × 15 cm		
				Weathering	Water	Acid	25°C to 0–300°C	Setting point	Annealed		Tempered		Annealed		
									Normal service °C	Extreme service °C	Normal service °C	Extreme service °C	3.2 mm Thick °C	6.4 mm Thick °C	12.7 mm Thick °C
(1) Soda lime	Clear	Lamp bulbs	I	3	2	2	93.5	105	110	460	220	250	65	50	35
(2) Potash soda lead	Clear	Lamp tubing	I	2	2	2	89.5	97	110	380	—	—	65	50	35
(3) Aluminosilicate	Clear	Electron tube	I	1	1	3	46	54	200	650	400	450	125	100	70
(4) High lead	Clear	Solder sealing	II	1	1	4	84	92	100	300	—	—	—	—	—
(5) Borosilicate	Clear	General	I	2	2	2	34	37	230	450	250	250	160	130	90
(6) 96% Silica	Clear	High temp.	I	1	1	1	7.5	5.5*	900	1200	—	—	—	—	—
(7) Fused silica	Clear	Optical	I	1	1	1	5.5	3.5*	900	1100	—	—	—	—	—
(8) Glass-ceramic	White	Missile nose cones	II	—	1	4	57	—	700	—	—	—	200	170	130

TABLE 10-13 (continued)

	Thermal stress resistance °C	Knoop hardness KHN$_{100}$	Density g/cm^3	Young's modulus multiply by 10^3 Kg/mm^2	Poisson's ratio	Log$_{10}$ of volume resistivity ohm-cm			Dielectric properties at 1MHz, 20°C			Refractive index
						25°C	250°C	350°C	Power factor %	Dielectric constant	Loss factor %	
(1)	16	465	2.47	7.1	.22	12.4	6.4	5.1	.9	7.2	6.5	1.512
(2)	20	382	3.05	6.0	.22	17.+	10.1	8.0	.12	6.7	.8	1.560
(3)	26	514	2.64	8.8	.24	17.+	13.5	11.3	.16	6.3	1.0	1.547
(4)	21	—	5.42	5.6	.28	17.+	10.6	8.7	.22	15.	3.3	1.86
(5)	52	442	2.24	6.3	.20	17.	9.4	7.7	.18	4.5	.79	1.473
(6)	220	487	2.18	6.9	.19	17.+	9.7	8.1	.04	3.8	.15	1.458
(7)	286	489	2.20	7.4	.16	17.+	11.8	10.2	.001	3.8	.0038	1.459
(8)	16	657	2.6	12	.24	16.7	10.0	8.7	.30	5.6	1.7	—

Column 4
^{2}Since weathering is determined primarily by clouding which changes transmission, a rating for the opal glasses is omitted.
^{3}These borosilicate glasses may rate differently if subjected to excessive heat treatment.

Column 6
Normal Service: No breakage from excessive thermal shock is assumed.
Extreme Limits: Glass will be very vulnerable to thermal shock. Recommendations in this range are based on mechanical stability considerations only. Tests should be made before adopting final designs.
These data approximate only.

Column 7
These data approximate only.
Based on plunging sample into cold water after oven heating. Resistance of 100°C (212°F) means no breakage if heated to 110°C (230°F) and plunged into water at 10°C (50°F). Tempered samples have over twice the resistance of annealed glass.

Column 9
These data subject to normal manufacturing variations.

Column 10
Determined by revised ASTM standard: number of standard not yet assigned.

Column 14
^{6}at 10 kHz

(Corning Glass Works)

Module 10   Appendix of Tables   469

10.3 HARDNESS/TENSILE STRENGTH CONVERSION

HARDNESS CONVERSION CHART FOR HARDENABLE CARBON AND ALLOY STEELS
APPROXIMATE RELATIONSHIP BETWEEN HARDNESSES AND TENSILE STRENGTH
(DATA FROM 1966 SAE HANDBOOK)

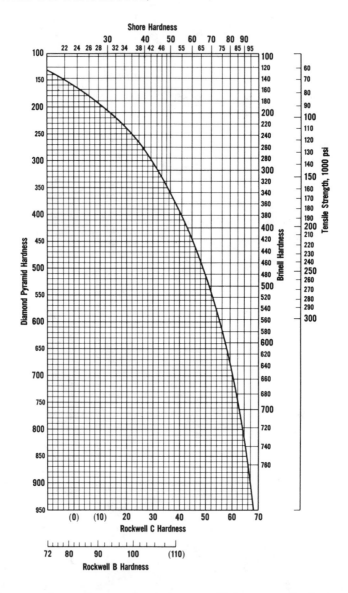

Conversions from one scale to another are made at the intercepts with the curve crossing the chart. For example, follow the horizontal line representing 400 Diamond Pyramid Hardness to its intersection with the conversion curve. From this point follow vertically downward for equivalent Rockwell C Hardness (41), horizontally to the right for Brinell Hardness (379) and Tensile Strength (187,000 psi), and vertically upward for the equivalent Shore Hardness (55).

INDEX

Air for polymers, 238, 242
AISI (American Iron and Steel Institute), 170
  steel designation, 213
Alcohol in plastics, 250
Allotropes (polymorphs) of carbon, 313
Allotropy, 69–70, 180
Alloy, 72–73
  chemical composition, 171
  steel, 183, 214–15
Alloying, 122
Alpha ferrite (ferrite), 181
Al-Si alloy, 177
Alumina (Al_2O_3) (aluminum oxide) (see Ceramics,
      alumina)
Alumina trihydrate (ATH), 251
Aluminum, 8, 216, 218–19, 228
  in ceramics, 310
  chromium substitute, 228
  numbering system, 220–21
  refining flow chart, 219
  rivets, 157, 202
Aluminum Association (AA) numbering system,
      220–21
American Iron and Steel Institute (AISI) (see AISI)
American Lumber Association, 279–80
American Plywood Association, 281–83
American Society for Metals (see ASM)
American Society for Nondestructive Testing (see
      ASNT)
American Society for Testing and Materials (see
      ASTM)
American Society of Mechanical Engineers (see ASME)
Amorphous, 42, 58
  glass, 327
  metals, 82–83
  silicon (aSI), 419
  structure, 243
Ampere (A), 401
Angle of twist, 107
Anion, 24–25, 117
Anisotropic, 277–78
Anisotropy, 132, 191
Annealing:
  glass, 329
  polymers, 245
  steel, 186
Anodizing, 222
Antimony, 224
Applications of materials (see specific materials; *Pause
      and Ponder* and *Applications and
      Alternatives* sections in each module)
Aramid, 340, 359,
Area defects, 80
Aromatic polyamide (aramid) fiber, 359
Artificial aging, 201
ASM, 182
  metals handbooks, 217
ASME, 214
ASNT, 146, 147
ASTM, 16
  adhesives defined, 291

concrete classifications, 338 (*table*)
elastomer standards, 265, 272
glass defined, 326
plastics defined, 247
plastics standards, 240, 249, 255, 264
structural adhesives defined, 295
Athletic equipment, 1 (see also Recreational industry)
Atom:
  bonding, 33 (*table*)
  defined, 22
  mass number, 23, 62, 436–37
  model, 23, 24
  packing factor (PF), 70–71
  planes, 64
  radius, 164
  size, 23, 28
  structure, 22
Atomic packing factor (APF), 70–71
Austempering, 198–99
Austenite, 164, 181
Autoclave, 240, 383
Automation, 349
Automobile design and materials applications, 156,
      220, 225
  body panels, 157
  bumpers, 248, 273
  Chrysler XL car, 22
  composites, 20–21, 350, 371, 392–93
  Ford, 392–93
  GMC Corvette, 21, 392–93
Average molecular weight, 242

Bainite, 194–96
Bakelite™, 381
Bamboo, 348
Barium (Ba), in plastics, 250
Basic oxygen furnace, 209–10
BCC (see Body-centered cubic unit cell)
BCT (see Body-centered tetragonal unit cell)
Bending strength, 109
Beryllium (Be), 226
BH product, 135
Biasing, 411
Bifunctional ethylene, 239
Bifunctional monomers, 240
Bioengineering, 16
Biological attack, 287
Biological resistance, 122
Biological polymers, 234
Birefringence (in plastics, 257)
Birefringency, 143
Bismuth, 224
Bisque, 325
Bit, 431
Block copolymer, 270–71
Block molecules, 270
BMC (see Bulk molding compound)
Body-centered cubic unit cell (BCC), 62, 163
Body-centered tetragonal unit cell (BCT), 63
Bonding, 17, 28
  chemical, 28